Introduction to Biomaterials

This textbook gives students the perfect introduction to the world of biomaterials, linking the fundamental properties of metals, polymers, ceramics, and natural biomaterials to the unique advantages and limitations surrounding their biomedical applications.

- Clinical concerns such as sterilization, surface modification, cell–biomaterial interactions, drug delivery systems, and tissue engineering are discussed in detail, giving students practical insight into the real-world challenges associated with biomaterials engineering.
- Key definitions, equations and concepts are concisely summarized alongside the text, allowing students to quickly and easily identify the most important information.
- Bringing together elements from across the book, the final chapter discusses modern commercial implants, challenging students to consider future industrial possibilities.

Concise enough to be taught in a single semester, and requiring only a basic understanding of biology, this balanced and accessible textbook is the ideal introduction to biomaterials for students of engineering, materials science, and medicine.

C. Mauli Agrawal is the Vice President for Research at the University of Texas at San Antonio (UTSA), and the Peter Flawn Professor of Biomedical Engineering. Previously, he served as the Dean of the College of Engineering at UTSA. He specializes in orthopedic and cardiovascular biomaterials and implants and his inventions have been licensed to various companies. He is a member of the International College of Fellows of Biomaterials Science and Engineering, a Fellow of the American Institute for Medical and Biological Engineering, a former President of the Society for Biomaterials, and was awarded the 2010 Julio Palmaz Award for Innovation in Healthcare and the Biosciences.

Joo L. Ong is Chair of the Department of Biomedical Engineering and the USAA Foundation Distinguished Professor at the University of Texas at San Antonio. His research focuses on modification and characterization of biomaterials surfaces for dental and orthopedic applications, tissue engineering ceramic scaffolds, protein-biomaterial interactions, and bone–biomaterial interactions. He is a Fellow of the American Institute for Medical and Biological Engineering.

Mark R. Appleford is an Assistant Professor of Biomedical Engineering at the University of Texas at San Antonio, focusing on tissue–biomaterial interactions, cellular engineering, reconstructive tissue engineering, and biocompatibility.

Gopinath Mani is an Assistant Professor of Biomedical Engineering at the University of South Dakota, focusing on surface modification and characterization of biomaterials, nanomaterials and nanomedicine, biodegradable metals and drug delivery systems. He is the Program Chair for the Surface Characterization and Modification Special Interest Group of the Society for Biomaterials, and has developed and taught numerous graduate-level programs in biomaterials engineering.

CAMBRIDGE TEXTS IN BIOMEDICAL ENGINEERING

Series Editors
W. Mark Saltzman, Yale University
Shu Chien, University of California, San Diego

Series Advisors
Jerry Collins, Alabama A & M University
Robert Malkin, Duke University
Kathy Ferrara, University of California, Davis
Nicholas Peppas, University of Texas, Austin
Roger Kamm, Massachusetts Institute of Technology
Masaaki Sato, Tohoku University, Japan
Christine Schmidt, University of Florida, Gainesville
George Truskey, Duke University
Douglas Lauffenburger, Massachusetts Institute of Technology

Cambridge Texts in Biomedical Engineering provide a forum for high-quality
textbooks targeted at undergraduate and graduate courses in biomedical engineering.
It covers a broad range of biomedical engineering topics from introductory texts to
advanced topics, including biomechanics, physiology, biomedical instrumentation,
imaging, signals and systems, cell engineering, and bioinformatics, as well as other
relevant subjects, with a blending of theory and practice. While aiming primarily at
biomedical engineering students, this series is also suitable for courses in broader
disciplines in engineering, the life sciences and medicine.

"This is a book that is destined to be a classic in biomaterials education. Written by leading bioengineers and scientists, it can serve not only as a textbook to support a semester-long undergraduate course, but also as an introduction to graduate-level classes. It is a well-written, comprehensive compendium of traditional and also modern knowledge on all aspects of biomaterials, and I am sure that both students and instructors will embrace it and use it widely."

Kyriacos A. Athanasiou
University of California, Davis

Introduction to Biomaterials

Basic Theory with Engineering Applications

C. Mauli Agrawal
University of Texas at San Antonio

Joo L. Ong
University of Texas at San Antonio

Mark R. Appleford
University of Texas at San Antonio

Gopinath Mani
University of South Dakota

CAMBRIDGE
UNIVERSITY PRESS

CAMBRIDGE
UNIVERSITY PRESS

University Printing House, Cambridge CB2 8BS, United Kingdom

One Liberty Plaza, 20th Floor, New York, NY 10006, USA

477 Williamstown Road, Port Melbourne, VIC 3207, Australia

314-321, 3rd Floor, Plot 3, Splendor Forum, Jasola District Centre, New Delhi - 110025, India

79 Anson Road, #06-04/06, Singapore 079906

Cambridge University Press is part of the University of Cambridge.

It furthers the University's mission by disseminating knowledge in the pursuit of
education, learning and research at the highest international levels of excellence.

www.cambridge.org
Information on this title: www.cambridge.org/9780521116909

First published 2014
3rd printing 2017

A catalogue record for this publication is available from the British Library

Library of Congress Cataloging in Publication data
Introduction to biomaterials : basic theory with engineering applications /
C. M. Agrawal . . . [et al.].
   p.   cm. – (Cambridge texts in biomedical engineering)
Includes bibliographical references.
ISBN 978-0-521-11690-9 (Hardback)
I. Agrawal, C. Mauli (Chandra Mauli)   II. Series: Cambridge texts in biomedical engineering.
[DNLM: 1. Biocompatible Materials.   2. Biomedical Technology.   QT 37]
R857.M3
610.28´4–dc23   2013016363

ISBN 978-0-521-11690-9 Hardback

Additional resources for this publication at www.cambridge.org/agrawal

I dedicate this work to my parents who taught me to love excellence, and to my wife and children (Sue, Ethan and Serena), who have always supported my pursuit of it.

C. Mauli Agrawal

To my family, who have put up with me all these years.

Joo L. Ong

I express my deepest appreciation for my wife Lindsey, best friend, greatest love, supplier of green limes and good joss.

Mark R. Appleford

I dedicate this work to my wife Priya Devendran, my daughter Manushri Gopinath, and my parents Mani and Bagyam Mani.

Gopinath Mani

Contents

Preface

Biomaterials have helped millions of people achieve a better quality of life in almost all corners of the world. Although the use of biomaterials has been common over many millennia, it was not until the twentieth century that the field of biomaterials finally gained recognition. With the advent of polymers, new processing and machining processes for metals and ceramics, and general advances in technology, there has been an exponential growth in biomaterials-related research and development activity over the past few decades. This activity has led to a plethora of biomaterials-based medical devices, which are now commercially available.

For students in the area of biomaterials, this is an especially exciting time. On the one hand, they have the opportunity to meet and learn from some of the stalwarts and pioneers of the field such as Sam Hulbert, one of the founders of the Society for Biomaterials (SFB). Other greats include Allan Hoffman and Buddy Ratner (biomaterials surfaces), Robert Langer (polymers and tissue engineering), Nicholas Peppas (hydrogels), Jack Lemons (orthopedic/dental implants), Joseph Salamone (contact lenses), and Julio Palmaz (intracoronary stents). Most of these individuals are still active in research and teaching. The authors of this book have been privileged to interact and learn from them in various forums, and students today have the same opportunities. On the other hand, with the current availability of sophisticated processing and characterization technologies, present day students also have the tools to take the field to unprecedented new levels of innovation.

This book has been written as an introduction to biomaterials for college students. It can be used either at the junior/senior levels of undergraduate education or at the graduate level for biomedical engineering students. It is best suited for students who have already taken an introductory course in biology. We have felt the need for a textbook that caters to *all* students interested in biomaterials and does not assume that every student intends to become a biomaterials scientist. This book is a balance between science and engineering, and presents both scientific principles and engineering applications. It does not assume that the student has a background in any particular field of study. Therefore, we first cover the basics of materials in Chapters 1 and 2 followed by basic biological principles in Chapter 3.

After presenting various techniques for the characterization of biomaterials in Chapter 4, we dedicate a chapter each to the discussion of metals, polymers, ceramics, and natural biomaterials (Chapters 5–8). Surface modification methods are presented in Chapter 9, followed by sterilization techniques in Chapter 10. The success of any biomaterial depends on the biological response to it and so protein chemistry, cell–biomaterial interactions, and the effect of biomaterials on tissue response are addressed in Chapter 11. The last three chapters (Chapters 12–14) cover the application of biomaterials in the clinical world; specifically drug delivery systems, tissue engineering, and clinical applications are presented and discussed.

This book has been designed to present enough material so that it can be comfortably covered during a regular length semester-long course. It should provide the student with a concise but comprehensive introduction to biomaterials and lays the foundation for more advanced courses.

The authors would like to thank the following individuals for assisting in a variety of ways in compiling this book: Jordan Kaufmann, Ethan Agrawal, Serena Agrawal, Tim Luukkonen, Amita Shah, Steve Lin, Angee Ong, Kevin Ong, Lisa Actis, Marcello Pilia, and Stefanie Shiels.

1 Introduction

Goals

After reading this chapter the student will understand the following.
- History of implants and biomaterials.
- Various definitions for biomaterials.
- Different types of chemical bonds.
- Basic families of materials.
- Future directions for the progress in biomaterials.

The Rig Veda, one of the four sacred Sanskrit books of ancient India that were compiled between 3500 and 1800 BC, relates the story of a warrior queen named Vishpla, who lost a leg in battle and was fitted with an iron leg after the wound healed. There is also mention of lacerated limbs treated with sutures. Sushruta, a renowned Indian physician from *circa* 600 BC, wrote a very comprehensive treatise describing various ailments as well as surgical techniques. His technique for nose reconstruction using a rotated skin flap is still used in modern times. Sutures made of vegetable fibers, leather, tendons, and horse hair were commonly used in his time. There are also reports of the use of linen sutures in Egypt 4000 years ago.[1]

These ancient records show that, since time immemorial, humans have tried to restore the function of limbs or organs that have ceased to perform adequately due to trauma or disease. Often, this was attempted through the use of materials either made or shaped by humans and used external to the body. These were the earliest form of biomaterials. Although examples of successful external prosthetic devices can be found in history, materials placed inside the body, also known as implants, were usually not viable due to infection. This changed in the 1860s with the advent of aseptic surgical techniques introduced by Dr. J. Lister. The discovery of antibiotics in the mid 1900s also reduced the incidence of infections related to surgery. Today, implants are very successful and are used in a wide variety of applications in the practice of medicine, improving the quality of life for millions and saving countless lives. However, the successes of today have come after a long history of trial and error and scientific endeavor.

Perhaps the most common implants in ancient times were dental in nature and thus, these implants provide an interesting history of progress over the millennia. Human remains from the first century AD recovered from a Gallo-Roman necropolis in France show the use of an iron implant to replace a tooth. In 1923, an archeological dig in Honduras uncovered pieces of shells that were used as a dental implant for a young woman in AD 600. In the Middle East, ivory implants have been discovered with skeletons from the Middle Ages. In more modern times, in the early 1800s, gold posts were placed in sockets immediately after tooth extraction with limited results. In the years following, other metals such as platinum were also investigated. In the 1940s, the Strock brothers from Boston tested Vitallium dental implants. A major breakthrough came in 1952 when Ingvar Branemark from Sweden tried titanium implants and found them to attach well to bone. Such implants provide a post onto which artificial teeth or crowns can be attached. These implants are in use even today.

Box 1.1

- In coronary artery disease, blood vessels supplying blood to the heart tissue are clogged leading to heart attacks.

- In 1977, Andreas Gruentzig introduced a procedure called balloon angioplasty where a balloon mounted on a catheter is inserted into the artery and inflated at the site of the blockage to compress the plaque against the arterial wall thus restoring blood flow. The procedure was very successful but a large percent of the patients suffered a subsequent narrowing of the artery called restenosis.

- In 1978, after hearing a lecture by Gruentzig, Julio Palmaz conceived the idea of a stent – a metal scaffold inserted into the artery and expanded using a balloon to keep the artery propped open.

- He initially experimented with wires wrapped on pins inserted into pencils, and wires soldered together. He finally got his inspiration from a tool left behind in his garage by a worker.

- Working in San Antonio he partnered with a cardiologist, Richard Schatz, and a restaurateur and investor named Phil Ramono to patent and develop the stent.

- Johnson & Johnson licensed the technology, which was introduced into the market in 1991.

- Today the stent is used in more than two million procedures annually and is credited with saving numerous lives.

Table 1.1 Significant developments in the history of biomaterials

Year	Individual(s)	Development
Prehistoric		Various sutures, metal wires, pins
1860s	J. Lister	Aseptic surgical technique
1886	C. Hansmann	Plates, screws for fracture fixation
1887	Adolf Fick	Glass contact lens
1912	W. D. Sherman	Vanadium steel plates and screws
1930–31	A. S. Hayman, M. C. Lidwill	Portable pacemaker
1936–37	A. E. Strock	Vitallium for dental implants
1938	Philip Wiles	Total hip replacement
1949	Sir Harold Ridley	Intraocular lens
1951	G. K. McKee, J. Watson-Farrar	Biomechanically successful total hip design
1952	Ingvar Branemark	Osteointegration of metal implants
1952	A. B. Voorhees	Prosthetic vascular graft
1958	Earl E. Bakken	Wearable pacemaker
1959	Sir John Charnley	Use of polymer in total joints
1960	L. Edwards, A. Starr	Mitral valve replacement
1981	W. Kolff and others	Implantable artificial heart
1980s	Julio Palmaz	Balloon expandable stent

Adapted from reference 2.

Ever since the introduction of aseptic surgical techniques by Lister in the 1860s, there has been rapid progress in the development of biomaterials and implants for a variety of applications in the body (Table 1.1) including dental implants, artificial total joints for hips (Figure 1.1), shoulders (Figure 1.2), and knees (Figure 1.3), spinal implants (Figure 1.4), fracture fixation rods and plates, cardiac pacemakers, stents to keep blood vessels open, and endovascular grafts to repair aneurysms. Although historically metals have been extensively used as biomaterials, there has been a significant increase in the use of ceramics and polymers over the past 40–50 years, thereby leading to a plethora of implants now available for clinical applications.

Figure 1.1

An implant for total hip replacement. The long metal stem is inserted into the medullary cavity of the femur. The metal hemisphere is lined with a polymer and is fixed to the acetabulum on the pelvic side. (Courtesy of Exactech, Inc.)

Figure 1.2

An implant for total shoulder replacement. The long metal stem is inserted into the medullary cavity of the humerus. The polymer component serves as a bearing surface. (Courtesy of Exactech, Inc.)

Figure 1.3

An implant for total knee replacement. The polymer component is made out of ultrahigh molecular weight polyethylene and serves as a bearing surface (Courtesy of Exactech, Inc.)

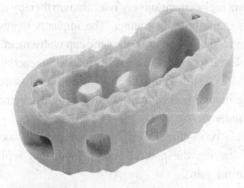

Figure 1.4

A spinal implant made of the polymer polyetheretherketone (PEEK). The fenestrated design facilitates the introduction of bone graft. (Courtesy of Exactech, Inc.)

1.1   Definitions

Biomaterials do not necessarily have to be natural materials as the name may suggest. In 1974, at the 6th Annual International Biomaterials Symposium held at Clemson University, a biomaterial was defined as

... a systemically, pharmacologically inert substance designed for implantation within or incorporation with a living system.[3]

This definition reflected the understanding of the use and function of implants at that time and it imposed the requirement of inertness on the material. Through the years, as science has progressed and resulted in a better understanding of the interaction between biology and materials in the body, the definition of biomaterials has changed as well.

In 1986, at a consensus conference of the European Society for Biomaterials, a biomaterial was defined as

a nonviable material used in a medical device, intended to interact with biological systems.[4]

Perhaps an even more complete definition was provided by Williams[5] as:

a material intended to interface with biological systems to evaluate, treat, augment, or replace any tissue, organ, or function of the body.

Box 1.2

- Today, hip joint replacement surgery is a common therapy to treat hip joints affected by acute arthritis or trauma. The implants represent a ball and socket joint and often comprise a polymeric cup with a metal or ceramic ball moving in it. But it took a lot of experimentation to reach this point.
- In 1925, M. N. Smith-Petersen, a Boston-based surgeon, tried to re-surface the natural ball of the joint using a glass hemisphere but the glass failed under stress.
- In the 1950s a procedure called hemiarthroplasty was popular. This consisted of leaving the natural cup in place but replacing the ball component of the joint.
- In the 1960s, John Charnley from England developed a full hip replacement implant with a Teflon cup and a metal ball attached to a metal stem that was inserted into the marrow cavity. When the Teflon failed, he tried polyethylene with success. The polyethylene–metal combination is still used to this day.

Thus, the definition of biomaterials has changed over the years as progress in science and technology has made it possible to:

- use implants to rapidly restore organ and/or tissue function,

- influence the long term viability of implants by better designing the biomaterial–biology interface, and
- drive the inevitable biological response in desired directions.

1.2 Changing focus

Not only has the definition of biomaterials been changed and refined over the years, but so has the emphasis on biology – the *bio* aspect of biomaterials. From the earliest days of biomaterials through the middle of the twentieth century, work on biomaterials concentrated on basic material properties such as strength, stiffness, ductility, fracture resistance, and corrosion. Over the ensuing decades, there was growing realization that biology plays a significant role in the success of an implant and should be taken into account in its design. In more recent years, emphasis has turned to the manipulation and tailoring of the biological response using principles rooted in the natural sciences. Biomaterials do not necessarily have to be inert in the body, but rather should interact with it at the cellular and molecular levels to ensure the success of the implant.

Although the application of biomaterials is unique, they still fundamentally belong to the same families of materials as those used for various other industries including construction, aerospace and sports equipment. In this respect, biomaterials are no different from most other materials. In the next few sections of this chapter, we will cover some of the basics types of chemical bonds that bind atoms to form materials as well as the major types of materials. An understanding of the various bonds is important, as bonds are responsible for a variety of material properties.

1.3 Types of bonds in materials

At a high level of classification, materials can be divided into two classes: natural and synthetic. At a more fundamental level, materials can be separated into different general categories based on their molecular structure and the type of bonding between their atoms. Since the latter plays a primary role in determining the properties of a biomaterial, it is important to first gain an understanding of the different types of bonds.

1.3.1 Ionic bonds

Ionic bonds are based on one atom donating an electron to form a cation and another atom accepting the electron to form an anion. The charged anion and

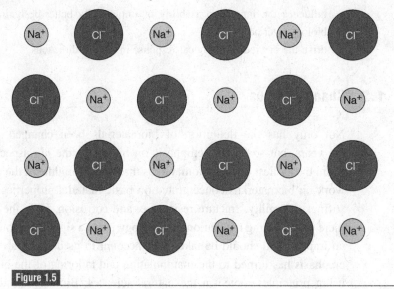

Figure 1.5

Ionic bonding between Na and Cl atoms.

cation are then held together by strong electrostatic attraction to create an ionic bond. A common example of such bonding is in sodium chloride (NaCl), where the sodium (Na) atom transfers one of its valence electrons to the chlorine (Cl) atom (Figure 1.5). After the transfer of an electron, the Na atom becomes a net positively charged ion and is represented as Na^+. The transfer of an electron to the Cl atom renders it a negatively charged ion, which is represented as Cl^-. The Na^+ and Cl^- ions are then held together by Coulombic forces.

To reduce the anion–anion or cation–cation repulsion, each anion is surrounded by as many cations as possible and vice versa. In general, ionic bonding is non-directional and the bond has equal strength in all directions. These properties give rise to highly ordered structures, which result in solids that have high strength and stiffness but are relatively brittle due to the inability of atoms to move in response to external forces. Additionally, the electrons are closely held in place and are not available for charge transfer, making the materials bad conductors of electricity and heat. Bonding energies for ionic bonds are generally high and usually range between 600 and 1500 kJ/mol, which is manifested in high melting temperatures.

1.3.2   Metallic bonds

Metal atoms are good donors of electrons and metallic bonds are characterized by tightly packed positive ions or cores surrounded by electrons. As shown in

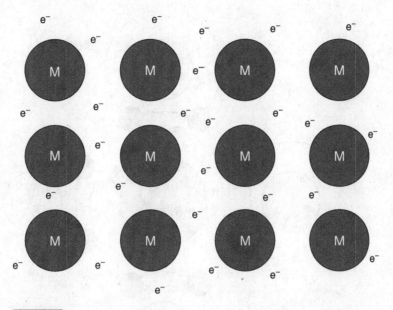

Figure 1.6

Metallic bonds showing positive ions surrounded by electron cloud.

Figure 1.6, the valence electrons are not firmly attached to any one atom but form an electron cloud to float and drift through the material.

The cores are highly organized structures, which consist of the nuclei and the remaining non-valence electrons. Having a net positive charge, these cores would repel each other were it not for the electron cloud surrounding them and working as an "adhesive." Additionally, the loose electron cloud allows for good charge transfer, making the metals good conductors of electricity and heat. Bonding energies can vary in metallic bonds, as shown by the examples of mercury (0.7 eV/atom) and tungsten (8.8 eV/atom). The non-directional nature of this bond gives planes of these ions the ability to slide on each other, thereby allowing the materials to deform under applied forces.

1.3.3 Covalent bonds

Some atoms have the ability to share electrons in their outer shells with other atoms. As shown in Figure 1.7, bonds formed by this sharing of electrons are called covalent bonds. Many molecules comprising dissimilar elements use covalent bonds. Examples include organic materials such as ethylene (C_2H_6), which is a gas, or its polymerized form polyethylene, which is a solid. Other non-metallic

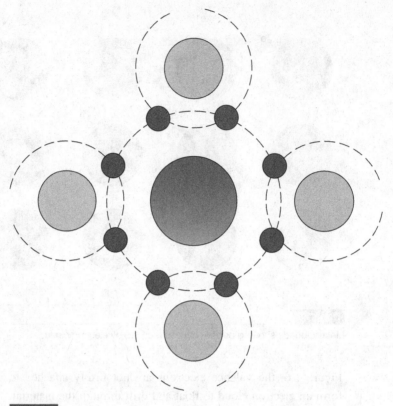

Figure 1.7

Covalent bonds showing atoms sharing electrons in their outer orbits.

elements also form molecules (Cl_2, F_2) using covalent bonds. Carbon and silicon use covalent bonds to form molecules such as silicon carbide.

Covalent bonds are strong and highly directional. However, the rotational ability of atoms around bonds based on single-electron sharing (single bonds) generates flexibility, and materials with such bonds usually have good deformational characteristics. As with ionic bonds, the electrons in these bonds are held in place, and these materials are generally not good conductors of electricity and heat.

1.3.4 Secondary bonds

In addition to the three primary bonds described above, there are weaker bonds such as *van der Waals* and *hydrogen bonds*. Such bonds are based on the attraction between atomic or molecular dipoles. Dipoles are formed when there is uneven or asymmetrical charge distribution, causing separation between the negative and positive portions of the molecule. These dipoles result in electrostatic attraction between adjacent atoms or molecules.

Hydrogen bonds are a special type of bonding between polar molecules and occur when hydrogen is covalently attached to elements such as oxygen, nitrogen or fluorine. Sharing of a single hydrogen electron results in a molecule which is positively charged on one end and thus forms a dipole. Hydrogen bonds are the strongest of the secondary bonds and can have bonding energies as high as 0.52 eV/atom, whereas other secondary bonds typically have bonding energies *circa* 0.1 eV/atom.

- Ionic bonds are the strongest, followed by metallic, covalent, and secondary bonds in descending order. These differences are reflected in the mechanical and physical properties of the materials.
- *Ionic bonds* are formed by the exchange of electrons between atoms. The bonds are non-directional and the materials formed by such bonds have uniform properties in all directions. The bonds are very structured and so the materials are usually strong but brittle.
- The valence electrons in *metallic bonds* form a cloud surrounding the cores of the atoms. Materials with metallic bonds are good conductors of heat and electricity.
- Atoms forming *covalent bonds* share electrons in their outer shells. The bonds are very strong and highly directional. Since the electrons are held in place, the materials are not good conductors.
- *Secondary bonds* are formed due to the attraction between dipoles – asymmetrically charged atomic or molecules. The bond strength is weak.

Secondary bonds do not involve the exchange or sharing of electrons, are less directional, and possess less than 10% of the strength of covalent bonds. However, they can significantly influence the properties of materials, especially polymers.

Now that we have learned about the basic types of bonds that hold atoms together, we will explore the different types of materials that these bonds form in the following sections.

1.4 Types of materials

1.4.1 Ceramics

The word ceramic is derived from the Greek word, "keramos," which stands for pottery. It has its roots in the ancient Sanskrit term "ker" or to burn. These linguistic roots refer to the fact that subjecting these materials to high temperatures often yields desirable properties. Ceramics are solid materials characterized by ionic bonds or combinations of covalent and ionic bonds. Carbons are included in this class. Usually ceramics are compounds of metallic and non-metallic elements and often are nitrides, carbides and oxides. Ceramics may be crystalline or amorphous (non-crystalline) glasses.

As a consequence of the nature of their bonds, ceramics are difficult to deform and have low ductility. Thus, in general they are extremely hard and brittle. Owing to this lack of ductility and hence devoid of the ability to blunt crack tips effectively, ceramics are more susceptible to the presence of sharp microcracks. These sharp cracks act as sources of stress concentration and as stress risers. The effective stress at these crack tips can be several fold higher than the applied stress, thereby resulting in material failure. Ceramics usually have a low resistance to crack propagation and possess low fracture toughness. They have high elastic modulus, high compressive strength but low toughness and tensile strength. If processed into fibers, they can possess superior mechanical properties and are often used as the reinforcing component in composite materials.

In general, ceramics are good insulators of both heat and electricity. They can withstand high temperatures and challenging chemical environments better than most other classes of materials. If properly polished they can also exhibit good wear properties.

As a biomaterial, ceramics are used for total joint prostheses, and tissue engineering scaffolds in the orthopedic field, as crowns for dental implants, and in heart valves among other applications (Table 1.2).

1.4.2 Metals

Metals are inorganic materials with atoms held together by non-directional metallic bonds. These atoms are in a closely packed configuration causing metals to have a high density and be readily visible on X-rays. The cloud of free electrons surrounding the atoms makes metals good conductors of electricity and heat.

The strong nature of the bonds and the close packing renders metals strong and they have high elastic modulus, strength, and melting points. In addition they are ductile and can be formed into complex shapes using a variety of techniques such as machining, forging, casting, and forming. Metals, however, can be susceptible to corrosion, especially in media with chloride ions. This can be of concern when they are used as biomaterials, and so, proper passivation of the metallic biomaterials is required.

Table 1.2   Examples of biomaterials and their applications

Biomaterial	Applications
Ceramics	
Alumina, Zirconia	Total joint prostheses, dental implants, implant coatings
Calcium phosphates	Tissue regeneration scaffolds, cements, drug delivery systems
Metals	
Stainless steel	Fracture fixation devices, stents, wires, dental implants
Cobalt chrome alloy	Total joint prostheses, stents
Titanium and alloys	Total joint prostheses, dental implants
Nitinol	Stents
Polymers (synthetic)	
Polyethylene	Total joints for hips, knees etc.
Polymethylmethacrylate	Bone cement
Polytetrafluoroethylene	Vascular grafts
Polyethylene terephthalate	Vascular grafts, hernia repair meshes
Polydimethyl siloxane	Contact lenses, breast implants
Polyhydroxyethyl methacrylate	Contact lenses
Polyurethane	Coatings for long term implants
Polylactic acid	Resorbable screws
Poly(lactic-co-glycolic) acid	Resorbable sutures
Polycaprolactone	Drug delivery devices
Polymers (natural)	
Collagen	Tissue engineering scaffolds, cosmetic bulking agent
Chitosan	Wound dressings

Of all materials, metals have the longest history as biomaterials and are commonly used as load-bearing implants and fixation devices in the dental and orthopedic fields (Table 1.2). They are also used in a variety of other applications such as guide wires, stents, heart valves, and electrodes.

1.4.3 Polymers

Polymers are organic materials with long molecular chains held together by covalent bonds along their backbone. Usually, they are based on carbon, hydrogen and other non-metallic elements. A single sample of polymer may contain molecular chains of a variety of lengths. These chains may be tangled and held together by covalent bonds (cross-linked) or weaker bonds such as van der Waals and hydrogen bonds. Polymers may be crystalline, amorphous, or a combination of the two. Usually, they are not good conductors of heat or electricity.

Under an applied load, the molecular chains in a polymer have the ability to slide past each other (unless cross-linked) or may have rotational capability about their bonds. This flexibility renders the polymer highly deformable and ductile. However, compared to metals and ceramics, polymers have low elastic modulus, strength, and thermal transition temperatures. From a manufacturing perspective, they can be easily and inexpensively produced and then shaped using a variety of techniques including melt molding.

Polymers have seen an increasing role as biomaterials over the past 50 years and are used extensively as sutures, coatings for leads, and implantable devices such as pacemakers, bearing surfaces for total joints, resorbable meshes, fixation plates and screws, contact lenses, and tissue engineering scaffolds (Table 1.2).

1.4.4 Composites

Composites are materials that comprise two or more different phases or constituents with distinct interfaces. Typically, one of these phases is a "filler," and the other phase is a "matrix." Although these phases can be theoretically identical in composition, composites usually are combinations of different materials. The properties of composites are determined by the shape, size, orientation, and distribution of the constituent materials as well as their relative proportions. Acting as fillers, fibers of a material with high mechanical strength are often dispersed in a matrix of a polymeric material to make a composite material with high strength and toughness. If the reinforcing filler is in the form of fibers, the resulting fiber-reinforced composite can be anisotropic if the fibers, with identical properties, are aligned in a certain direction. Particulate-reinforced composites are more likely to have isotropic properties, that is, the reinforcing particulates having identical properties but are not aligned in a certain direction.

For composites to be successful in enhancing mechanical properties there is a need for effective stress transfer at the interface of the filler and the matrix. Weak interfaces result in separation of the constituents and failure.

Bone is a composite at the microstructural level with ceramic particles (apatite) acting as the filler distributed in a matrix of collagen, which is a natural polymer. This combination gives bone its strength and toughness. However, since the collagen fibers are organized in aligned bundles, bone is anisotropic with different properties in different directions.

1.5   Impact of biomaterials

Biomaterials have played a very significant role in improving the quality of human life, especially for the elderly. A good example is the development of total joint prostheses to replace natural hip and knee joints that have become non-functional due to trauma or conditions such as arthritis. Although surgical experiments with total hip replacements started in the 1890s, it was not until the early 1960s that a successful total hip replacement was developed by Sir John Charnley using stainless steel and ultrahigh molecular weight polyethylene. This was followed by the development of total knee replacements in the period from 1968 to 1972. Today, joint-replacement surgeries are immensely successful in restoring function and relieving pain for thousands of patients who usually are walking within a few days after their surgical procedure. In the USA alone, there were 193 000 total hip arthroplasties and 381 000 total knee replacements in 2002.[6] Compared to 1990, this signified a 62% increase in total hip implants and a 195% increase in total knee implants placed in patients. In the area of orthopedics, success is not limited to just the hip and knee implants, but replacements have also been developed for shoulders, elbows, ankles, and finger joints. Additionally, the use of intramedullary nails has significantly improved the outcomes of surgeries for long bone fractures.

- The main types of bonds that hold atoms together to form materials are:
  - ionic
  - metallic
  - covalent
  - secondary.
- Common materials can be classified as:
  - metals
  - polymers
  - ceramics
  - composites.

Dental implants with good bone integration have had a major influence on dentistry. These implants are used as posts on which porcelain crowns can be fixed to yield artificial teeth that restore function and are also esthetically pleasing.

Another area where biomaterials have improved the quality of life is eye care. Contact lenses have revolutionized the treatment of impaired eyesight. Soft contact lenses have reduced the discomfort associated with the earlier hard lenses and have led to the development of long-wear contacts.

Biomaterials have had a very significant impact on not just improving the quality of life but also prolonging life. In the 1980s Dr. Julio Palmaz, then on the faculty at the University of Texas Health Science Center at San Antonio, developed the balloon expandable coronary artery stent to keep arteries open after occlusions caused by plaque have been cleared using balloon angioplasty. This metal (stainless steel or cobalt chrome alloy) device is now used extensively around the world ·and is instrumental in reducing the incidence of heart attacks and saving countless lives.

Other life-prolonging implants based on successful biomaterials include pace-makers, heart valves, and polymer-based aortic grafts and endovascular stent grafts for treating life-threatening aneurysms.

Biomaterials-based implants and treatment strategies have become an integral part of healthcare today. It is anticipated that the use of biomaterials will continue to grow rapidly as scientists, engineers, and clinicians continue to attain a better understanding of the interaction between materials and biology in the body.

1.6   Future of biomaterials

The field of biomaterials is changing as science yields a more detailed understanding of human biology at the cellular and molecular levels. A large part of current biomaterials research is devoted to the investigation of the interaction between materials and proteins, cells and tissue. All biomaterials are immediately covered by deposits of proteins and other biological molecules upon insertion in the body. These deposits then play a significant role is determining the future response of the body to the implant. Efforts are underway in laboratories around the world to develop hybrid biomaterials where the surfaces of traditional biomaterials (such as those described in Table 1.2 above) are modified by using biochemical moieties such as proteins, peptides, and receptors for antibodies. These modified surfaces can actively interact with the body's molecules and cells to control and drive the biological reaction to an implant in desired directions with the goal of achieving better long-term outcomes. No longer is the goal to make implants simply "tolerable" to the body but the aim is to develop biomaterials that interact in an intelligent fashion with the host biology and bond with, repel, or evaluate their *in vivo* surroundings.

In the future the boundaries between biomaterials and naturally occurring biological entities and systems in the body will further diminish and scientists will be able to reproduce or grow tissue, organs, and biological molecules to restore function.

1.7 Summary

Biomaterials have been in use since ancient times, but their progress was severely limited by lack of aseptic surgical techniques. In recent years, there has been an emphasis on ensuring the long-term viability of implants and how they interact with the biology of the body. To begin to study biomaterials, it is important to first understand the type of chemical bonds that hold atoms together. These basic bond types include ionic, metallic, covalent, and secondary bonds. Each type of bond imparts characteristic properties to the materials it forms and renders them hard or soft, brittle or ductile, and good or bad conductors of electricity and heat. The basic families of materials are ceramics, metals, polymers, and composites. Each of these material types have distinct properties, which determine their behavior and make them more or less appropriate for different applications as a biomaterial. In recent years, scientists and engineers have been working to make biomaterials more responsive to the conditions in the body so that they can interact with its complex biological milieu in an intelligent fashion and to ensure a better outcome for the patient.

References

1. Duraiswami P. K., Orth, M. and Tuli, S. M. (1971). 5000 Years of Orthopaedics in India. *Clinical Orthopaedics and Related Research*, **75**, 269–280.
2. Bhat, Sujata, V. (2002). *Biomaterials*. The Netherlands, Kluwer Academic Publishers, ISBN 0–7923–7058–9.
3. Park, J. B. (1984). *Biomaterials Science and Engineering*. New York, Plenum Press.
4. Williams, D. F. (1987). *Definitions in Biomaterials, Proceedings of a Consensus Conference of the European Society for Biomaterials*. Chester, England, 1986, Volume **4**. New York, Elsevier.
5. Williams, D. F. (1999). *The Williams Dictionary of Biomaterials*. Liverpool, UK, Liverpool University Press.
6. Kurtz, S., Mowat, F., Ong, K. *et al.* (2005). Prevalence of primary and revision total hip and knee arthroplasty in the United States from 1990 through 2002. *J. Bone and Joint Surgery*, **87**, 1487–1497.

Problems

1. What is the main difference between the old definitions for biomaterials and those from more recent years?

2. Which of the following would be considered to be made from a biomaterial according to the modern definitions for a biomaterial:
    (a) pacemaker for the heart,
    (b) walking stick,
    (c) stethoscope,
    (d) vascular graft,
    (e) toothbrush,
    (f) suture?

3. Why are polymers not good conductors of electricity in general?

4. Which of the basic bonds are the most directional in nature? Which are the least?

5. Of the different families of materials which are generally the most hard and brittle? Which are the least?

6. Of ceramics, polymers, and metals, which have low resistance to crack propagation and fracture?

7. If you had to produce an artificial artery, what are the essential properties you will incorporate in your design? Which type of material would you first consider? Why?

8. You have been asked to design the stem component of a total hip joint. This component is inserted into the medullary cavity of the femur and serves as a major load-carrying unit. The load on a hip joint can exceed twice the body weight during normal walking and can be even higher when running or jumping. What family of materials would be a good candidate for this application?

9. Why are the surface properties of a biomaterial very important?

10. Is it preferable to make implants bioactive or inert? Why?

2 Basic properties of materials

Goals

After reading this chapter the student will understand the following.

- The mechanical properties and behavior of materials.
- Material failure under ductile or brittle conditions.
- The time-dependent mechanical behavior of materials.
- Corrosion and its various forms.
- Concepts related to the basic surface properties of materials.

In everyday life, we often define materials using relative terms such as soft or hard, flexible or rigid, strong or weak, tough or brittle, and in a variety of other qualitative ways. What do these terms really mean in the world of engineering? Is such qualitative categorization sufficient for the design and manufacture of a product? The answer is definitely a no, especially when human health or lives may depend on the product. For example, you certainly would not want engineers who are building bridges to pick materials based on such relative and qualitative descriptions! Choices based on much more rigorous, scientific, and *quantitative* characterization would be expected. The same is true when selecting biomaterials.

Material properties can be characterized quantitatively using standardized tests under defined conditions. Once characterized, these properties can be used in conjunction with engineering design techniques to predict the behavior of the engineered product under the expected operating conditions and to ensure that it would function safely. This is important because properties may change based on independent variables such as temperature or rate of application of force. Often a variety of material properties need to be considered for each product.

Some material properties fall in the category of mechanical properties, which predict the deformation, failure behavior, and fracture of materials under the action of forces. As an example, mechanical properties would be very important for a hip-replacement implant because it would be expected to withstand heavy loads generated during walking, and such loads can be as high as several times a person's body weight. For some materials and operating conditions, the electrochemical properties, which characterize corrosion behavior, may be important. This would be true for metals used for a stent that is placed in contact with blood inside an artery. Surface properties such as surface energy, hydrophilicity, or hydrophobicity can be important for implants because these characteristics influence whether cells would attach to the material or determine how proteins will interact with the surface. Since the *in vivo* environment is very complex, the careful selection of appropriate biomaterials based on material properties is essential. In this chapter, we will discuss some of the basic material properties and how they are measured.

2.1   Mechanical properties

The response of a material to deforming forces is characterized by its mechanical properties. These forces can be of various types such as tensile, compressive, torsional or combinations of these forces (Figure 2.1). In general, the type of response varies based on the kind of force applied. Other factors that also play a role in changing the response include the rate of application of force, the temperature of the material, and the surrounding environmental conditions.

To determine the mechanical properties of a material, force versus deformation tests are conducted. In these tests, samples of a material are loaded at a constant rate and both the deformation and the force required to cause that deformation are measured at various time points. To avoid complexities that may otherwise arise, most mechanical properties reported are based on unidirectional and uniaxial loading. However, multi-axial load tests can be conducted if needed. Also under dynamic loading conditions, the direction of the applied force can be reversed and varied to give sinusoidal, saw-tooth, square-wave or other loading patterns.

Materials can also be subjected to time-dependent tests. For example a material sample may be loaded to a certain deformation and that amount of deformation is held constant while the behavior of the force is measured over a period of time. Conversely, a material may be subjected to a constant force and the deformation is recorded as a function of time.

Properties measured under various conditions and testing modes have distinct standard definitions and nomenclature. Standardized testing methods are made available by agencies, such as the American Society for Testing and Materials International (ASTM) and the International Organization for Standardization (ISO), so that properties measured at different test facilities can be meaningfully compared.

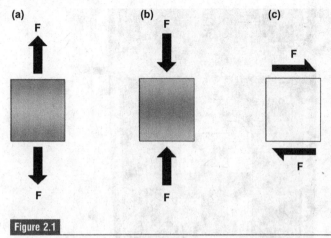

Figure 2.1

Different types of loading: (a) tensile; (b) compressive; (c) shear.

2.1.1 Tensile testing

In the classic form of tensile testing, both ends of a material specimen are clamped between a pair of jaws. The lower jaw or holder is held fixed in place. As shown in Figure 2.2, the upper jaw is attached to the moving crosshead of a tensile tester via a load cell, which measures the applied force. The material specimens tested can be bar shaped with straight sides and uniform rectangular or round cross-sections. Alternatively, the test specimens can have varying cross-sections and be shaped into "dog bone" (rectangular cross-section) or "dumbbell" shapes (circular cross-section), as seen in Figure 2.3. An extensiometer is used to measure the change in the dimension of the test sample along the axis of loading. Using this test, information on force versus deformation is collected and used to develop a stress–strain plot. Shown in Eqs. (2.1) and (2.2) are the formulae for stress and strain, where stress has units of force per unit area, and strain is dimensionless and has no units:

$$\text{stress }(\sigma) = \frac{\text{force}}{\text{original cross-sectional area}}, \tag{2.1}$$

$$\text{strain }(\varepsilon) = \frac{\text{change in length}}{\text{original length}}. \tag{2.2}$$

The gage section of the test sample is measured to determine both its cross-sectional area and the length (Figure 2.4). For straight-edge samples, the gage section is the portion of the sample between the upper and lower jaws. For the shaped dog bone and dumbbell samples, the gage section is the mid-section, which has a uniform but reduced cross-section.

Figure 2.2

A tensile testing machine.

- It is imperative to understand that force and stress are not one and the same.
- A force of the same magnitude can generate very different stress levels based on the area over which it is applied.
- Example: a force applied on an apple through the flat side of a knife blade (large area) may cause a small depression but when the same force is applied through the edge of the blade (small area), it will cut the fruit because of the very high stress generated.

In the seventeenth century, Robert Hooke showed that when a solid material is subjected to a tensile force, its deformation is proportional to the load applied. This is known as Hooke's law and is represented by the linear region of the material's stress–strain curve. An example of a stress–strain curve is shown in

(a)

(b)

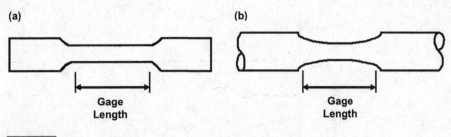

Gage
Length

Gage
Length

Figure 2.3

Standard types of specimens for testing the mechanical properties of a material using a tensile testing machine: (a) dog bone and (b) dumbbell.

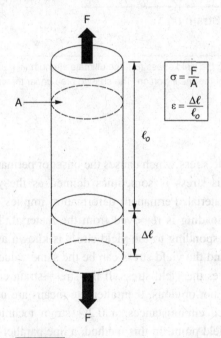

$$\sigma = \frac{F}{A}$$

$$\varepsilon = \frac{\Delta \ell}{\ell_o}$$

Figure 2.4

A rod under tensile loading showing stress and strain.

Figure 2.5. In the linear region of the stress–strain curve, the material is said to be elastic and behaves like a spring. The slope of this linear region of the curve is known as the *elastic modulus*, *E*. Equation (2.3) defines the relationship between elastic modulus, stress, and strain:

$$\sigma = E\varepsilon. \tag{2.3}$$

The *proportional limit* is the point on the stress–strain curve corresponding to the highest stress at which the stress is linearly proportional to the strain.

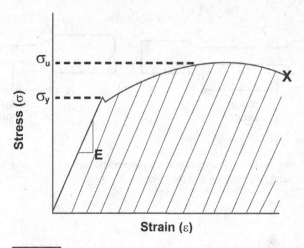

Figure 2.5

A stress–strain curve showing the yield stress (σ_y), the ultimate strength (σ_u), and the elastic modulus (E) given by slope of the linear portion. The shaded area under the curve represents the toughness of the material.

The yield stress, σ_y, is the stress which causes the onset of permanent deformation in the material, and this stress is sometimes defined as the *yield strength* or the elastic limit of a material. Permanent deformation implies deformation that persists even when all loading is removed from the material. The point on the stress–strain curve corresponding to the yield stress is known as the *yield point*. The proportional limit and the yield stress can be the same value, but the proportional limit often precedes the yield stress. If the stress–strain curve is not linear or a clear yield point is not obvious, then alternate means are used to determine the yield stress. In such circumstances, a 0.2% strain technique is generally used to determine the yield point. In this method, a line parallel to the linear part of the stress–strain plot is drawn to intersect the strain axis at 0.2% strain. The point where this line intersects the stress–strain curve is designated as the yield point.

Under elastic conditions, the material reverts to its original dimension when unloaded. On the other hand, if the stress is increased beyond the elastic limit, the material may either undergo failure or become permanently or plastically deformed. Plastic deformation is made possible by the movement of large numbers of atoms or as in the case of polymers, movement of molecular chains. Brittle materials, such as ceramics, do not exhibit plastic deformation or yield points. Other mechanical properties can also be obtained from the stress–strain curve: the maximum stress reached prior to fracture is defined as the *ultimate tensile strength;* the area under the curve up to the failure point is a measure of the work required to cause fracture and is a measure of the *toughness* of the material.

For many ductile metals the stress levels continue to increase after the yield stress is reached and plastic deformation is initiated. This is due to a phenomenon known as *strain hardening*, which is the resistance to further deformation. This resistance is caused by the decreased mobility of atomic planes within the material due to the interaction of multiple dislocations (imperfections in the lattice of atoms in a material). As a result, the stress continues to increase until it reaches the ultimate strength, beyond which the material deforms rapidly until it fractures. It is interesting to note that materials with very few dislocations, such as single crystals, or materials with large number of dislocations tend to have high strength.

- The stress value calculated for most engineering design considerations is called the *engineering stress* and is defined as $\sigma = F/A$, where A represents the original cross-sectional area of the sample.
- In reality the cross-sectional area changes with loading as the sample deforms.
- The *true stress* in the sample can be calculated by using the instantaneous area.

In general, the elastic modulus of polymeric materials is low and they tend to undergo large plastic deformations. The linear region of their stress–strain plot is usually short or may be non-existent. In some polymers, the slope of the stress–strain plot may increase after initial loading due to the alignment of molecular chains. Most polymers have relatively low ultimate strength but deform significantly before fracture and possess high toughness (Figure 2.6). Metals have higher elastic modulus, a clear linear region and often a well-defined yield point, beyond which they undergo plastic deformation. They have high ultimate strength. Ceramic materials have very high elastic modulus values and very little deformation. They usually have high strength but are brittle with very low toughness.

As a material is loaded, it undergoes deformation along the axis of the load. However, since its volume does not change, the longitudinal deformation necessitates a corresponding transverse deformation. For example, if a cylindrical specimen is loaded in tension and undergoes an increase in length, it will also exhibit a simultaneous decrease in diameter in order to maintain a constant volume. This is known as the Poisson Effect. The ratio of the transverse strain

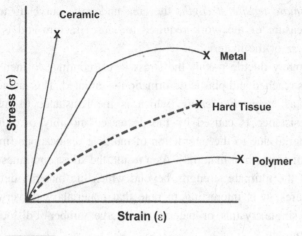

Figure 2.6

Representative stress–strain curves for ceramics, metals, hard tissue, and polymers.

divided by the longitudinal strain is called the *Poisson's ratio*, as denoted by v (nu). For most isotropic simple materials, the Poisson's ratio lies in the range 0.2–0.5. Steels, when loaded within their elastic limits (before yield), usually exhibit a v value of approximately 0.3. Rubber has a v value of nearly 0.5, whereas cork has very little transverse deformation and has a v value close to 0. A negative value of Poisson's ratio indicates a material which, when stretched in one direction, becomes thicker in the perpendicular direction. Such materials are called *auxetic* and are usually composed of macroscopic elements with hinge-like structures. Some polymer foams behave in this manner.

2.1.2   Compressive testing

In compressive testing, block-shaped samples are used instead of the samples described for tensile tests. The same type of test equipment may be used as in the tensile tests but with modified holders or fixtures. Depending on the type of the fixture, the machine crosshead can be lowered or raised to impose compressive loading on the specimen. Often, the failure modes as well as elastic properties differ between tensile and compressive testing modes. For example, ceramics are usually stronger in compression than tension.

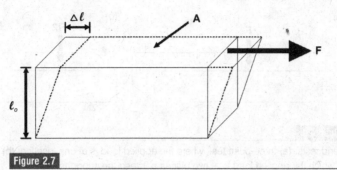

Figure 2.7

Deformation under shear loading.

2.1.3 Shear testing

In shear tests, the load is applied parallel to the surface on which it is acting (Figure 2.7). The deformation is measured in the direction of the force applied. The shear stress is usually denoted by τ, the shear strain by γ, and the shear modulus by G. Hooke's law applies in the elastic region for shear, just as in the case of tension and compression. The relationship between shear stress, shear strain, and shear modulus is represented in Eq. (2.4):

$$\tau = G\gamma. \tag{2.4}$$

For an *isotropic* material, the mechanical properties are uniform in all directions, irrespective of the orientation of the test specimen or direction of testing. For such materials, only two constants, E and G, are needed to fully characterize their stiffness. However, for an *anisotropic* material, the properties differ in different directions, and thus, the results of the tests described above would vary with orientation. Most natural tissues, such as bone, ligaments, skin, and cartilage, are anisotropic in nature.

2.1.4 Bend or flexural tests

Samples can also be tested under a bending load to determine their mechanical properties. In such cases, the samples have a uniform cross-section, which can be rectangular or circular. In the case of a three-point test (Figure 2.8a), the sample is laid on two supports that are placed at a distance, L, apart. A load is imposed at the mid-point of the span, causing the sample to bow. In this test configuration, the material on the inner side of the bow is in compression while that on the outside is under tensile loading (the reader is encouraged to sketch a bent rod to

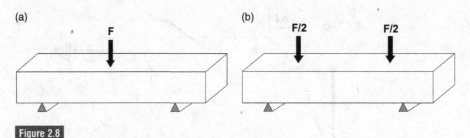

Bend tests: (a) three-point test, where the applied load is at one location; (b) four-point test, in which the applied load is at two places between the supports.

see why this happens). For a beam with a rectangular cross-section, Eqs. (2.5) and (2.6) are used to determine stress and strain, respectively:

$$\sigma_f = 3FL/2bd^2, \tag{2.5}$$

$$\varepsilon_f = 6\Delta d/L^2, \tag{2.6}$$

where σ_f = stress in the outer material at the mid-point (MPa),
ε_f = strain of the outer material at the mid-point,
Δ = maximum deflection of the mid-point under the load (mm),
d = depth or thickness of the beam (mm),
b = width of the beam (mm),
L = length of the span between the two supports (mm), and
F = load applied at mid-point (N).

In a four-point test (Figure 2.8b), the load, F, is delivered through two points ($F/2$ at each point) within the span, L, between the two support points. In this test configuration, the stress is constant between the two loading points. While the four-point test is preferable for determining material properties, the three-point test is more useful for determining the overall strength of a sample or a structure.

2.1.5 Viscoelastic behavior

The properties described above assume that the deformation in a material is instantaneous at the application of a force. However, in some materials, there may be an *additional* time-dependent deformation component due to viscous flow within the material. These materials are known as viscoelastic materials and can be modeled using a combination of a spring and a dashpot, as shown in Figure 2.9.

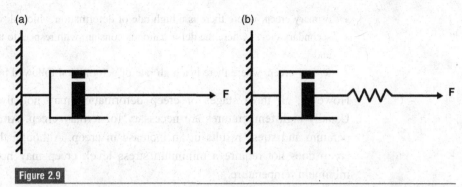

Figure 2.9

Models for a viscoelastic material: (a) cylinder and piston or dashpot model for viscous deformation; (b) dashpot and spring model for combined viscous and elastic deformation.

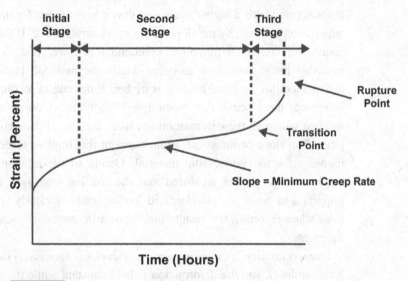

Figure 2.10

The different stages of creep deformation. Upon loading of a viscoelastic material there is initial elastic deformation followed by an increasing rate of strain in primary creep. This is followed by a linear creep strain–time relationship during secondary creep. Lastly, under tertiary creep, there is rapid deformation leading to failure.

Viscoelasticity is manifested in creep and stress relaxation behavior. Creep occurs when a viscoelastic material is loaded quickly to a certain stress, followed by that stress being held constant over time. Under this type of loading, the viscoelastic material will exhibit an instantaneous deformation (the equivalent of the spring in the model), followed by an additional deformation that is proportional to time (viscous dashpot). Creep deformation usually occurs in the following three stages (Figure 2.10):

- primary creep, where there is a high rate of deformation which slows down rapidly,
- secondary creep, where the deformation is constant with respect to time on a log scale, and
- tertiary creep, where there is a high rate of deformation followed by failure.

However, all three stages of creep deformation may not always be evident. Usually high temperatures are necessary for tertiary creep. An increase in temperature and stress results in an increase in creep. Although the occurrence of creep does not require a minimum stress level, creep may not occur below a minimum temperature.

When a viscoelastic material is subjected to a load during a test, the deformation that is measured is a combination of the instantaneous elastic deformation and the viscous deformation that occurs over the duration of the test. This combined deformation yields a higher strain and thus a lower value for the elastic modulus, which is represented by the slope of the stress–strain curve. If the rate of loading is faster, the viscous deformation component is lower due to the shorter time available for viscous flow to occur within the material. Hence, the calculated elastic modulus for rapid loading is higher. If the rate of loading is progressively increased, the viscous deformation will continue to decrease until a limit is reached where all the deformation is elastic and the viscous component is negligible. The slope of the stress–strain curve at this limit represents the true elastic modulus for the viscoelastic material. Owing to the relationship between the mechanical properties measured and the loading rate used for the test, it is important to either use standardized loading rates or clearly specify the loading rates when reporting test results for viscoelastic materials such as polymers and ligaments.

Stress relaxation is observed when a viscoelastic material is rapidly deformed to a certain level, and the deformation is held constant while the stress is measured over time. In this case, the stress will decrease as a function of time. Viscoelasticity is exhibited in most polymers, in some metals, but rarely in ceramics.

2.1.6 Ductile and brittle fracture

Failure in a material can be defined in a variety of ways. For example, the onset of plastic deformation, reaching a certain percent strain, or final fracture can all be considered failure based on the functional constraints placed on a device. Fracture itself may be defined as the separation of a body into multiple pieces and occurs when the cohesive strength between adjoining atoms is exceeded by the applied stress. The failure mode at fracture can be divided into two categories:

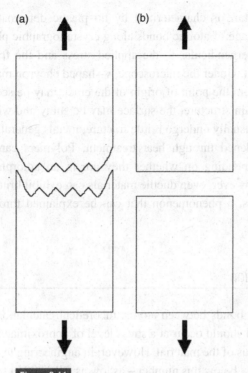

Figure 2.11

Types of fracture: (a) moderately ductile fracture showing a cup–cone failure surface; (b) brittle fracture usually presents a macroscopically flat surface.

ductile or brittle. Whether an engineering material undergoes ductile or brittle fracture depends on its intrinsic material properties as well as its temperature and rate of loading.

Ductile fracture occurs after a material undergoes plastic deformation. As the stress level increases, microdefects, such as voids or cavities, form inside the bulk of the material or on its surface. These defects then coalesce to form a crack, which grows by the assimilation of more defects. This crack usually travels perpendicular to the applied load. Once the crack attains a certain length, it reaches a critical size and starts to propagate rapidly. In the case of a test specimen with a round cross-section, as the fracture approaches the surface, failure occurs at 45° to the applied load along the plane of maximum shear stress. This gives the appearance of a cone shape to one failed section and a cup shape to the mating section, leading to a "cup and cone" fracture (Figure 2.11). When examined under the high magnification of a scanning electron microscope, the fracture surface of a ductile failure reveals a dimpled texture. These dimples represent the microdefects that coalesced to form the crack.

True brittle fracture is characterized by no plastic deformation, and failure occurs by the cleavage of atomic bonds along crystallographic planes. The brittle crack propagates perpendicular to the applied stress and the fracture surface is macroscopically flat. Under the microscope, v-shaped chevron marks or a fan-like pattern radiating from the point of origin of the crack may be seen. For materials with a very fine grain structure, the surface may be shiny and without texture.

While ceramics usually undergo brittle fracture, metals generally exhibit ductile failure unless hardened through heat treatment. Polymers can undergo either type of failure, depending on whether they are in an amorphous glassy or a crystalline state. However, even ductile materials can exhibit brittle fracture under some circumstances, a phenomenon that can be explained through the field of fracture mechanics.

2.1.7 Stress concentration

Based on cohesive bonds between atoms, theoretical calculations show that the failure of a material should occur at a stress level of approximately $E/10$, where E is the elastic modulus of the material. However, in engineering tests, failures occur at stress levels vastly below this number – as low as $E/10\,000$ in some cases. In the first half of the twentieth century, this discrepancy was brought to the forefront by some spectacular failures of ships and airplanes. In a few cases, unexpected, rapidly propagating cracks broke entire ships into two while under normal sailing conditions. In other cases, jet aircraft lost their wings while in flight due to similar cracks. These systems had been designed properly using the engineering standards of that time and yet underwent catastrophic failures.

These failures led to renewed interest and research in the area of fracture of materials. In the 1920s, A. A. Griffith postulated that most materials contain inherent flaws in the form of microscopic voids or cracks. These flaws serve as points of stress concentration or as stress risers. Consequently, the applied stress is magnified at the tips of these flaws, and the degree of stress amplification increases as the radius of the tip of the flaw decreases. Thus, the sharper the crack, the higher will be the stress. The flaws do not have to be microscopic in size to cause stress concentration. In fact, any discontinuity, such as a screw hole in a fracture fixation plate or a change in cross-section in the stem of a total hip prosthesis, can cause stress amplification. The local stress at these stress risers can exceed the theoretical cohesive strength of the material and cause fracture, even though the applied stress or the stress at locations away from the stress riser are below this level. As shown in Eq. (2.7), the stress at the edge of a defect is proportional to the applied stress and the shape and size of the defect:

$$\sigma_t = k \cdot \sigma, \tag{2.7}$$

where σ = applied stress,

σ_t = stress at edge of defect, and

k = factor determined by shape and size of flaw or defect; k has a value of 2 for a hemispherical defect but can be very high for other cases depending on the sharpness (radius) of the crack tip.

As a crack propagates during fracture, it has a sharp tip (small radius) which causes a stress concentration and a high stress ahead of the crack. In the case of a ductile material, this stress causes plastic deformation of the material at the tip, thus blunting the crack and reducing the stress. However, since a brittle material does not exhibit plastic deformation, the crack tip in it remains sharp causing high stresses and rapid failure. Brittle materials are usually stronger under compressive loading compared to other modes of loading because the crack tips are compressed, decreasing the rate of crack propagation.

2.1.8 Fracture toughness

The stress distribution near a flaw can be described in terms of the stress intensity factor, K. This factor helps to determine the stress at any point situated at a distance, r, and at an angle, θ, from the crack tip. For a given stress σ,

$$K = Y\sigma(a)^{1/2}, \tag{2.8}$$

where Y = a constant based on the geometry of the specimen and the crack,

σ = the stress at (r, θ), and

a = the crack length.

A critical value of K defines the conditions for brittle fracture and is known as the *fracture toughness* K_c. Based on the applied stress and the size of the existing crack, brittle fracture will occur when the value of K in Eq. (2.8) exceeds K_c. Thus, fracture toughness is a material property that is a measure of the resistance of the material to brittle fracture in the presence of an existing flaw. During the engineering design of a medical device, the ultimate or yield strengths need to be taken into consideration and an ample factor of safety should be built into the design. However, it is also prudent to consider the fracture toughness and design to prevent brittle fractures. Such analysis is either based on the maximum-sized defect that may potentially exist within a material without detection, or on the size of a known notch, hole, or slit in the designed structure.

2.1.9   Fatigue

When subjected to a single application of load, materials fail at their ultimate strength. However, if they are subjected to multiple load cycles, they can fail at much lower stress levels. This is known as fatigue failure. The stress cycles do not necessarily have to completely unload the specimens but can only partially reduce the load. The loading profile can take a variety of forms including sinusoidal, saw-tooth, or otherwise uneven. Fatigue failure is a serious issue for implants because a variety of devices, such as total joint prostheses, dental implants, and heart valves, are subjected to cyclic loading.

Fatigue failure has several stages and the first stage of fatigue-induced failure is crack initiation. Cracks usually initiate either at existing sites of stress concentration or as the result of an increase in the number of dislocations within the crystal structure of the material, which leads to imperfections and crack formation. The second stage of fatigue failure involves crack growth or propagation which occurs because the tip of the newly initiated crack acts as a stress riser, leading to sharply elevated stresses and material failure at the tip. The cyclic nature of the stress causes the crack growth to occur intermittently with propagation taking place every time the material is loaded. The third stage is the final fracture, and this takes place once the crack grows to a critical size and the critical stress intensity factor is reached.

- Both strength and fracture toughness are material properties.
- Stress intensity is related to fracture toughness in the same manner as stress is related to strength.
- Failure occurs when stress exceeds strength or stress intensity exceeds fracture toughness.
- Fracture toughness should not be confused with toughness, which is a different material property and a measure of the amount of energy required to cause failure.

The fracture surfaces of materials that have failed under fatigue are characteristic in nature and exhibit striations or wave-like patterns emanating from the point of initiation. The distance between these striations is a measure of crack growth for each loading cycle.

As shown in Figure 2.12, the fatigue properties of a material are graphed as a stress (S)–number of cycles to failure (N) curve. The number of cycles is plotted on a logarithmic scale and every point on the curve denotes a 50% probability of failure at that given stress–cycle combination.

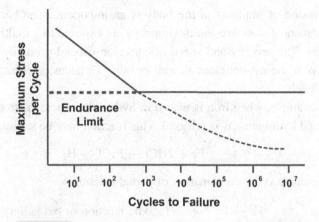

Figure 2.12

Examples of fatigue *S–N* curves. Solid lines show the behavior of a material with an endurance limit (below a certain load the material will not undergo fatigue failure). Dotted line shows typical behavior for a material which does not exhibit a fatigue limit.

In most materials, there is a minimum stress level below which fatigue failure does not occur irrespective of the number of loading cycles. This stress level is known as the *endurance limit*, and the *S–N* plot becomes horizontal at this point. However, certain metals, such as aluminum and its alloys, do not have an endurance limit, and as a result, their *S–N* curve shows a steady decrease in the failure stress with increasing loading cycles. The use of such materials should be avoided where cyclic loading is anticipated.

2.2 Electrochemical properties

2.2.1 Corrosion

Corrosion may be defined *as the chemical deterioration of a material as a result of interaction with its environment*. Although this broad definition covers environment-related damage in all materials, including both ceramics and plastics, this section will be limited to corrosion in metals, while the other material types will be addressed in later chapters. Corrosion may be broadly divided into the following two categories:

- wet corrosion, which takes place in the presence of a liquid, and
- dry corrosion, which occurs due to corrosive gases.

Of the above two different categories of corrosion, only wet corrosion is relevant in the case of biomaterials.

Corrosion of implants in the body is an important issue because the *in vivo* environment is complex and the implants are exposed to a multitude of chemical moieties. The severity and speed of corrosion depend primarily on the chemical makeup of the environment as well as other variables, such as temperature and stress levels.

For example, when iron is placed in hydrochloric acid, a strong reaction takes place and hydrogen gas is released. This reaction may be summarized as below:

$$Fe + 2HCl \rightarrow FeCl_2 + H_2. \tag{2.9}$$

This outcome is the combination of partial reactions:

$$Fe \rightarrow Fe^{+2} + 2e \text{ (anodic reaction or oxidation)}, \tag{2.10}$$

$$2H^+ + 2e \rightarrow H_2 \text{ (cathodic reaction or reduction)}, \tag{2.11}$$

$$Fe^{+2} + 2Cl^- \rightarrow FeCl_2. \tag{2.12}$$

In the above equations, e represents an electron. The net result is the loss of iron (Eq. (2.10)) due to corrosion and its conversion to ferrous chloride (Eq. (2.12)). In the corrosion process, both oxidation and reduction processes take place simultaneously and at equal rates.

Another example is the corrosion of iron in the presence of an aqueous medium with dissolved oxygen. The oxidation and reduction of this corrosion phenomenon is represented in the following equations:

$$2Fe \rightarrow 2Fe^{+2} + 4e \text{ (oxidation)}, \tag{2.13}$$

$$O_2 + 2H_2O + 4e \rightarrow 4OH^- \text{ (reduction)}, \tag{2.14}$$

$$2Fe^{+2} + 4OH^- \rightarrow 2Fe(OH)_2. \tag{2.15}$$

Since ferrous hydroxide [$Fe(OH)_2$] is unstable, this by-product is further oxidized to form ferric hydroxide [$Fe(OH)_3$], which is commonly known as rust:

$$2Fe(OH)_2 + H_2O + \frac{1}{2}O_2 \rightarrow 2Fe(OH)_3. \tag{2.16}$$

Equation (2.17) can be used to describe the general oxidation reaction of a metal (M):

$$M \rightarrow M^{+n} + ne. \tag{2.17}$$

The reduction reaction can take several forms:

- hydrogen evolution, as described in Eq. (2.11),
- oxygen reduction in netural or basic solutions, as described in Eq. (2.14), and

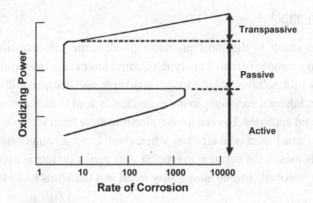

Figure 2.13

Corrosion behavior of an active–passive material as a function of electrode potential.

- oxygen reduction in acidic solutions, as described in Eq. (2.18):

$$O_2 + 4H^+ + 4e \rightarrow 2H_2O. \tag{2.18}$$

Passivity

Metallic passivity is an interesting phenomenon that is complex in nature, and its reasons are difficult to explain. Usually, as the oxidizing power of a solution is increased, the rate of corrosion increases exponentially. This is known as the active zone. However, in certain metals, such as iron, chromium, titanium, and their alloys, an increase in the oxidizing power beyond a certain critical point results in a decrease in corrosion rates (Figure 2.13). This decrease in corrosion rate can be significant and metals tend to behave as inert noble metals. The active region is followed by the metal reaching a stable passive region wherein increasing the oxidizing power of the solution has no effect on the corrosion rate. However, further increase in the oxidizing power beyond the stable passive region results in an increase in the corrosion rate once again. This zone is known as the transpassive region.

2.2.2 Types of corrosion

Wet corrosion can be categorized into several different forms. Some of these forms of corrosion are more relevant to metallic implants than others, but all forms need to be considered when designing products.

Uniform attack

Uniform attack is the most prevalent of all forms of corrosion, and its most common example is rust. This type of corrosion occurs uniformly over a surface exposed to a solution which makes oxidative and reductive chemical reactions possible. Inherent variations in the material may lead to different sections acting as anodes and cathodes. In such a case, electrons flow from the anode to the cathode. Uniform attack occurs in all cases where there is not an equilibrium concentration of metals ions in the surrounding media. The corrosion rate is usually slow, and in the case of cobalt chrome alloys, may result in a thickness loss of 0.1 μm per year.

Galvanic corrosion

Metals have a characteristic electrical potential. Under ideal conditions, these potentials can be measured and referenced against the hydrogen electrode. The potential for the hydrogen electrode is defined as zero. When two metals with different potentials or electromotive force (EMF) are connected, there is a possibility of current flow. Immersing this pair of metals in an ionic solution leads to the completion of the electrical loop and corrosion ensues. The electrode potentials for various different metals are listed in Table 2.1. However, it is not possible to find these potentials for alloys with reactive components. In addition to the electrical potential, the environment also has an effect on the corrosion of the metals. Thus, a practical grouping of metals has been developed using seawater and is known as the galvanic series (Table 2.2). There is not a significant difference between the positions of metals on the EMF and the galvanic series. At the top of the list are the noble metals or cathodic materials, while the anodic materials are listed at the bottom. If the metals or alloys are in their passive state, they are listed further up in the series and closer to the noble metals.

The farther apart the two metals or alloys are in the galvanic series, the greater the difference will be in their potential. The greater the potential difference, the higher the rate of corrosion will be when these two metals are in contact or connected. On the other hand, materials that are very close to each other in the galvanic series pose little danger of galvanic corrosion. The degree of corrosion also depends on the distance from the junction of the two metals, with the maximum corrosion occurring at the junction site and decreasing at spots farther away. The relative surface area of the anode and cathode also influence the amount of corrosion, with increased corrosion observed when the anode is much smaller than the cathode.

Implants made of more than one metal, or implanted metal devices of different metals but are in contact with each other can undergo galvanic corrosion.

Table 2.1 Standard EMF series for metals

Metal	Metal ion	Electrode potential (volts) vs. normal hydrogen electrode at 25 °C	
Gold	Au^{+3}	+1.498	
Platinum	Pt^{+2}	+1.2	
Silver	Ag^+	+0.799	
Mercury	Hg_2^{+2}	+0.788	Noble or cathodic
Hydrogen	H^+	0.000	
Lead	Pb^{+2}	−0.126	Active or anodic
Nickel	Ni^{+2}	−0.250	
Cobalt	Co^{+2}	−0.277	
Iron	Fe^{+2}	−0.440	
Chromium	Cr^{+3}	−0.744	
Zinc	Zn^{+2}	−0.763	
Aluminum	Al^{+3}	−1.662	
Magnesium	Mg^{+2}	−2.363	
Sodium	Na^+	−2.714	
Potassium	K^+	−2.925	

For additional information consult Mars G. Fontana and Norbert D. Greene, *Corrosion Engineering*, ISBN-13: 978-0-071-00360-5, McGraw Hill Higher Education, 1986.

This may occur in implants such as modular hip joints, dental implants, fracture fixation plate and screw combinations, or an intramedullary nail in contact with a cerclage wire holding pieces of bone together. Sometimes, an inclusion or impurity introduced into a metal implant during fabrication can act as the second metal and set up a galvanic corrosion pair.

Crevice corrosion

Crevice corrosion occurs in restricted areas where there is fluid stagnation, such as in cracks, modular press-fit joints, beneath screw heads or underneath gaskets. The corrosion is driven by reactions described in Eq. (2.13) and Eq. (2.14).

Table 2.2   Galvanic series for select metals

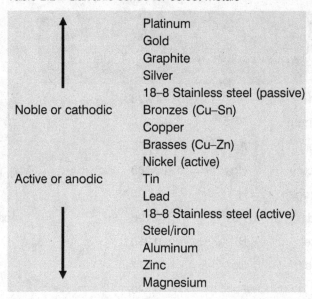

Noble or cathodic

Platinum
Gold
Graphite
Silver
18–8 Stainless steel (passive)
Bronzes (Cu–Sn)
Copper
Brasses (Cu–Zn)
Nickel (active)

Active or anodic

Tin
Lead
18–8 Stainless steel (active)
Steel/iron
Aluminum
Zinc
Magnesium

With time, the oxygen in the crevice gets depleted due to limited fluid flow and the reduction reaction slows down *in that locale*. The reduction reaction continues unfettered in areas external to the crevice, and the number of metal ions going into solution does not decrease significantly to maintain balance between the two processes in the crevice. This leads to a buildup of positive metal ions in the crevice, attracting negatively charged chloride ions. The presence of chloride ions in the crevice, for reasons not fully understood, further accelerates the dissolution of metal ions in the crevice, causing even more chloride ions to migrate, and setting up an autocatalytic process. Crevice corrosion is very localized and causes severe localized damage while the surrounding areas of the metal remain relatively untouched.

Pitting corrosion

Pitting corrosion is similar to crevice corrosion in its mechanism and can be initiated by scratches, inclusions or other damage. Usually, it starts on horizontal surfaces, and its growth follows gravity. Once initiated, the corrosion quickly forms surface pits and can grow at ever increasing rates, causing holes in the metal. However, since the corrosion is on a surface and not protected within a crevice, the initiation sites can be unstable and easily disrupted or stopped by any change in convective flow which may alter ion concentration. The pits are usually numerous in number and small in size. Often the pits change the surface texture of the implant giving it a matte-finish look. Improved manufacturing processes have led to fewer inclusions in implants and a reduction in pitting corrosion.

Intergranular corrosion

Cast or forged metals and alloys contain continuous regions known as grains. The size and structure of these grains depend on the material and also on heat treatment. During manufacturing, the size of the grains is manipulated intentionally by varying heating/cooling rates to obtain desired properties such as increased hardness or ductility. The areas at the edges of the grains are disordered and are known as grain boundaries. These boundaries often contain material compositions different from the grains and may contain more impurities. For example, the chromium content is depleted at the grain boundaries in stainless steel. These variations, coupled with the high surface area of the grain boundaries, render them more reactive and hence subject to corrosion. Such localized corrosion at or near the grain boundaries with limited corrosion in the grains is known as *intergranular corrosion* and can lead to a decrease in mechanical properties. In extreme cases, the metal can disintegrate as the grain boundaries are destroyed, and the grains fall out. The incidence of intergranular corrosion can be decreased by using proper heat treatment and by reducing impurities and inclusions.

Leaching

Leaching is the selective dissolution of a constituent of an alloy from the grains. This may happen for a variety of reasons: the device or implant may be immersed in a medium that selectively attacks one component; or more than one phase may exist within the alloy. These phases, although having the same constituents, may have different relative compositions with different dissolution properties. Leaching is often a concern when nickel-based alloys are used as biomaterials, due to documented adverse responses to nickel in a segment of the patient population.

Erosion corrosion

Erosion corrosion is defined as the acceleration in corrosion damage when there is relative movement between a metal surface and the corroding media. The actual corrosion may involve any of the mechanisms described above, but the relative movement may remove the corrosion products or protective passive layers thereby exposing new metal surface to attack.

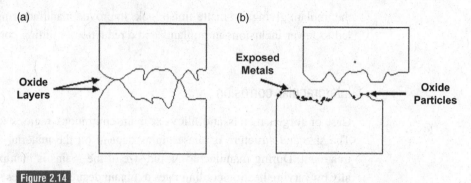

Figure 2.14

Fretting corrosion showing the breakdown of oxide layers due to relative motion caused by repetitive movement or vibration.

Fretting corrosion

In many materials, the surface reacts with oxygen to form oxides. These oxides then form a protective layer, preventing the metal underneath from further chemical attack. Fretting corrosion is a form of erosion corrosion, which occurs at contact areas between metal components subject to vibration or slip. The relative movement causes mechanical wear of the components and repeatedly removes the protective oxide layers, which form every time new metal surface is exposed and reacts chemically. This cyclic process can cause rapid damage. Fretting corrosion can be observed at the interface of screws with fracture fixation plates and intermedullary rods, and between press-fit components of modular hip prostheses.

Stress corrosion

Stress corrosion can occur when there is a tensile force applied to a metal. The applied stress can change the electrochemical nature of the metal, thus setting up galvanic corrosion sites. For example, when a metal implant is loaded in bending *in vivo*, one side is subjected to tensile forces while the opposite side is loaded in compression. This difference in loading can establish a difference in the electrochemical potential between these two sides leading to corrosion.

Stress corrosion can also lead to the initiation and rapid progression of cracks. This may be related to changes in the electrochemical potentials in areas near the crack tip, which are subjected to very high stresses due to stress concentrations. Alternatively, these high stresses may rupture the protective oxide films formed at the crack tips, causing more corrosion damage and contributing to crack growth in a synergistic manner. The factors influencing stress corrosion include stress, composition of fluid, and type of metal. Areas of implants that are subject to stress concentration such as screw holes, sudden changes in cross-section, and scratches are susceptible to stress corrosion.

2.3   Surface properties

The surface of a material differs from the bulk material in many respects. The atoms comprising the surface layer may not have all their bonds satisfied, which lead to a net force field. Surfaces should be less favorable energetically than the bulk. Otherwise, there would be a driving force to create more surfaces. Consequently, surfaces have more energy than the bulk, and *surface energy* is a measure of this extra energy. Surfaces of materials are highly reactive, and this leads to oxidation and other chemical changes which then render the surface properties different from those of the bulk. Moreover, the arrangements of the atoms at the surface may be significantly different from that in the bulk due to the effects of the adjacent environment and foreign atoms to which they are exposed. Additionally, surfaces often contain contaminants that they acquire from the surrounding environment. In general, the surface properties of a material may differ from those of the bulk. Since the surface is usually very thin compared to the bulk for most medical devices, the techniques used for characterizing the surface are usually different and need higher sensitivity.

The physical nature of surfaces may vary in many aspects. For example, the surface roughness of the material may vary from smooth to rough. In addition, surface morphology could vary from smooth to a stepped or pockmarked roughness. The surface may contain different types of atoms, and their spatial arrangement may vary leading to different chemistries. The arrangement of atoms or molecules in the depth direction may also vary. There may be different domains or phase separations on the surface. The surface may have regions that are crystalline and others that are amorphous or disordered. These variations are all important considerations when designing an implantable device because the body's response to the implant is mediated by the surface and its properties. In fact, upon implantation, the outer surface of a device is rapidly coated by proteins and other molecules. These protein-coated or molecule-coated surfaces then influence the cellular and immune response. Designing and controlling surface properties can have a significant role in the long-term success of an implant. Presented here are brief descriptions of contact angle and hardness measurement techniques as examples of how surface properties are measured. Methods for characterizing surfaces and surface modification techniques are covered in more detail in Chapters 4 and 9, respectively.

2.3.1 Contact angle

The principles of contact angle measurements are provided in Chapter 4. Briefly, when a drop of liquid is placed on a surface, it may either spread out or remain in the form of a droplet. This outcome depends on whether the attraction between the liquid molecules (cohesive force) is greater or lower than the attraction between the liquid and the substrate atoms (adhesive force). This phenomenon is exemplified by the behavior of a water droplet on a car surface before and after it has been waxed. Using Young's equation shown in Eq. (2.19), a force balance can be used to calculate the surface energy of the substrate:

$$\gamma_{sg} = \gamma_{sl} + \gamma_{lv} \cos \theta, \tag{2.19}$$

where θ is the angle between the liquid–solid interface and the liquid–vapor interface, γ_{sg} is the surface energy, γ_{sl} is the solid–liquid interfacial tension, and γ_{lv} is the liquid–vapor surface tension. Complete wetting occurs when $\theta = 0°$, and a complete non-wetting surface is characterized by $\theta = 180°$.

2.3.2 Hardness

Hardness is a measure of the ability of a material to resist localized plastic deformation. This property is most relevant at the surface of the material. As we have discussed above, the surface of a material can have very different properties compared to the bulk. Thus, although hardness can be inferred from the E and σ_y of the bulk material, it is best to measure it directly on the surface. Hardness is an important property when contact between two different materials is anticipated during use.

Hardness is often measured using an indentation technique. There are various scales for indentation hardness. However, these different scales use variations of the same measuring protocol, wherein a standard material of a known geometry is used to indent the surface of the test material under a known standard load. The hardness is determined either from the depth or area of the indentation, depending on the scale used. For example, the Rockwell scales (R_A, R_C, etc.) use a variety of indentors (spheres or cones) of known dimensions, and the hardness number is related to the inverse of the depth of the indentation. Thus, the more shallow the indentation, the higher the hardness. Other scales that use indentation include the Vickers test which uses a square pyramidal indentor, and the Brinell Hardness test that uses spheres for indentation. The Brinell test covers a relatively large area and is not affected as much by grain boundaries.

The Mohs Hardness test is one of the oldest (*circa* 1812) hardness tests and is based on observing whether the test material can be scratched by another material of known hardness. The hardness is given a numerical value or a rank based on a scale related to ten known minerals. This test is only a relative test, and for most engineering materials, the other tests described above are used.

2.4 Summary

As biomaterials are fashioned into useful devices or implants, it is important that during the design process, their behavior under operating conditions is carefully predicted and deemed to be safe and efficacious. Whether the device will be able to withstand the mechanical loads, whether the implant surface will adversely reacts to the electrochemical conditions in its immediate environment, or how the implant surface interacts with the biological milieu surrounding it, will all depend on the various intrinsic material properties of the biomaterials comprising the implant. By using standard tests to measure the material properties of a biomaterial, one can predict to a large degree how the material will behave, thus minimizing the risk of failure in use. This is the basis of all engineering design. For biomaterials that are commercially available, the material properties are usually available on specification sheets provided by the source. If the material has been self-produced, then it is important to conduct standard tests to fully characterize its relevant properties.

Suggested reading

- Black, J. (1988). *Orthopedic Biomaterials in Research and Practice*. Churchill Livingstone Inc., ISBN 0–443–08485–8.
- Callister, W. D. and Rethwisch, D. G. (2007). *Materials Science and Engineering: An Introduction*. John Wiley & Sons, ISBN-13: 978–0–471–73696–7.
- Fontana, M. G. and Greene, N. D. (1986). *Corrosion Engineering*. McGraw Hill Higher Education, ISBN-13: 978–0–071–00360–5.
- Hertzberg, R. W. (1995). *Deformation and Fracture Mechanics of Engineering Materials*. John Wiley & Sons, ISBN 0–471–63589–8.

Problems

1. You have two metal specimens both of which are cylindrical with a diameter of 20 mm and a length of 200 mm. One is made of a titanium alloy (elastic modulus, $E = 100$ GPa) and the other out of stainless steel ($E = 200$ GPa). Both of them are subjected to a tensile force of 500 N. Which of them will have a higher stress? Which will develop a higher strain? Given the information provided can you tell which of the specimens will be stronger?

2. The figure below shows the stress–strain curves for two different materials A and B. Which of these materials:
    (a) is stronger,
    (b) is more ductile,
    (c) will absorb more energy prior to failure?

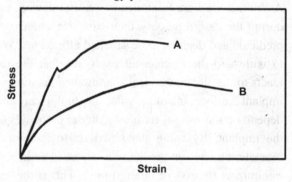

3. What is the difference between isotropic and anisotropic materials?
4. A bar made of a brittle material is tested in a three-point bend test. Where is the failure likely to occur – on the inside (loading side) or outside? Why?
5. Describe the three stages of creep.
6. Creep and stress relaxation are both manifestations of viscoelasticity. What is the difference between them?
7. In one study the critical stress intensity factor for human bone was calculated to be 4.05 MN/m$^{3/2}$. If the value of Y in Eq. (2.8) is 1.2 and there may be a 2 mm crack present in a bone specimen, what would be the maximum tensile stress that can be applied before fracture occurs?
8. You have been asked to examine a broken fracture fixation plate and determine the reason for failure. If the fracture surface shows a dimpled structure what kind of failure would you suspect? If the surface had a wave-like pattern? If the surface was smooth?
9. An engineer is designing a total hip implant. She intends to make the femoral stem out of titanium because it forms a good interface with bone. However, since titanium does not have optimum wear properties she intends

to make the ball component out of stainless steel and attach it to the stem using a press fit. Should she be worried about corrosion? If so, what types of corrosion could take place?

10. A polymer surface was treated using an oxygen based gas plasma technique. Contact angle measurements prior to treatment showed $\theta = 120°$. Post treatment θ was 43°. Did the treatment make the surface more or less hydrophilic?

11. A dumbbell test specimen with a gage length of 100 mm and circular cross-section with a diameter of 10 mm is loaded in tension in a tensile testing machine. Upon application of a tensile force of 1000 N, the gage length extends to 100.5 mm. Calculate the stress and strain in the gage section. If the deformation is all within the linear elastic range, what is the elastic modulus of the material?

3 Biological systems

Goals

After reading this chapter, students will understand the following.

- Key terms such as homeostasis, reductionism, flow, and flux in terms of biological systems.
- Fick's first law of diffusion and how it is useful for cellular interaction with biomaterials.
- The principal functions of the plasma membrane.
- The major classes and operations of cell junctions.
- Cell signaling pathways and secondary messengers.
- Commonly used biological testing techniques in biology–material interactions.

3.1 The biological environment

One of the most basic principles of systems biology is the concept of *reductionism* where behavior of a "whole" can be explained by the corresponding behavior of its "parts." By successively deconstructing the organism and studying its components, biologists can explain the function of the body in terms of its organ systems, cells, subcellular organelles, macro- and biochemicals, and so forth until we reach fundamental particles. This concept of reductionism is among the most common techniques used to study biomaterials as we explore their influence at the biochemical, protein, cellular, and whole-organism levels. However, great care must be taken as reductionism leads to more and more missing information with each step taken. As an example, *in vitro* experimentation often lacks the cellular complexity and hormonal control of *in vivo* behavior and can lead to false or

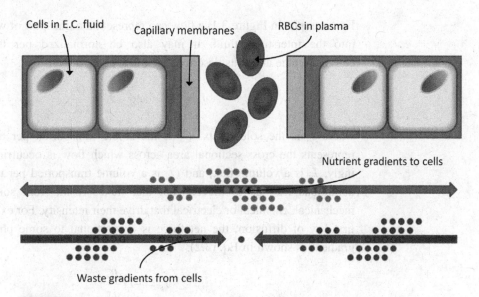

Cells in E.C. fluid   Capillary membranes   RBCs in plasma

Nutrient gradients to cells

Waste gradients from cells

Figure 3.1

Schematic drawing of cell transport of waste and nutrients across the cell membrane, into the
interstitial fluid, crossing other barriers such as the capillary membrane and entering the plasma with
binding to red blood cells. Note that such transport occurs across chemical gradients and may
be quantified with knowledge of the specific surface areas, properties of the membranes,
and relative quantities of the substances.

contradictory conclusions. Despite the limitations of reductionism, it is among
our best tools for understanding and testing biomaterials when care is taken to
identify the assumptions of each experiment and result. In this chapter, we will
focus on the communication systems at the cellular level and its implications
for biomaterials.

In addition to breaking complex biological behavior down into its constituent
parts, physical properties and relationships can also be used in parallel to help
describe the biological system. A central premise of the cell-based organism is that
a constant internal environment must be maintained. This constant state is known
as *homeostasis*. A basic representation of this is shown in Figure 3.1 where
nutrients and waste follow gradients to and from the cell, passing into the
interstitial fluid, crossing other barriers such as a capillary membrane, entering
the blood plasma, and potentially binding with a red blood cell. Both active and
passive processes mediate these transports, and all can be modeled using physical
and chemical systems.

Transport is essential to the maintenance of cell homeostasis and is described as
either a *flux* or *flow*. Transport flow is often symbolized by the variable Q and
represents material transported per unit time as either a fluid volume or a mass of
solute. Flow can be normalized by dividing it by the area across which it occurs.

For example, in Figure 3.1, a flow can represent the transport of waste from a cell into the interstitial fluid. It may also be normalized per the area of the cell membrane and thus described as a flux defined by

$$J_s = \frac{Q_s}{A} \quad \text{and} \quad J_v = \frac{Q_v}{A}, \tag{3.1}$$

where J_s is the solute flux, Q_s is the solute transported per unit time and A represents the cross-sectional area across which flow is occurring. Correspondingly, J_v is a volume flux, and Q_v is a volume transported per unit time. These definitions lead us to the concept that fluxes result from some force, often mechanical, chemical or electrical that drive their intensity. For example, in Fick's first law of diffusion, the net force is proportional to some phenomenological gradient as shown in Eq. (3.2):

$$J_s = -D\frac{dC}{dx}. \tag{3.2}$$

For simplicity, the above equation assumes one direction rather than the true vector definition of a flux where both magnitude and direction are accounted for. In Fick's first law, J_s denotes the solute flux, D is the diffusion coefficient defined below and dC/dx is the concentration gradient. The general form of this equation should look familiar as it mirrors the Fourier's law of heat transfer, Ohm's law, and also works for pressure driven flow. Each of these laws can be applied to biological systems and can define fluxes depending on a calculated coefficient such as D. For example, in Eq. (3.3) D is defined by the Stokes–Einstein equation for diffusion of spherical particles through fluids with relatively low Reynolds numbers:

$$D = \frac{kT}{6\pi\eta a_s}, \tag{3.3}$$

where k is Boltzmann's constant, T is the absolute temperature, π is a geometric ratio, η is the fluid viscosity, and a_s is the radius of the spherical particle. Note that Eq. (3.3) describes how a flux moves downhill, from high to low concentrations, and is valid when the concentration difference is the only driving force. However, these types of fundamental equations may also be used to sum the effect of multiple flow processes on a single system and can be further developed by accounting for both passive and active transport mechanisms which alter the overall transport fluxes. These diffusion equations can help us understand the rates and behaviors of solute, ion, and volume transport through membranes. They are among the first tools used in quantifying transport behavior in cell–biomaterial interactions, whether they model the release of ions from a surface, the transport of proteins into a hydrogel, or the degradation rate of a biodegradable material. However, they are not the only method and much of this chapter will focus on the various active communication pathways which cells use to interact with their environment.

3.2 Genetic regulation and control systems

While this chapter aims to focus on cell communication pathways that are relevant for biomaterial design, a brief introduction to the genetic systems is necessary to understand how this communication is ultimately processed by the cell. Cells display an outward appearance known as a *phenotype* which is often distinct for a given behavior or function but which may or may not be reflective of the cells genetic information or *genotype*. From fundamental biology, we know that the genome of nearly all cells in the body consists of deoxyribonucleic acid (DNA), which is organized into paired chromosomes with specific genes to encode for all the proteins expressed in an organism. It is the regulation of this system that ultimately defines the type of cell and is responsible for the diversity of phenotypes observed despite the identical genotype.

The wealth of chemical, biomolecular, physical, and mechanical inputs the cell experiences are all interpreted as environmental signals to alter the rates of intracellular activities. At the genetic level, this involves alterations to the transcription process such as changes in chromatin remodeling, transcription factors, and regulatory sequences. Closely related are changes that occur post transcription, such as mRNA processing, compartmentalization, degradation rates, and translation, with each processing possessing its own distinct control system.

3.3 The plasma membrane

It would be logical to assume that the boundary between a cell and its environment would necessitate a robust and impenetrable barrier to protect the delicate compartmental workings of the cell. However, in reality, this barrier, the plasma membrane, is permeable and only 5–10 nm in thickness. This thin structure cannot even be observed by using traditional light microscopy, and was first seen by J. D. Robertson who used an electron microscope to resolve its three layered design known as the lipid bilayer. The main functions of the membrane are to:

• create a selectively permeable barrier for communication, and substance exchange,
• provide transportation of solutes via the proteins and pores spanning the membrane,

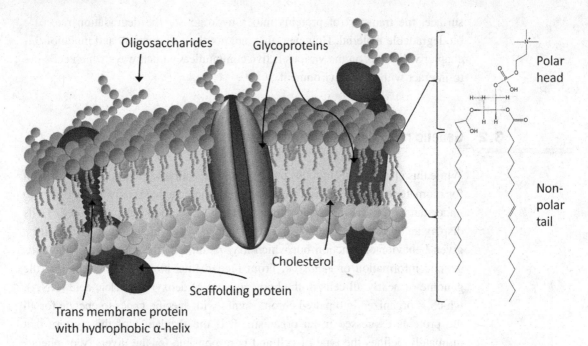

Figure 3.2

Drawing of the lipid bilayer membrane consisting of multiple phospholipids organized with their hydrophobic carbon tails pointed toward the interior of the membrane and their polar hydrophilic head groups facing the extra- and intracellular regions. The drawing represents the diversity of the membrane structure with many varieties of lipids (regionally different), glycoproteins, cholesterols, and scaffolding proteins present in the structure. Chemical structure of the most common glycoprotein of the plasma membrane, phosphatidylcholine, with its polar hydrophilic head group and hydrocarbon hydrophobic chain tail is shown on the right.

- create compartmentalization, to not only separate internal cell systems from the surrounding environment but to establish internal frameworks for cell partitions,
- provide intercellular and signaling response via connections to other cells,
- provide intercellular and signaling response through surface receptors for molecules,
- provide a substrate for biochemical actions by linking structural scaffolds internal to the cell with local protein organization.

3.3.1   Membranes are phospholipid layers

Membrane structures are assembled as a core bimolecular layer of phospholipids with their polar, water-soluble head groups facing out and their hydrophobic fatty acid tails facing the interior (Figure 3.2). The first proposal of this lipid bilayer arrangement was made by E. Gorter and F. Grendel in 1925 who were able to extract the lipids from red blood cells. They postulated that because mammalian red blood cells do not contain nuclei or cytoplasmic organelles, the major source of lipid would be from the cell membrane. This assumption allowed Gorter and Grendel to conclude that the ratio of lipids to surface area was 2:1, and that the thermodynamically favored arrangement would be to have the polar groups directed outward to interact with the aqueous environments, both outside and inside the cell. This early model evolved as researchers identified other proteins, both on the membrane surface and spanning the membrane. By the early 1970s, the membrane structure was described by S. J. Singer and G. Nicolson as the *fluid–mosaic model* representing a much more dynamic and mobile system for semi-permanent interactions between proteins and movement of the lipid layers. This mosaic model illustrates the flexibility of the membrane where lipids may have lateral, rotational, flexional, and reversal motion within the plane. Also seen in Figure 3.2 are other lipid components of the membrane such as cholesterol, which has affinity for flexible phospholipids and creates local regions of rigidity in the membrane due to a less flexible ring structure.

3.4 Cytoskeleton and motility

The cell membrane represents the limits of the cell, and the interior portions of this membrane are coated with a variety of proteins to form networks within the *cytoplasm*. The cytoplasm is a mixture containing both the organelles of the cell, and the *cytosol*, which is a fluid of organic compounds and dissolved ions. The shape of the cell is defined by highly regulated series of filaments which also function as controls for cell motion, compartmentalization, cell division, and internal transport of organelles and substances. Three major types of filaments are termed:

- microtubules,
- intermediate filaments, and
- actin filaments (microfilaments).

Microtubules are approximately 25 nm in diameter, hollow, have directionality, and are assembled from heterodimers of tubulin. Microtubule-associated proteins such as kinesin and dynein are found with these filaments and represent a family of motor proteins. These proteins are responsible for moving vesicles along the track of the microtubules in + and − directions, depending on the directionality of the tubule. Additionally, the microtubules are also responsible for large structures such as the cilia which span the plasma membrane into the extracellular environment, acting as sensors and to conduct local fluid behavior.

Unlike microtubules, intermediate filaments are approximately 9 nm in diameter, assembled in long thin rods without directionality, and thus can assemble from either end or in the middle. They are assembled from a heterogenous group of proteins. Keratin, lamin proteins, nestin, and desmin are a few of their members and are often found in load bearing applications.

In comparison to microtubules and intermediate filaments, actin filaments are amongst the smallest intracellular filaments. With a diameter of approximately 8 nm, actin filaments are assembled from globular actin (G-actin) monomers into a filamentous non-covalent dimer chain (F-actin). These filaments, that like microtubules have directionality, have a + and − end. They are responsible for much of the local membrane architecture as they control both the shape and motion of structures, like the microvilli in epithelial cells. They also control the dynamic processes of cell locomotion and cytokinesis. Just as microtubules have the motor proteins kinesin and dynein for transporting vesicles, actin filaments have the myosin super family of proteins. The most well-studied myosin protein belongs to the conventional type II class. First identified in muscle, myosin proteins are involved in the sliding filament model of muscle contraction. However, this class also has numerous functions with non-muscle applications such as splitting the cell during division and for creating tensile forces at focal adhesion locations.

The sliding filament model of muscle fiber contraction continues to be one of the best models to study actin–myosin mechanics. However, this model does not reflect the complexity of non-muscle cellular motility. Contractility at the cellular level is far less ordered than in the sarcomere in the muscle, and operates in a tightly controlled and localized region of the plasma membrane often termed as the *cortex* in biological literature. This region has a wealth of actin-binding proteins which are responsible for directing the pattern and assembly of microfilaments for locomotion.

Most research concerning cell locomotion has traditionally been performed on two-dimensional (2D) substrates in migration assays. Only recently have we begun to explore more complex three-dimensional (3D) assays and analysis of cells as they navigate more native architectures *in vitro*. Despite this limitation, cell locomotion has been well documented for cells crawling over a substrate. Cells first display a leading edge with flattened appearance that protrudes in the direction of motion with temporary "foot-holds" of anchorage. The body of the cell is then pulled forward via the cytoskeleton to the new anchorage spot while the "tail" of the cell breaks its contact with the substrate to move forward. This process is repeated as the cell explores a surface as shown for a fibroblast in Figure 3.3. The motion of this process is not linear but rather intermittent and halting. The flattened protrusion is known as the lamellipodium and can produce substantial traction forces. Some early experiments with biomaterials unintentionally demonstrated this process when

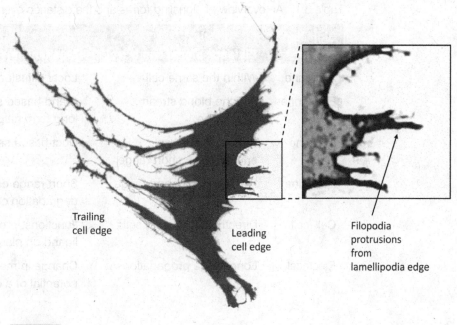

Trailing
cell edge

Leading
cell edge

Filopodia
protrusions
from
lamellipodia edge

Figure 3.3

Multiple leading and trailing edges of a fibroblast as it explores a two dimensional substrate showing the lamellipodia, a flattened region usually devoid of major cell organelles.

cells in local contact with the materials exerted powerful local strains and caused unanticipated material deformation.

From our study of cell locomotion, the question of how the cell starts moving along a direction could be asked. Biochemical gradients and signaling systems can be received by cells along the lamellipodium borders. In the case of hematopoietic cells, a family of proteins called the Wiskott–Aldrich syndrome proteins (WASP), activates a signaling system to begin the polymerization of the actin filaments. Other proteins mediate this signaling system in non-hematopoietic cells and start the motion. Cytoskeletal structure and cell locomotion are closely linked with cell communication pathways and signaling which will be covered in depth in the next sections.

3.5 Cell to cell communication pathways

During the processes of tissue repair, regeneration, and early development, cells employ a unique set of communication tools to organize themselves and signal their neighbors. Our knowledge of these molecules and pathways has been greatly advanced with the ability to knockout or knockdown selected molecules and

Table 3.1   An overview of signaling forms and the distance over which they work

Signaling system	Distance signal can travel	Mechanism
Autocrine	Within the same cell	Local diffusion and transport
Endocrine	Into the blood stream, long range	Gland-based secretion and long term response
Exocrine	Directly to external environment, short range	Duct-based secretion
Paracrine	Into the ECM, short range	Short range due to degradation or ECM hindrance
Cell–cell	Directly neighboring cells	Junctions, presentation of ligand on plasma membrane
Electrical	Long range propagation	Change in membrane potential of a cell

observe the effects in model organisms. Classifications of how cells signal vary considerably but one of the most common systems organizes signaling based on the distance between cells. Cells touching each other can communicate directly through gap junctions, which are small pores in the membrane, or by presenting a ligand on their plasma membrane and making it available for a neighbor cell to bind with using a receptor. With added distance, cells can signal by releasing molecules into the extracellular space. These molecules can be chemically or electrically transmitted. An overview of signaling forms is shown in Table 3.1.

A diffusible signal released by a cell to affect itself is termed *autocrine* signaling, whereas a diffusible signal released by a cell and affecting a second cell is known as *paracrine* signaling. Both mechanisms share similar characteristics in that:

- they are highly localized signals since they are quickly degraded or captured at a cell membrane, and
- they rely on simple diffusion transport mechanisms.

Why would a cell seek to communicate with itself? An example of autocrine pathways is best observed in feedback systems. At many stages of development, cells need to reinforce their expression of proteins. Autocrine feedback provides a continual message to stay on task in activities such as differentiation into more specialized phenotypes. An example of paracrine signaling has recently been observed in mesenchymal stem cell transplants to prevent scar tissue formation in regions such as the heart following myocardial infarction. Local signaling has been implicated in tasking implanted cells to improve angiogenesis – the formation of new blood vessels around an injured location.

In contrast to the short range interaction of autocrine and paracrine signaling, is the long range consequence of signaling molecules which are released by cells into the blood stream and affect cells at great distances. Such signals are known as *endocrine* signals. These far reaching products, such as insulin, are often hormones with diffuse effects across a range of target cells. In the case of insulin, which is an endocrine signal, the response selectivity comes from a target cell's number of receptors available to receive the signal. In contrast, *exocrine* signals are secreted into ducts such as from the lining of the small intestine where digestive enzymes and proteins are released to the lumen. The lumen of the intestine is considered an exterior surface as it is continuous with the external environment. These duct-based systems have relatively short pathways as most are at or near the exterior of the body.

Electrical signal conduction is one of the fastest and longest range methods of signaling. In neural cells, an electrical potential can travel across the extensive length of the plasma membrane and end at a terminal where transmission of a signal can continue either by electrical means through a gap junction or chemically by the release of a neurotransmitter received at a receptor in the target cell. This is predominantly observed in the region of neurons called the axon.

3.6   Cell junctions

When cells are in direct contact with one another, straight cell to cell signaling is possible using specialized junctions. A schematic representation of these junctions is shown in Figure 3.4. These specialized communication pathways provide near direct transmission of a signal from cell to cell, bypassing membrane signaling. Many of these junctions are routinely found in epithelial cells or other "compartmentalized" tissues of the body. These are typically regions where one face of the cell is in contact with an environment that must be separated from its opposite face. This orientation is known as an apical–basolateral organization.

3.6.1   Tight junctions

From our discussion of the plasma membrane, we identified the key features of this protective layer that prevent molecules from diffusing across it. However, the space between cells would be freely accessible for diffusion if cells lacked *tight junctions*. As seen in Figure 3.5, these localized "seams" found close to the outer

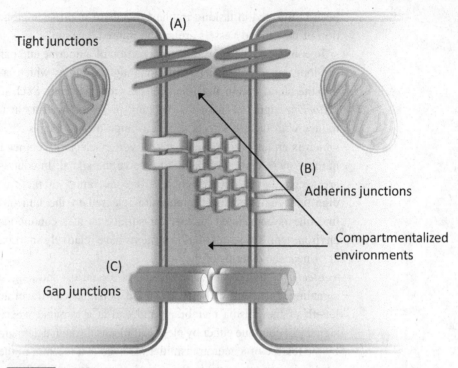

Figure 3.4

Schematic visualization of cell junctions. Tight junctions (A) are meetings of the cell membrane at very specific points of contact that run in long strands to seal out solute molecules. Adherens junctions (B) are made possible through the assembly of cadherin proteins on neighboring cells through a calcium-dependent linkage. In gap junctions (C) a more specialized communication is mediated between cells without direct contact using proteins called connexins that are clustered in groups of three in neighboring cells.

surface of adjoining cells prevent random diffusion by forming long arrangements of interconnected strands made from integral proteins. This type of protein has been given the specialized name "occludin." This protein entirely surrounds the cell, thus occluding the extracellular space from one compartment to another. In addition to the occludin family of proteins, tight junctions are also composed of claudins, tricellulium, junctional adhesion molecules, and Crumbs proteins.

As with all integral-based linkages, variety is found in the affinity of the linkage. In tight junctions, there is an observable scale of how well the seal operates in different tissues. In fact, a variety of strong and leaky tight junctions exist. For example, in the capillary beds of the brain, the junctions are extraordinarily tight and dense to form the fundamental "blood brain barrier." In contrast, many carcinomas, including those that can originate at the epithelium such as breast, prostate, and lung, initially result from a detachment of the epithelium from

APICAL

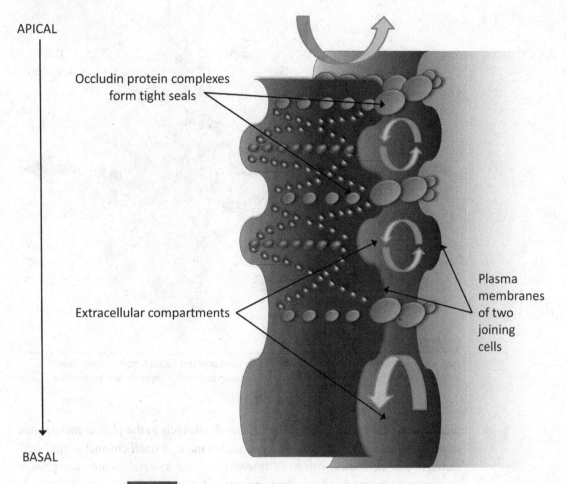

Occludin protein complexes
form tight seals

Plasma
membranes
of two
joining
cells

Extracellular compartments

BASAL

Seams of tight junctions function to join cells at specialized structures termed occludins to join local extracellular spaces into compartmental configurations.

its neighboring extracellular matrix. In the field of pharmacokinetics, the motion of drug molecules is studied as they move across these barriers, and modern pharmacology has implemented strategies to open tight junctions to gain access to the brain circulation. Nevertheless, from a biomaterials perspective, modifying junctions such as this should be considered with caution.

3.6.2 Gap junctions

One of the most specialized intercellular communication pathways is mediated by *gap junctions*. These sites represent regions where cells can pass ions, metabolites,

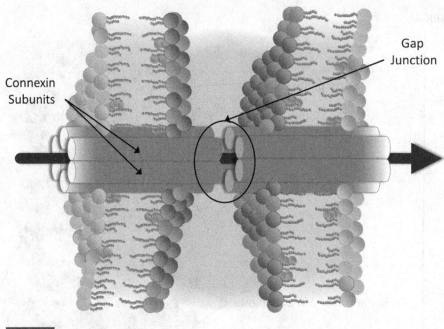

Connexin
Subunits

Gap
Junction

Figure 3.6

Gap junctions cluster 6 connexin subunits into a functional unit called a connexon to allow molecules as large as 1.2 kDa to directly pass between cells without entering the extracellular space.

and second messengers directly through small channels in the plasma membranes of neighboring cells. While no direct contact is made, a small channel is "framed" through the membrane using an assembly of the integral membrane protein connexin. These proteins are clustered into a 6 connexin functional unit known as the **connexon** (Figure 3.6). Connexin proteins have extremely short half-lives, on the order of a few hours. Such short half-lives of connexin complement their role in helping to signal dynamic events such as growth, proliferation, and differentiation. The size of the annulus or passageway in the center of a connexin has been estimated at 1.5 nm, and molecules as large as 1.2 kDa molecular mass have been documented crossing the channel.

Gap junctions have been implicated as major transport signalers especially during embryonic development and wound repair. Because they allow potent signaling molecules to pass across the membrane, gap junctions directly allow clusters of cells to behave as a functional unit. During embryogenesis, the connexin 43 (Cx43) has been linked with cardiogenic as well as osteogenic differentiation. Like all of the cell-to-cell communication pathways, there exists a range of connexin affinities to correspond with varied tissue-specific distributions. Over 20 mammalian connexin types have been identified with adjustable docking between subtypes. The most prominent example of a connexin type is the expression of Cx43,45 proteins found primarily in atrial contractile ventricular myocytes. Another example is the Cx40 which is expressed in atrial myocytes and His–Purkinje fibers, allowing the systems to operate as near independent signaling pathways. In addition, signaling through the gap junction can be rapidly stopped by a rise in calcium ion (Ca^{2+}) or hydrogen ion (H^+) concentration in the cell as well as through protein phosphorylation events.

3.6.3   Adherens and desmosomes

In addition to the signaling channels discussed in the prior sections, there are two distinct types of adhesion junctions which blend the concept of a cell communication with adhesion. Adherens junctions are functionally very similar to the aforementioned tight junctions. They are often found in similar locations such as at the epithelial lining to encircle each cell and use a family of 30+ glycoproteins known as the *cadherins*. Traditional classifications of this family were denoted by tissue defining prescript letters, i.e. E-cadherin found in the epithelia, N-cadherin in neural tissue, and P-cadherin in the placenta. These glycoproteins help join similar cells to one another and have a subcellular inter-action with catenin proteins to transmit information across the membrane and anchor themselves into the actin cytoskeleton. On their surface, cadherins have a large extracellular domain with five distinct units in series. Between these units is a space with high affinity for calcium which has been linked with the structural cohesion of the protein. Neighboring cells create an interlock of these proteins that provides the adhesive strength to this type of junction. This family of proteins is thought to have an essential role in helping cells find "like" cells in tissue aggregates and have both a diffuse presence in the membrane as well as a localized expression at the adherens junction.

Like adherens junctions, desmosomes also use cadherin membrane spanning proteins to create a junctional complex between cells with the same extracellular calcium binding mechanism. But unlike the adherens, desmosomes link to inter-mediate filament loops on the subcellular membrane in distinct plaques on the scale of 1000 nm in diameter. This linkage has strong correlation with absorbing the energy of mechanical deformation or strain energy density that can be clearly observed in tissues such as the dermis where deformation can be pronounced. Displacement of these stretching forces can be easily transferred through desmo-somes to entire networks of cells, allowing them to receive a similar "mechanical" signal and react uniformly. Each of the signaling systems and junctions also lead directly to intracellular signaling pathways.

3.7   Cell signaling pathways

As evidenced in the prior sections, cells have an assortment of membrane spanning systems to connect the external environment with intracellular processes. The process of cell signaling begins at the membrane where some form of stimulus must be received, converted into a usable form, and cause a change in cell behavior as shown in Figure 3.7. Cell signaling pathways are of vital interest to the biomaterials engineer since their behavior directly alters cell phenotype. Many novel biomaterials have immobilized therapeutics on their surface or bulk structure, which when released, can cause a physiological change. Examples of such novel implant devices can be orthopedic implants that can be designed to increase the expression of secreted collagen for bones or vascular stents coated with anti-inflammatory agents to prevent the receptor based binding of lymphocytes.

There is substantial variety in each of the steps in cell signaling, and a few general points can be made before we explore the machinery of signaling. A signal arriving at a cell can activate multiple internal cascades, and correspondingly there are many cascades that require multiple external signals. Quantity, both in the

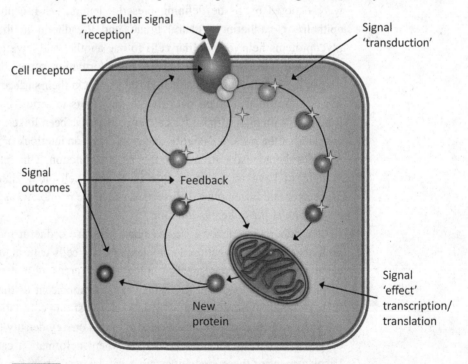

Figure 3.7

Cell signaling behavior can be classified as a "reception" of an external signal at or near the plasma membrane, its "transduction" into a useful signal inside the cell and finally its "effect" on a process in the cell, i.e. its conversion into a messenger to effect gene expression.

amount of signal and the number of receptors, may be needed to achieve a response. When we consider that most cells are barraged with signals such as hormones, cytokines, and lipids that are present in the blood stream and extracellular fluid, the effect of selectivity must be considered to allow cells to receive the correct signal based on their expression of receptors. This fundamental concept means that every cell cannot respond to every signal, but rather only to a defined set of signals, specific to a tissue type. Cells accomplish this selectivity by changing the type and number of receptors at their surface.

3.7.1   Receptors as signaling sensors

There are three distinct cases of cell signaling. In one case, chemical compounds (typically referred to as ligands) can diffuse freely across the plasma membrane, whereas in the second case, the chemical compounds cannot diffuse because of their size or charge. In both cases, these compounds are able to signal by binding to either intracellular receptors or membrane bound receptors, respectively. In the third case, a physical stimulation such as light or stress can alter binding of a receptor, causing a conformational change in its structure and initiating a chain of reactions to propagate the signal. *Signal transport* occurs if the ligand is taken in or "endocytosed" with the activated receptor or brought inside through the creation of a membrane pore as occurs in ion channels. In the case of *signal transduction*, the ligand is not brought into the cell but changes the enzymatic activity of the receptor and often activates "second messengers" to pass the signal downstream of the receptor–ligand complex. Both pathways are schematically shown in Figure 3.8, demonstrating multiple ways to communicate a signal.

Receptor activity can be understood through several key features which include its affinity, selectivity, reversibility, and saturation kinetics. Typically receptors have reasonably high affinity with their ligands. Receptors have an affinity constant, K_a, averaged on the order of 10^8 and a dissociation constant, K_d, with high variability. Binding is moderately selective so that a receptor binds with its native ligand or a closely related ligand. However, this selective binding has become a point of discussion since many receptor–ligand bindings can occur with a gradient of activation or inactivation. *Reversibility* refers to a ligand's ability to freely dissociate from a receptor rather than permanently active it, but once again many mechanisms can exist in this classification. For example, opiods such as morphine can bind to a specific μ type opiod receptor in the central nervous system and limit the receptor's ability to be internalized, thus maintaining signaling at a high rate but ultimately leading to tolerance and signal desensitization. Finally, all

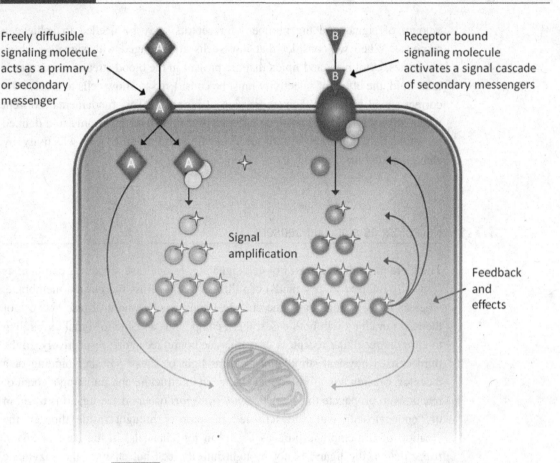

Freely diffusible signaling molecule acts as a primary or secondary messenger

Receptor bound signaling molecule activates a signal cascade of secondary messengers

Signal amplification

Feedback and effects

Figure 3.8

Two distinct examples of cell signaling across a membrane. Signal transport with compound A freely diffusible across the membrane and signal transduction with compound B altering the receptor's activation of a second messenger. It can be observed in both pathways that amplification of the signal occurs downstream with multiple effects.

these phenomena are influenced by *saturation kinetics*, where the number of receptors on a cell and the concentration of ligand are key players. Receptor number can vary from just a few to thousands and even millions per cell, but a fixed number of receptors can be saturated by high ligand numbers.

3.7.2 Receptor classes

Several superfamilies or categories are used to group receptors based on their structure and similar function. Five traditional superfamilies encompass the majority of receptors: ion channel, G-protein, enzyme linked, tyrosine kinase, and

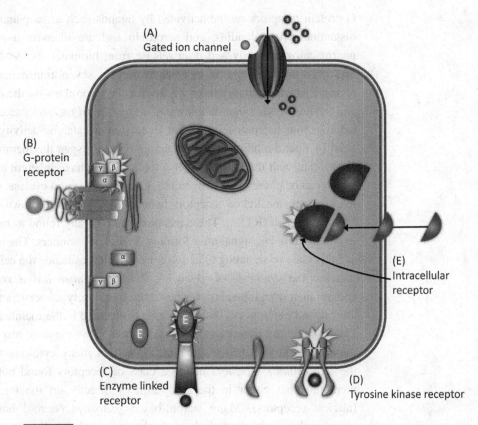

(A)
Gated ion channel

(B)
G-protein
receptor

(E)
Intracellular
receptor

(C)
Enzyme linked
receptor

(D)
Tyrosine kinase receptor

Figure 3.9

Representations of the structure and activation of the various receptor families; ion-channel based (A), G-protein (B), enzyme linked (C), tyrosine kinase (D), and intracellular (E), and their mechanisms of action on subcellular signaling.

intracellular. While these families represent the starting point for a study of receptor signaling, there is substantial heterogeneity in structure and function even among each class. An overview of these different receptor groups is observed in Figure 3.9

The *ion-channel receptors* are among the most well-studied receptors, and can be found in most cells including those of the neurotransmission system. Ligand binding alters the permeability of ions across the plasma membrane by changing the receptor structure. As an example, the nicotinic acetylcholine receptor allows a variety of positively charged ions such as sodium ions (Na^+), calcium ions (Ca^{2+}), and potassium ions (K^+) into the cell. A major characteristic of these receptors is that conformation changes occur with great speed. *G-protein linked receptors* are named due to their activation of sub-membrane trimeric G-proteins, α, β, and γ, which activate major signaling pathways in the cell. A large diverse family, the

G-protein receptors can be activated by ligands such as; dopamine, epinephrine, histamine, prostaglandins, and serotonin and are likewise a major target for pharmacological study and drug release from biomaterials. A shared feature of this class of receptors is their characteristic seven transmembrane spanning regions, which configure their affinity for the G-proteins on the inner membrane and ligands on the outer membrane. *Enzyme linked receptors* are one of the most heterogenous receptor families but share similar catalytic activity which is regulated by ligand-binding. These receptors typically span the membrane only once, contrasting with the G-protein receptors, and can have inherent enzyme capacity which can be based on a phosphatase, kinase or guanylyl cyclase. One of the most relevant enzyme linked receptors belongs to a subclass known as the *receptor tyrosine kinases* (RTKs). These receptors are typically found as monomers before ligand binding but signal after forming dimers or tetramers. The insulin receptors use this class of signaling to activate protein kinases inside the cell and ultimately increase the synthesis of glycogen. *Tyrosine kinase linked receptors* do not contain their own inherent tyrosine kinase but closely associate and activate them after ligand binding. As such, they are also located in the membrane but recruit a kinase from the cytoplasm upon activation. This class is responsible for the majority of growth factor signaling as well as many cytokine signals. Finally, the *intracellular receptors* define a class of receptors found not in the plasma membrane but either in the cytoplasm or directly on the nuclear membrane (nuclear receptors). Many steroid-based pathways (steroid hormone receptor superfamily) signal through these mechanisms and directly alter gene regulation by effecting gene transcription. These so-called nuclear receptors can undergo major conformational changes to pass through nuclear pores and directly bind specific sequences of DNA, altering transcription rates and thus can also be called transcription factors.

3.7.3 Second messengers and their activation/deactivation

In the description of the receptor classes in the previous section it was noted that most membrane receptors bind a ligand on the outside of the cell and change some activity on the inside of the cell. Many pathways utilize second messengers to propagate a signal. A major reason to use a second messenger is to keep the machinery of signal reception at the membrane where it can be reused continuously. This accounts for the incredible efficiency of signaling behavior in the transduction pathway. Second messengers can be water soluble compounds such as cyclic AMP (cAMP), ions such as Ca^+, and lipid soluble compounds like diacylglycerol (DAG). In most varieties, a second messenger passes the message

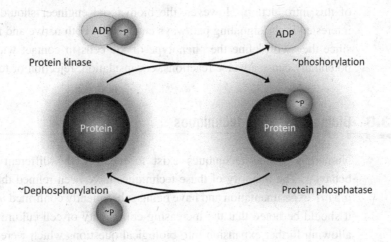

Figure 3.10

Representation of the role of second messenger pathways and regulation that occurs in the events of phosphorylation/dephosphorylation carried out by kinases and phosphatases.

down a chain, and ultimately to a target protein which alters cell behavior. This series of chain activations is coordinated by a central premise of *phosphorylation* events. Proteins can either be inactive or active, and many of these on–off switches are made possible by adding or removing a phosphate group through the process of *phosphorylation or dephosphorylation*. There are several mechanisms to facilitate this change including

- a shift in pH,
- a change in the redox environment of a protein, and
- the presence or absence of a phosphate group in the amino acid structure of the protein.

The inherent attraction of phosphorylation control is its reversibility, speed, and efficiency. Rather than having the cell manufacture new proteins for every incoming or outgoing signal, the intracellular messenger can be quickly reused with minimal energy demands. This process is mediated by two distinct classes of enzymes: *kinases* (also known as phosphotransferases) which transfer phosphates from high-energy molecules such as adenosine-5′-triphosphate (ATP) to a target protein, and *phosphatases* which remove the phosphate group. A schematic representation of this system is shown in Figure 3.10. In the section on receptor classes, we have already described a receptor with intrinsic kinase abilities, the RTK, where auto-phosphorylation of the receptor complex opens active docking sites for second messenger signaling.

There are many distinct classes of second messengers, kinases, and phosphatases that result in the complexity of cellular signaling and are far beyond the scope of this introduction. However, the biomaterial engineer should be exceedingly interested in the signaling pathways engaged by both native and foreign materials since they will define the phenotype of the cells in contact with materials, and potentially cause adverse reactions, encapsulation, rejection or toxicity.

3.8   Biological testing techniques

Numerous testing techniques exist to measure the different aspects of cell behavior. The majority of these techniques have been refined through the use of *in vitro* experimentation and have being subsequently confirmed *in vivo*. However, it should be noted that the increasing complexity of cell-culture based studies is allowing further expansion into biological questions which were only previously possible in animal studies. This point is strongly made in the shift toward three-dimensional (3D) cell-culture techniques described in more detail in Chapter 13. The advent of 3D tissue culture with extracellular matrix mimics has opened a vast resource for the biomaterial engineer to study multi-cellular responses.

Techniques in the study of cell structure, genetic regulation, motility, and communication signaling behavior have grown at a rapid pace in the last century. A progression has been realized from observations of the whole organism to tissue and cellular level knowledge, and now to subcellular and genetic levels of understanding. The field of molecular biology has most recently contributed a great amount of understanding to our traditional concepts of cellular processes, and a few of the key techniques and experiments are listed in the following sections.

3.8.1   Probe and labeling technologies

Developed from the field of immunology, labeling systems have become one of the most remarkable means to observe all levels of biological events. Observation of fluorescence and luminescence compounds by using microscopes designed to capture their excitation have helped resolve major questions in receptor biology, cell signaling, membrane architecture, and genetic regulation. Confocal microscopy has become a mainstay tool for the study of cell and molecular biology where the unique focusing characteristics of the lenses provide a researcher with the ability to observe real-time 3D images of cell components at high magnification. In earlier sections we discussed cell membranes, the cytoskeleton, cell junctions and internal signaling mechanisms. Each of these cellular components can be visualized by using antibodies that recognize the unique proteins to each system and both visualization and quantification can be accomplished using the techniques outlined in Section 11.4.

3.8.2 Examination of gene expression

For the study and measurement of cell behavior, isolating changes in gene expression are among the most essential techniques used by the biomedical engineer. Reverse transcription polymerase chain reaction (RT-PCR), complimentary DNA (cDNA) library screens, and cDNA microarray hybridization are among the common techniques used to find minute and/or remarkable changes in mRNA levels in response to biomaterial surfaces or elution products. Following the creation of a cDNA library for control and biomaterial-treated cells, microarray hybridization can quickly identify regions of the genome that do not overlap and demonstrate a change in expression between the control versus treated groups. The great advantage of microarray technology permits the analysis of more than 10 000 genes at the same time. Additional advantages to microarray technology allow for quantification of *time* as well as *intensity* based changes in gene expression.

3.8.3 The plasma membrane

Common assays to evaluate the mechanics of the plasma membrane include single cell techniques such as micropipette aspiration. Using a glass micropipette to apply a quantified amount of suction on a single cell, a small region of cytoskeleton can be drawn into the pipette. Measurement of the required force, the deformed shape, and motion over time can help resolve questions regarding membrane viscoelasticity. More recently, the technique called a "cell poker" has been used to measure the deformation in the membrane using a small indentation tip and has become relevant to observe local changes in membrane architecture, as seen in Figure 3.11. Both of these mechanical techniques provide valuable information, especially when coupled with biochemical and molecular approaches. In addition to micromechanical techniques, one of the first and still best analyses of membrane composition is the use of freeze–fracture. A block of frozen tissue can be impacted with a knife blade and form a fracture plane that runs through the lipid bilayer. Both new faces of this plane can be coated with a thin layer of metal to form a replica, or stained for chemical and protein reactivity. Since the fracture plane passes through the middle of the lipid bilayer, it is possible

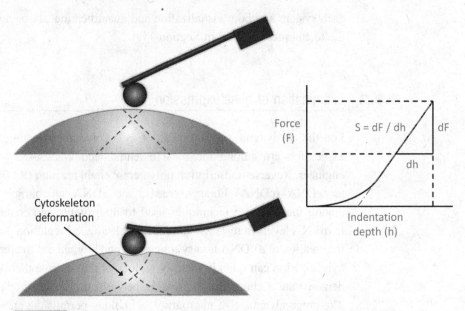

Force
(F)

S = dF / dh | dF

dh

Indentation
depth (h)

Cytoskeleton
deformation

Figure 3.11

Representation of the mechanics of membrane force measurement or cell poker design used to
identify the local strain behavior of membranes. This system has also found new applications by
the immobilization of proteins and antibodies onto the cell tip to measure membrane
physical–chemical interactions.

to identify any transmembrane protein or other proteins that are embedded at
least half way through the membrane.

3.8.4 Cytoskeleton and motility

Highly relevant to the biomaterials engineer are the mechanisms by which a
cell will attach, move, and secrete ECM proteins onto a material surface. The
cytoskeleton, composed of microfilaments, intermediate filaments, and
microtubules, participates in the generation of the force needed for cell motion.
Studies of the cytoskeleton include the use of electron and fluorescence microscopy, as shown in Figure 3.12, where it becomes possible to resolve the spatial
and temporal distribution of its components. Like the biochemical studies of the
plasma membrane, imaging techniques of the cytoskeleton become more relevant
when coupled with micromechanical quantification. In the case of the cytoskeleton,
displacement analysis is commonly used to assess micromotion of a cell or group
of cells. This technique creates a vector map, a measurement of both direction and
magnitude, using fluorescent tags that are time resolved to quantify motion.

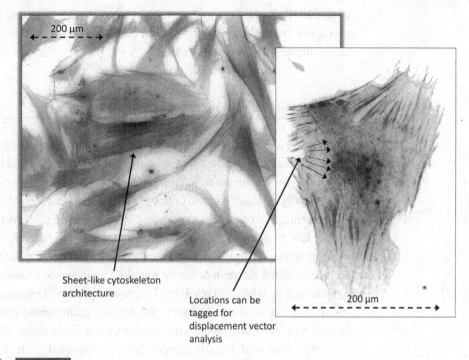

200 µm

Sheet-like cytoskeleton
architecture

Locations can be
tagged for
displacement vector
analysis

200 µm

Figure 3.12

Cell cytoskeleton fluorescence imaging demonstrating the microfilament structure of a fibroblast which can be used in time-lapse imaging displacement analysis.

3.8.5 Communication between cells

Among the most relevant techniques to identify cell communication are (1) dye transfer and (2) electrophysiology studies. Dye transfer uses small labeled molecules in the sub-kilodalton (sub-kDa) range to determine the "leakage" of the molecule through gap junctions. Cells are loaded with tracers (typically one large tracer to identify the labeled cells and one smaller tracer to reveal communication) via microinjection and then cultured with non-labeled cells for a period of time to establish gap communication with new cells. Alexa 488 (mol wt 643 Da) is among the most popular choices for these studies due to its stability and ease of imaging by direct fluorescence microscopy. However, tracer size is not the only factor in crossing a gap junction, and care must be taken also to consider the tracer's charge and structure. In addition, many intercellular communication pathways are ion based and fall into the category of electrophysiology studies. By far, the most historical and prevalent of these techniques is the patch clamp system that targets ligand and voltage-gated channels. This classical system utilizes glass microelectrodes patched onto the surface of the cell by using giga-ohm (GΩ) resistance so that the ionic currents inside the cell can be measured.

3.8.6   Mapping intracellular signaling

Quantifying and mapping the progress of signal transport and transduction pathways have generated remarkable understanding in abnormal biological function, however the techniques for the quantification and mapping have only recently been designed. Traditionally, researchers have utilized two-dimensional (2D) gel electrophoresis for the quantification of protein kinases and other signaling proteins but were hindered by the low availability of these signals as well as the inherent time limitation of measuring one or few signals per experiment. Recent advances have applied a technique called the protein kinase multiblot analysis to quantify 50–100 proteins simultaneously while still resolving variation in protein phosphorylation. Similar to multiblot analysis, the technique of bead-based multiplexing can improve analyte detection of 500–1000 proteins or signals within a single assay. The ability to detect multiple active and inactive signals has become vital to mapping intracellular pathways. It has become evident from the literature that most signaling pathways cannot be studied in a "linear" fashion with each pathway clear and independent, but rather as large integrated "networks" of signals with multiple feedback systems in play. For this reason, a new field has emerged that uses mathematical modeling of these interactions to provide organization to the complexity. The modeling approach can help to better identify the role of distinct pathways that operate in cooperation with a network, and in locating the major regulatory hubs so that they can be targeted for pharmaceutical intervention.

3.9   Summary

The cellular environment encompasses a diverse population of control systems that span from biomolecular phenomena to a remarkably complex coordination of signaling pathways. Of particular note for the biomaterials engineer are the communication processes that the mammalian cell uses to sense and respond to its environment. A vast amount of regulation occurs at the plasma membrane, at the junctional and receptor complexes expressed, and within the cell itself via second messengers and cascades that alter the phenotype and behavior of the cell. These regulation systems are distinct from genetic control and may be targeted by the engineer to alter both the functions of the cell and its response to a biomaterial.

Suggested reading

- Dunlop, J., Bowlby, M., Peri, R., Dmytro Vasilyey, D. and Arias, R. (2008). *Nature Reviews Drug Discovery* **7**, 358–368.
- Fleming, T. (2002). *Cell-Cell Interactions*. Oxford University Press.
- Terrian, D. (2002). *Cancer cell signaling: methods and protocols, Methods in Molecular Biology*, vol. **218**. Totawa, NJ, Humana Press Inc.
- Eungdamrong, N. J. and Iyengar, R. (2004). *Modeling Cell Signaling Networks, Biology of the Cell*, **96** (5), 355–362.

Problems

1. Define the flow and flux terms and identify the characteristic equations regarding diffusion in biological membranes.
2. Name the primary functions of the plasma membrane and describe their role in transport mechanics of solutes and liquids.
3. Define the term fluid mosaic model of membrane behavior and describe its evolution from a rigid structure.
4. Define the three types of mammalian cellular filaments, which have polarity and what motor proteins are associated with them?
5. Describe a model for cell motility and offer a hypothesis for the lack of organelles in the lamellipodium.
6. Actin has been identified as among the most evolutionarily conserved proteins. Explain what this implies about the key features of this protein and where it is found.
7. Compare and contrast autocrine, paracrine, and endocrine signaling. How would each be involved in local tissue trauma following damage to connective tissue?
8. What would be the implications if the body lacked tight junctions? What are the implications for biomaterials regarding the foreign body reaction with respect to tight junctions?
9. Identify how spatial cell motion studies are performed and explain how this information would be relevant to material integration with local tissues; give an example.

4 Characterization of biomaterials

> **Goals**
>
> After reading this chapter the student will understand the following.
> - The principles underlying various instruments typically used for the characterization of biomaterials.
> - The use of different instruments for biomaterials characterization.

During the development phase of any medical devices or implants, it is essential that the biomaterials used are thoroughly characterized. The material's surface properties are important because it is the implant's surface where the body's biology and the material first interact. On the other hand, the bulk properties of the material determine its mechanical and physical behavior and its long-term viability. Of course, the material's chemical properties influence its stability, biocompatibility, and the body's reaction to it. Since a material possesses different physical and chemical properties, the analytical instruments used for characterizing these properties can be broadly classified under the following three categories:

- surface characterization,
- bulk characterization, and
- chromatographic analysis.

The use of multiple characterization techniques is essential for complete analysis of the surface and bulk properties of a biomaterial. When a material is implanted in the body, protein adsorption almost instantaneously occurs on the material surface. Hence, the cells do not directly interact with the material surface but only through the layer of proteins adsorbed on the surface. This adsorption of proteins is governed by a combination of different surface characteristics of the material including surface energy, surface chemistry, surface texture, and surface roughness. The use of contact angle measurement provides information about the material's surface energy, whereas Fourier transform infrared spectroscopy (FTIR), X-ray photoelectron spectroscopy (XPS), and secondary ion mass spectrometry (SIMS) provide information on the material's surface chemistry at different depths. Scanning electron microscopy (SEM) provides details about the material's surface texture, whereas atomic force microscopy (AFM) allows for measurement of the material's surface roughness. Thus, data collection using different analytical instruments provides in-depth information regarding the material's surface and assists in predicting how surface properties would influence the biological outcome. Additionally, the physical properties of a biomaterial such as strength, hardness, ductility, and toughness are vital for its successful use as an implant. These physical properties are strongly influenced by the microstructure and crystallinity of the material, which can be assessed by using transmission electron microscopy (TEM) and X-ray diffraction (XRD), respectively. Advances in technology have also allowed biomaterials to be used for delivery of growth factors or other medically relevant biomolecules. The separation and analysis of therapeutic drug molecules, biomolecules, and other organic molecules are crucial for drug and protein delivery applications. High performance liquid chromatography (HPLC) and gel permeation chromatography (GPC) are commonly used for such separation and analysis. The latter is also useful for characterizing polymers.

In the following sections, various instruments and techniques used for the characterization of materials are discussed in detail. An example of a material's characterization is provided at the end of each section on a characterization technique to demonstrate the biomedical application of the analytical instrument.

4.1   Contact angle

Wettability is one of the vital parameters that determine cell–biomaterial interactions. Contact angle measurement is a simple technique that is commonly used to determine the wettability of any material surface. The contact angle, θ, is the quantitative measure of the angle maintained by a liquid at the boundary where liquid, solid, and gas phases intersect. A contact angle value of 0° indicates complete wettability of the surface, whereas a value of 180° indicates a completely non-wettable surface.

When a drop of water is placed on a clean glass surface, the water spreads and immediately wets the surface. On the contrary, when a drop of water is placed on Teflon, the water beads up and does not wet the surface. In the case of the water drop on the glass surface, the force between molecules of water (cohesive force) is

(a)   (b)

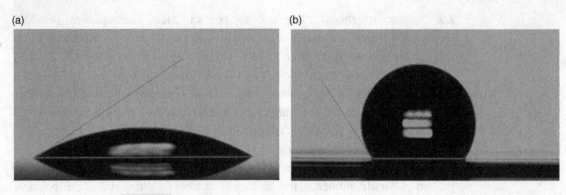

Figure 4.1

Photomicrograph of water contact angle measurements showing (a) hydrophilic surface, and (b) hydrophobic surface.

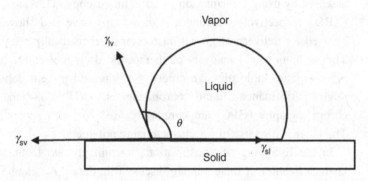

Figure 4.2

Schematic depicting the relationship between three interfacial tensions (γ_{sv}, γ_{sl}, γ_{lv}) and contact angle (θ).

lower than the force between water molecules and molecules at the solid glass surface (adhesive force). Hence, the water drop collapses and spreads on the glass surface. This property is commonly referred to as *hydrophilicity* (Figure 4.1a). In the case of the water drop on the Teflon surface, the cohesive force between molecules of water (liquid) exceeds the adhesive force between water molecules and molecules at the Teflon (solid) surface. Hence, the water drop beads up on the Teflon surface. This property is commonly referred to as *hydrophobicity* (Figure 4.1b).

The contact angle observed is dependent on the balance between the surface tensions (force per unit length) of the solid–vapor, solid–liquid, and liquid–vapor interfaces (Figure 4.2).

As shown in Eq. (4.1), a relationship between the interfacial tensions and the contact angle is described by Young's equation:

$$\gamma_{sv} = \gamma_{sl} + \gamma_{lv} \cos \theta, \tag{4.1}$$

where

γ_{sv} = interfacial tension between the solid and the vapor,
γ_{sl} = interfacial tension between the solid and the liquid,
γ_{lv} = interfacial tension between the liquid and the vapor, and
θ = contact angle measured.

In Eq. (4.1), there are two known (γ_{lv} and θ) parameters and two unknown (γ_{sv} and γ_{sl}) parameters. The parameter γ_{lv} is available in the literature for several commonly used liquids. For example, the interfacial tension between liquid and the vapor (γ_{lv}) for water is 72 mN/m, whereas the γ_{lv} for methylene iodide is 51 mN/m. The parameter θ is the contact angle measured on the solid surface. The parameter γ_{sv} provides the surface energy of the material surface and needs to be determined. It can be determined approximately using either of the following techniques:

- Zisman method, or
- solving simultaneous equations.

The Zisman method requires a set of liquids with known values of γ_{lv} to be used for measuring the contact angles on a solid surface. By plotting the known values of γ_{lv} against the θ measured, the value for γ_{sv} can be determined by extrapolating the plot to the contact angle value of 0° (Figure 4.3). This determined γ_{sv} value is also known as the critical surface tension value (γ_c) of the material.

Unlike the Zisman method, solving simultaneous equations to determine γ_{sv} requires the use of two different liquids with known values of γ_{lv}. By measuring the contact angles, θ, on a solid surface with the two different liquids, the γ_{lv} values and the corresponding measured θ values can be substituted in Eq. (4.1), yielding two equations with two unknown variables, γ_{sv} and γ_{sl}. Solving these two simultaneous equations allows the value for γ_{sv} to be determined.

A common technique for measuring contact angles is the use of goniometry or a sessile drop method. In principle, *static contact angles* are assessed through a goniometer by measuring the angle formed between the solid and the tangent to the drop surface. Problems with this technique are the control of drop volume, the time over which the angle is measured, and the accuracy of the operator in measuring the angles. With advances in technology, the efficiency and accuracy of this technique have been improved by incorporating an image capturing device and a computer to measure the contact angles. Today's sessile drop system typically consists of (a) a platform to hold the sample; (b) a syringe to place a liquid drop on the sample surface; (c) a light source to illuminate the liquid drop; (d) a charge-coupled device (CCD) camera and lens to capture the image; and (e) a computer to process the image.

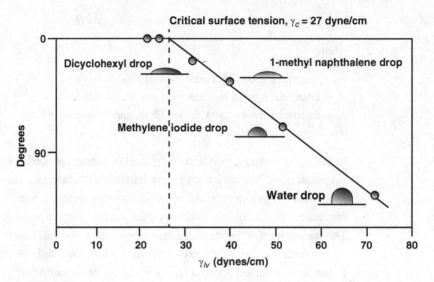

Figure 4.3

Zisman's plot to calculate the surface energy of a material surface. (Reprinted from *Biomaterials Science: An Introduction to Materials in Medicine*, 2nd edition, Ratner, B. D., Hoffman, A. S., Schoen, F. J. and Lemons, J. E., editors. Surface properties and surface characterization of materials, Ratner, B. D., page no. 45, Elsevier (2004), with permission from Elsevier.)

In measuring contact angles by using the sessile drop method, two different types of techniques are commonly employed. In the first method, contact angles are measured by simply placing a liquid drop on a material surface and taking measurements.

In the second method, a drop of liquid is first placed on a solid surface and the contact angle is measured. Then, additional droplets are gradually added to the original drop to maximize its volume without increasing the interfacial area between the solid and liquid phases. The contact angle is measured after the addition of each droplet. These measured contact angles reach a plateau after attaining a maximum value, and the angle at the plateau is called the *advanced contact angle* (Figure 4.4a).

Similarly, liquid from the expanded drop can be gradually removed to minimize its volume without decreasing the interfacial area between the solid and liquid phases. After each step, the contact angles are measured and they reach a plateau after attaining a minimum value. The angle measured at the plateau is called the *receded contact angle* (Figure 4.4b). The difference between advanced and receded contact angles is called *contact angle hysteresis* (Figure 4.5) and provides information regarding the homogeneity or heterogeneity of a material surface.

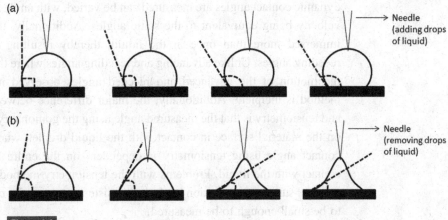

Figure 4.4

Schematic showing (a) advancing contact angle measurements, and (b) receding contact angle measurements.

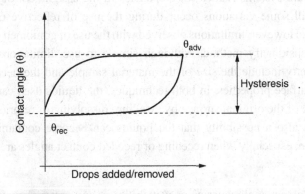

Figure 4.5

Plot of contact angle versus liquid drops, with contact angle hysteresis observed between advanced and receded contact angles.

A low hysteresis indicates that the surface is homogeneous while a high hysteresis indicates that the surface is heterogeneous.

Another common technique for measuring contact angles is the use of a tensiometer. Using this technique, the contact angle for a material is measured when the liquid, solid, and vapor boundaries are in motion. Using a microbalance, the principle of tensiometry is based on the change in forces when the solid comes in contact with the liquid. With known forces of solid–liquid interactions, known geometry of the solid, and known surface tension of the liquid, the contact angle for the material can be determined. The contact angle measured using this technique is also known as the *dynamic contact angle*. The speed at which dynamic contact angles are measured can be varied, with angles measured at a low velocity being equivalent to the static angles. Additionally, the material can be immersed more than once in the liquid, thereby resulting in advancing and receding angles. Unlike advancing and receding angles where the motion is actual, the motion of the advanced and receded angles observed in the sessile drop method is incipient. Additionally, the major difference between the goniometry and tensiometry is that the measured angle using the goniometry is dependent only on the material surface in contact with the liquid droplet, whereas the measured contact angle using tensiometry is dependent on the entire material that is in contact with the liquid. Problems with the tensiometry method are that the entire material surface composition has to be consistent, and that the material sample has to be small enough to be measured.

The advantages of contact angle goniometry are its simplicity and the use of inexpensive instrumentation. The limitations include the differences in the assignment of tangent and base lines of the drop by different operators. Although the use of software systems reduces discrepancies in the assignment of tangent and base lines, still some variations occur during the use of reflective or unconventional samples. However, limitations observed with the use of goniometry are not observed in the tensiometry method. Although simple to use, limitations with the use of tensiometry include the size of the material sample and the need for consistency of the surface properties. In both techniques, the liquids used can also change the structure of the original surface by swelling, dissolving or re-orienting the surface. There is also a possibility that the liquids can become contaminated during the operation, especially when receding or receded contact angles are measured.

- Contact angle determines wettability of any material surface.
- The technique is simple and straightforward.
- Advancing and receding contact angles are used to measure homogeneity and heterogeneity of a material surface.

4.2 Infrared spectroscopy

The electromagnetic spectrum consists of radiations varying from low to high energy as shown in Figure 4.6. All electromagnetic radiation is made of discrete packets of energy called photons, which travel in a wave-like pattern at the speed of light. The low-energy radiations have low-energy photons, low frequency, and long wavelength. The high-energy radiations have high-energy photons, high frequency and

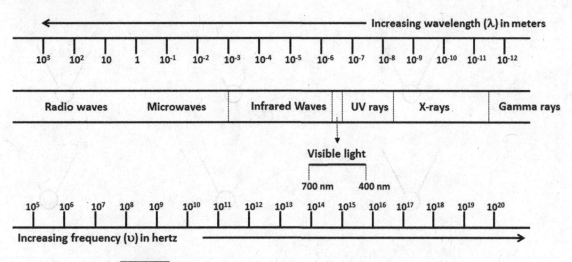

Figure 4.6

Electromagnetic spectrum.

short wavelength. In order of increasing energy, the spectrum consists of radio waves, microwaves, infrared waves, visible light, ultraviolet rays, X-rays, and gamma rays. When an electromagnetic radiation interacts with a molecule, the molecule undergoes distinct changes depending on the type of radiation. For instance, X-rays cause ionization (ejection of electrons) of molecules, and microwaves cause rotation of molecules. Similarly, infrared waves cause vibration of molecules.

Atoms in a molecule undergo vibrations, which can result in the stretching and bending of the bonds between these atoms. The stretching involves a change in the bond length, while bending involves a change in the bond angle. The different modes of stretching and bending vibrations are depicted in Figure 4.7. When a vibrating molecule is exposed to infrared radiation, the molecule will absorb that portion of the radiation whose frequencies are equal to the frequencies of the vibrations in the molecule. The analysis of the remaining portion of the radiation transmitted through the sample provides the information regarding the frequencies absorbed by the sample and hence information regarding the molecular structure of the sample.

Infrared spectroscopy consists of an infrared radiation source, a monochromator, and a detector. The infrared radiation emitted from the source covers a wide range of wavelengths. The radiation is allowed to pass through the sample and the monochromator. The monochromator disperses the radiation into its constituent wavelengths and processes one wavelength at a time. The intensities of different wavelengths are then measured using the detector. The main limitation of the conventional dispersive IR spectroscopy is the longer scan time. In modern day, Fourier transform infrared spectroscopy (FTIR), the monochromators have been replaced by interferometers which are known for their high speed and sensitivity.

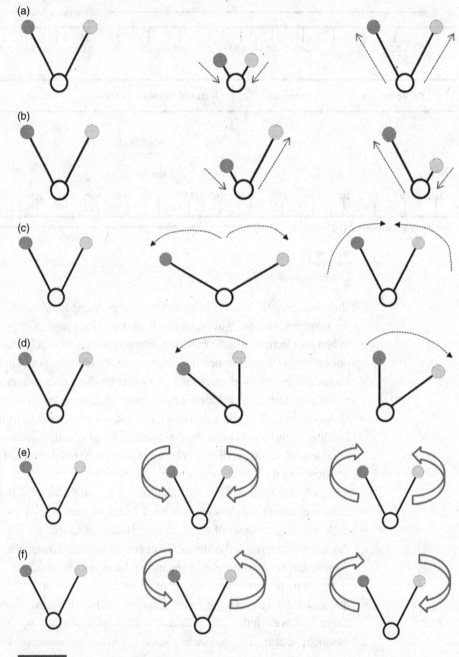

Figure 4.7

Schematics showing (a) symmetric stretching, (b) asymmetric stretching, (c) scissoring, (d) rocking, (e) wagging, and (f) twisting.

In FTIR, all wavelengths of the radiation are processed simultaneously and a Fourier transformation is applied to convert the obtained interferogram (a plot between signal intensity and time) to the final infrared spectrum – a plot between wavenumber and absorbance intensity or percentage transmittance. The wavenumber is defined as the number of waves per unit length. The unit commonly used for measuring wavenumbers in FTIR spectrum is cm^{-1}. As shown in Eq. (4.2), the wavenumber ($\bar{v}$) is inversely proportional to wavelength (λ):

$$\bar{v} = 1/\lambda. \tag{4.2}$$

The relation between frequency (v) and energy (E) is provided by the Planck–Einstein equation, shown in Eq. (4.3):

$$E = hv \tag{4.3}$$

where h is Planck's constant.

The relation between frequency (v), velocity (c), and wavelength (λ) is provided by Eqs. (4.4), (4.5), and (4.6):

$$v = c/\lambda \tag{4.4}$$

$$E = \frac{hc}{\lambda} \tag{4.5}$$

$$E = hc\,\bar{v}. \tag{4.6}$$

Based on the above equations, the wavenumber is directly proportional to the energy of the IR absorption. Hence, a low wavenumber corresponds to low energy absorption by a molecule and vice versa. Table 4.1 lists the C=O stretching vibration frequencies in different chemical groups.

In general, the FTIR offers very low signal-to-noise ratios. The advantage for using FTIR as a materials characterization tool is that it is a non-destructive technique. A variety of sampling modes are used in conjunction with the FTIR to analyze different types of samples, including:

- attenuated total reflection (ATR),
- specular reflectance,
- infrared reflection absorption spectroscopy (IRRAS), and
- diffuse reflectance infrared Fourier transform spectroscopy (DRIFTS).

4.2.1   Attenuated total reflection (ATR)

Attenuated total reflection (ATR) is one of the commonly used FTIR sampling modes for analyzing biomaterial surfaces (Figure 4.8). In ATR, the sample to be

Table 4.1 C=O stretching vibration frequencies in different chemical groups

Chemical groups	C=O stretching vibration frequencies in wavenumbers (cm^{-1})
Amide	1690
Carboxylic acid	1710
Ketone	1715
Aldehyde	1730
Ester	1735

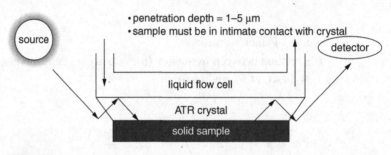

Figure 4.8

Schematic showing a setup for the attenuated total reflection (ATR). (Reprinted from *Biomaterials Science: An Introduction to Materials in Medicine*, 2nd edition, *Ratner, B. D., Hoffman, A. S., Schoen, F. J. and Lemons, J. E., editors. Surface properties and surface characterization of materials, Ratner, B. D., page no. 51, Elsevier (2004)*, with permission from Elsevier.)

analyzed is firmly pressed against a zinc selenide (ZnSe) or a germanium (Ge) crystal having high refractive index. The infrared radiation is allowed to enter the crystal and reflect through the crystal several times. During internal reflectance, evanescent waves are produced. Evanescent waves are the waves which undergo total internal reflection because they strike at an angle greater than the critical angle. These waves pass beyond the crystal and penetrate the sample to a depth of 1–5 μm. These evanescent waves then undergo attenuation depending on the amount of energy absorbed by the sample in the infrared region. The attenuated energy of evanescent waves is passed back to the IR radiation which comes out of the crystal and enters the detector. ATR is not an entirely surface sensitive technique because of the large penetration depth. However, it is capable of providing significant structural information about polymers and proteins.

4.2.2   Specular reflectance

In specular reflectance sampling mode, the angle of reflection of an infrared beam is equal to its angle of incidence. Specular reflectance is commonly used to study thin films on smooth and reflective surfaces. No sample preparation is needed exclusively for this sampling mode. When an infrared beam is allowed to pass through a thin film deposited on a polished surface, a combination of both absorption and reflection phenomena occur. The beam passes through the thin film through absorption, reflects at the polished surface, and subsequently passes out of the thin film. The absorption of infrared energy by the thin film at its characteristic wavelength is then used to obtain the final IR spectrum. A variety of organic and inorganic coatings on polished metal surfaces can be studied using this technique. An angle of 30–45° from normal is typically used in specular reflectance mode to study thin films, with thickness in the range of submicrometer to 100 μm.

4.2.3   Infrared reflection absorption spectroscopy (IRRAS)

In this sampling mode, the angle of incidence of the IR beam used is typically 80° or greater (Figure 4.9). Such high angle of incidence makes this technique highly surface sensitive. This technique is also known as grazing angle analysis. IRRAS is widely used to study self-assembled monolayers (1–3 nm thickness) on highly reflective gold surfaces or any films with thickness less than 10 nm on highly polished metal surfaces. The use of a polarizer in conjunction with IRRAS reveals information regarding the orientation of molecules on the metal surface.

4.2.4   Diffuse reflectance infrared Fourier transform spectroscopy (DRIFTS)

Diffuse reflectance infrared Fourier transform spectroscopy (DRIFTS) is the sampling mode that is commonly used for analyzing particles, powders, and rough surfaces (Figure 4.10). When the IR radiation is allowed to interact with a sample of particles, it can be either reflected from the surface of a particle or transmitted through a particle and reflected back from another particle. This transmission–reflection event occurs several times inside the sample. When the scattered radiation comes out of the sample, it is collected and focused on the detector using a large spherical mirror. During the sample–IR radiation

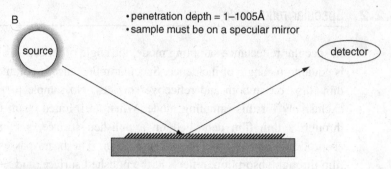

Figure 4.9

Schematic showing a setup for the infrared reflection absorption spectroscopy (IRRAS). (Reprinted from *Biomaterials Science: An Introduction to Materials in Medicine*, 2nd edition, Ratner, B. D., Hoffman, A. S., Schoen, F. J. and Lemons, J. E., editors. Surface properties and surface characterization of materials, Ratner, B. D., page no. 51, Elsevier (2004), with permission from Elsevier.)

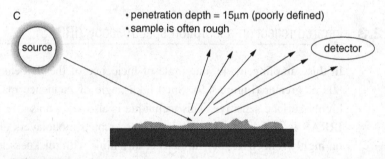

Figure 4.10

Schematic showing a setup for the diffuse reflectance infrared Fourier transform spectroscopy (DRIFTS). (Reprinted from *Biomaterials Science: An Introduction to Materials in Medicine*, 2nd edition, Ratner, B. D., Hoffman, A. S., Schoen, F. J. and Lemons, J. E., editors. Surface properties and surface characterization of materials, Ratner, B. D., page no. 51, Elsevier, (2004), with permission from Elsevier.)

interaction, some energy in the radiation is absorbed by the particles of the sample. This information is later used to generate an IR spectrum. The particle size is crucial in determining the efficiency of this technique. The size of particles is typically ground to 5 μm or less for effective analysis of the samples using DRIFTS. The sample powder is contained in an open cup and mixed with potassium bromide powder at dilutions of about 5%–10% to avoid any specular reflectance. In this sampling mode, the penetration depth of IR radiation into the sample is estimated to be about 100 μm.

- FTIR provides detailed information regarding the molecular structure of a sample.
- ATR is commonly used to study the structure of various polymers and proteins.
- IRRAS is used to study the structure of thin films on reflective surfaces.
- DRIFTS is used for analyzing particles, powders and rough surfaces.

4.3 X-ray photoelectron spectroscopy

X-ray photoelectron spectroscopy (XPS) is also known as electron spectroscopy for chemical analysis (ESCA). It uses an analytical instrument to obtain the chemical composition of a material surface and works on the principle of the photoelectric effect. When a beam of X-rays with photons of energy, hv, is irradiated on a solid surface, it ejects electrons from the inner shells of atoms present on the sample surface. The kinetic energy (KE) of the ejected electrons (photoelectrons) is measured. The binding energy (BE) of the photoelectrons is then calculated using the following relationship in Eq. (4.7):

$$BE = hv - KE. \qquad (4.7)$$

In the above equation, BE is the binding energy of the photoelectrons, hv is the energy of the photons (where h is Planck's constant (6.626×10^{-34} Js) and v is the frequency of the radiation), and KE is the kinetic energy of the photoelectrons. Since the binding energies of electrons coming from the inner shells of different atoms are readily available in the scientific literature, it is easy to identify the type of elements present on the surface.

XPS consists of an X-ray source, an ultrahigh vacuum chamber; an electron energy analyzer, and a computer to process the data (Figure 4.11). X-ray sources such as aluminum K_α with photon energy of 1486.7 eV and magnesium K_α with photon energy of 1253.6 eV are typically used in XPS. The samples to be analyzed are placed in an ultrahigh vacuum chamber to avoid potential contamination on the sample surface and to maximize the photoelectrons reaching the analyzer effectively. An electron energy analyzer measures the kinetic energy of the photoelectrons. Software in the attached computer then processes the data and converts the kinetic energy to binding energy.

XPS detects the presence of all elements except hydrogen and helium. It also provides information regarding the chemical bonding of an element. The binding

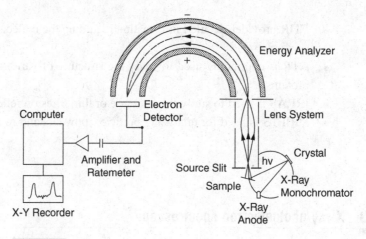

Figure 4.11

Schematic of an X-ray photoelectron spectrometer. (Sibilia, J. P., *A Guide to Materials Characterization and Chemical Analysis*, page no. 169 (1988). Copyright Wiley-VCH Verlag GmbH & Co. KGaA. Reproduced with permission.)

energies of electrons coming from an atom can be different depending on its bonding to the neighboring elements. For example, a carbon atom can bind to an oxygen atom through a single bond or double bond. Table 4.2 provides the binding energies of carbon at different bonding states.

Typically, the XPS spectrum is presented with BE on the y-axis and intensity of the peaks on the x-axis. As an example, Figure 4.12 shows the chemical structure of flufenamic acid (FA), an anti-inflammatory drug, and Figure 4.13 shows a high resolution XPS C 1s spectrum of FA. The high resolution C 1s spectrum is deconvoluted into five components, with the peaks of the components, C 1s (1), C 1s (2), C 1s (3), C 1s (4), and C 1s (5) observed at 285 eV, 286 eV, 289.5 eV, 291.2 eV, and 293.1 eV, respectively. The peaks of C 1s (1), C 1s (2), C 1s (3), C 1s (4), and C 1s (5) are assigned to carbon atoms in C—C, C—O, C=O, $\pi \to \pi^*$ shake-up satellite from the aromatic rings, and C—F_3 bonds, respectively.

The sampling depth of X-rays in XPS is 1–10 nm and so, it provides information regarding the top 1–10 nm of a material surface. The detection limit varies from 0.1 atom% to 1.0 atom%, and the minimum analysis area varies from 10 μm to 200 μm depending on the design of the instrument. Although XPS is a surface sensitive technique, *depth profiling* can provide elemental information regarding the material's subsurface. During depth profiling, a chemically inert beam of energetic ions is used to etch the sample surface and the new surface thus obtained is analyzed using XPS. This process is continued to obtain elemental composition for up to a depth of 1 μm. Thus, XPS depth profiling is very valuable in quantifying elements as a function of sample depth.

Table 4.2   Binding energies of carbon at different bonding states

Chemical species	Binding energy (eV)
C—C	285
C—O	286.5
C=O	289
C—F	290
C—F$_2$	292
C—F$_3$	293

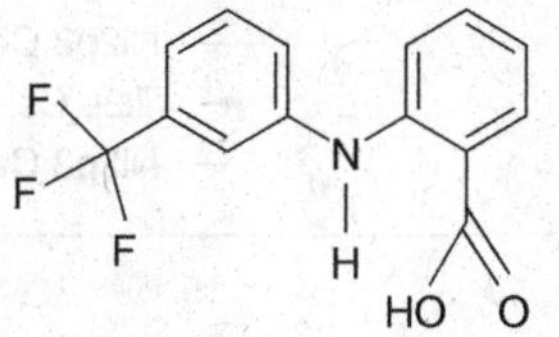

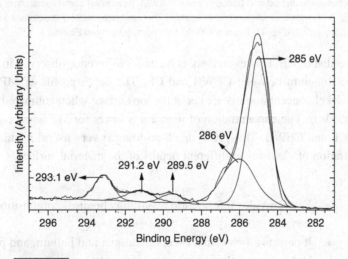

Figure 4.12

Chemical structure of flufenamic acid.

Figure 4.13

High resolution XPS C 1s spectrum of flufenamic acid.

Figure 4.14 shows an example of an XPS depth profile for a nitinol surface which was finished under different methods: (a) mechanically polished (MP); chemically etched (CE); chemically etched and boiled in water (CEWB).[1] The nickel concentration is slightly lower for CEWB at the top surface when compared

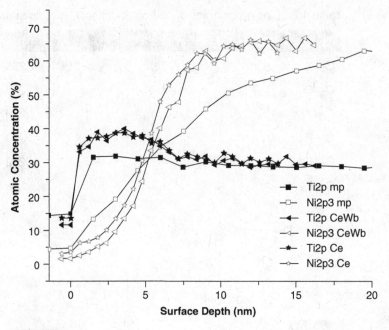

Figure 4.14

XPS depth profiles of Ni and Ti for mechanically polished (MP), chemically etched (CE), and chemically etched and boiled in water (CEWB). (Reprinted from *Biomaterials*, **30**, Shabalovskaya, S. A., Rondelli, G. C., Undisz, A. L. *et al.*, The electrochemical characteristics of native Nitinol surfaces, 3662–3671, Elsevier (2009), with permission from Elsevier.)

to that of CE. However, there is no major difference observed in the concentration of titanium between CEWB and CE. The depth profile of MP showed that the nickel concentration is greater at the top surface when compared to that of CE and CEWB. The concentration of titanium is lower for MP when compared to that of CE and CEWB. Thus, XPS depth profiling is very useful in studying the concentration of elements at different depths of the material surface.

- XPS provides information regarding chemical composition of a material surface.
- It detects all elements except hydrogen and helium, and provides chemical bonding information of an element.
- XPS depth profiling provides elemental information regarding the subsurface for up to a depth of 1 μm.

Other methods of evaluating the subsurface include the use of *angle-resolved* XPS. As grazing take-off angles provide more surface information compared to angles close to the surface normal, reduction in the photoelectron take-off angle allows XPS information to be obtained from decreasing depths. The use of angle-resolved XPS permits the estimation of the film thicknesses and thickness of the contaminated layer.

4.4   Secondary ion mass spectrometry

Secondary ion mass spectrometry (SIMS) is a highly sensitive analytical tool available to determine the atomic and molecular composition of a material surface. As shown in Figure 4.15, SIMS consists of (a) a primary ion gun; (b) an ultrahigh vacuum chamber; (c) an energy analyzer; (d) a mass analyzer; (e) an ion detector; and (f) a computer to process the data. The primary ions that are commonly used in SIMS are argon (Ar^+), xenon (Xe^+), oxygen (O_2^+), gallium (Ga^+), and cesium (Cs^+). The basic process involved in SIMS is sputtering. When a beam of primary ions is allowed to collide with a material surface, a stream of secondary particles is generated. The secondary particles consist of mainly neutral species and few

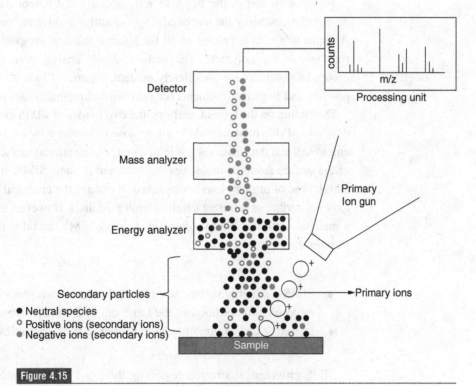

Figure 4.15

Schematic of the principle of secondary ion mass spectrometry.

positive and negative ion species (secondary ions). Approximately, 1% of the total secondary particles are secondary ions. The energy analyzer (located between the sample and mass analyzer) separates the secondary ions from the mixture of secondary particles, and passes them on to the mass analyzer. The mass analyzer separates the secondary ions based on their mass to charge ratio (m/z) and the detector counts the number of ions according to their m/z. Although different types of analyzers are available, the highly sensitive time of flight (Tof) is the most commonly used mass analyzer. Tof calculates the m/z of secondary ions based on the time taken by the ions to reach the detector. An ultrahigh vacuum chamber is essential in SIMS for the secondary ions to reach the detector with no significant loss. Additionally, the surface contamination due to adsorption of gas particles is prevented under vacuum conditions.

In graphical form, the data from SIMS is usually presented with m/z on the x-axis and intensity of the peaks on the y-axis. The elements can be identified by their exact m/z. The data interpretation in SIMS is relatively complex when compared to XPS. As shown in Figures 4.16 and 4.17, the peaks for a variety of combinations of atomic and molecular fragments can be obtained. The interpretation of SIMS spectra needs extensive training and experience.

Figure 4.16 shows the negative ion spectra for different methacrylate polymers,[2] demonstrating the use of SIMS in identifying different polymer structures. A signal at 85 Da is present in all the spectra and it is assigned to the backbone fragment of the polymers. The unique signals arising from the different side groups of polymers are also clearly evident. Figure 4.17 shows the characteristic positive and negative fragments for poly(methyl methacrylate) polymer.[2]

Depending on the type of analysis, the two modes of SIMS are static SIMS and dynamic SIMS. In static SIMS, a low dose of primary ions ($<10^{13}$ ions/cm^2) is employed, and thus involves minimal damage to a material surface. Less than 10% of the surface atoms or molecules are sputtered in static SIMS. In dynamic SIMS, a high dose of primary ions is employed. It obtains the chemical composition of a material surface at different depths (depth profiling). However, the degradation of a material surface is not avoidable in dynamic SIMS, and it is thus a destructive technique.

- SIMS – a highly surface-sensitive technique which provides atomic and molecular composition of the top 1 nm of a material surface.
- It needs extensive training and experience to interpret SIMS data.

SIMS provides information regarding the top 1 nm of a material surface and detects all elements including hydrogen. It is also capable of detecting the different

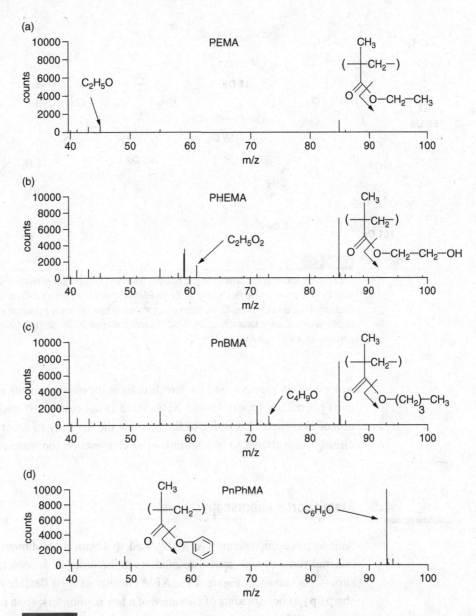

Figure 4.16

Negative ion fingerprint spectra for methacrylate polymers: (a) poly(ethyl methacrylate) (PEMA) (b) poly(hydroxyethyl methacrylate) (PHEMA) (c) poly(n-butyl methacrylate) (PnBMA), and (d) poly (phenyl methacrylate) (PPhMA). The key identifying fragment for each polymer has been highlighted. (Reprinted from *Biomaterials*, **24**, Belu, A. M., Graham, D. J. and Castner, D. G. Time-of-flight secondary ion mass spectrometry: techniques and applications for the characterization of biomaterial surfaces, 3635–3653, Elsevier (2003), with permission from Elsevier.)

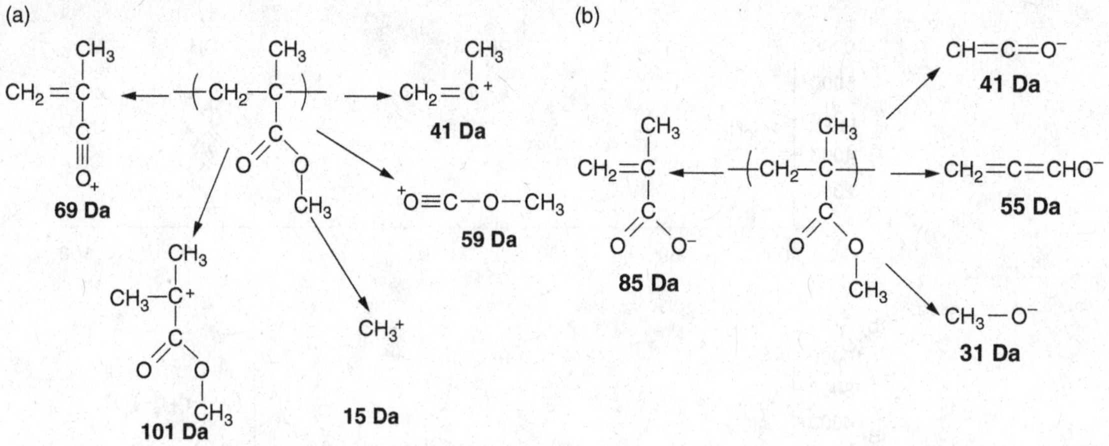

(a)

(b)

Figure 4.17

Schematics showing (a) characteristic positive fragments for poly(methyl methacrylate) (PMMA), and (b) characteristic negative fragments for PMMA. (Reprinted from *Biomaterials*, **24**, Belu, A. M., Graham, D. J. and Castner, D. G. Time-of-flight secondary ion mass spectrometry: techniques and applications for the characterization of biomaterial surfaces, 3635–3653, Elsevier (2003), with permission from Elsevier.)

isotopes of an element, and its detection limit for most elements is in the range of parts per million (ppm). Unlike XPS, SIMS is not commonly used for quantification of elements. This is due to the fact that the intensity of secondary ions is not directly proportional to concentrations of elements on the material surface.

4.5 Atomic force microscopy

Atomic force microscopy (AFM) is used to obtain three-dimensional images of material surfaces with spatial resolutions in the order of nanometers or even angstroms. As shown in Figure 4.18, AFM consists of (a) a flexible cantilever with a sharp tip (probe) of radius of curvature of a few nanometers; (b) a piezoelectric tube scanner capable of moving the sample in x, y, and z directions; (c) a laser source; (d) a photodetector; and (e) a computer to process the image. When the probe is brought close to the sample surface at a distance of few angstroms, the forces (electrostatic, van der Waals, or capillary) generated between the tip and the material surface deflect the cantilever. In order to maintain a constant force between the probe and the sample surface, the piezoelectric scanner moves the sample up and down, allowing the probe to follow the contour of the surface. In parallel, the deflection of the probe is measured using a laser beam reflected from the top surface of the cantilever into a

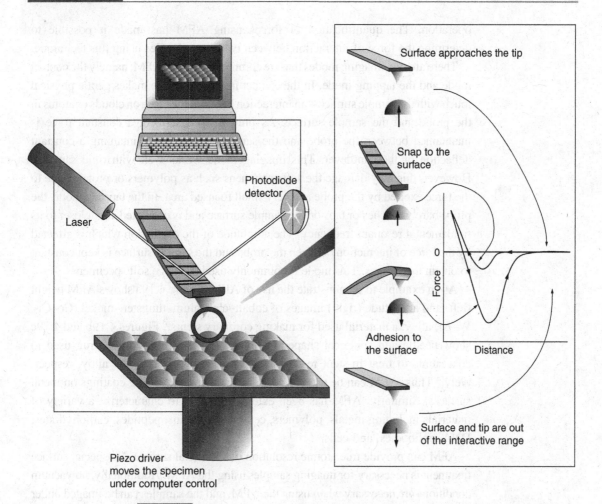

Laser

Photodiode detector

Piezo driver
moves the specimen
under computer control

Surface approaches the tip

Snap to the surface

0

Force

Distance

Adhesion to the surface

Surface and tip are out of the interactive range

Figure 4.18

Schematic illustrating the principle of atomic force microscopy. (Reprinted from *Biomaterials Science: An Introduction to Materials in Medicine*, 2nd edition, Ratner, B. D., Hoffman, A. S., Schoen, F. J. and Lemons, J. E., editors. Surface properties and surface characterization of materials, Ratner, B. D., page no. 53, Elsevier, (2004), with permission from Elsevier.)

photodetector. The detector receives signals continuously, and the acquired signals are converted into images using the computer.

Using Eq. (4.8), the force generated between the probe and sample surface can be quantified using Hooke's law:

$$F = -kx. \qquad (4.8)$$

The spring constant, k, is a known factor since cantilevers with a wide range of spring constants are commercially available. The cantilever deflection, x, is the difference between two signals received in the photodetector during the scanning operation. The quantification of forces using AFM has made it possible to determine the force of interaction between two biomolecules using this technique.

There are two imaging modes that are commonly used in AFM, namely the contact mode and the tapping mode. In the contact mode, the probe makes gentle physical touch with the sample surface – an interaction between the electron clouds of atoms in the probe and the sample surface. As previously described, a constant force is maintained between the probe and the sample surface by maintaining a constant deflection of the cantilever. This imaging mode works well with rigid samples. However, this may damage the soft specimens such as polymers or proteins due to the force exerted by the probe at a very small focused area. In the tapping mode, the probe barely touches or taps on the sample surface and is oscillated at or closer to its fundamental resonant frequency. The amplitude of the oscillation which is affected by the force of interaction between the probe and the sample surface is kept constant to obtain the images. This mode is commonly used to image soft specimens.

As an example to demonstrate the use of AFM, Figure 4.19a shows AFM height (left) and amplitude (right) images of cobalt–chromium–tungsten–nickel (Co–Cr–W–Ni) alloy, a material used for making coronary stents.[3] Figures 4.19b and 4.19c show needle and spherical shaped paclitaxel (an anti-proliferative drug used to coat stents to treat in-stent restenosis) crystals on Co–Cr–W–Ni alloy, respectively.[3] Thus, AFM can be used to characterize therapeutic drug coatings on metal surfaces. Similarly, AFM has been extensively used to characterize a variety of materials including metals, polymers, ceramics, proteins, peptides, carbohydrates, DNA, antibodies, and cells.

AFM can provide true atomic resolution of a material surface. No special surface treatment is necessary for imaging samples using the AFM. Additionally, no vacuum conditions are necessary when using the AFM, and the samples can be imaged under ambient laboratory conditions. Additionally, AFM is also capable of imaging biological specimens in aqueous environment. The limitations of AFM include its limited vertical scan range ($\sim$10 μm) and XY scan size ($\sim$100 μm $\times$ 100 μm). Artifacts are common in AFM images due to bad probes, scanner geometry, sample drifts, environmental vibrations, and surface contamination. Hence, care must be taken to ensure that the image is obtained from the actual sample surface and is free of artifacts.

- AFM provides 3D images of a material surface with spatial resolution in nanometers or ångströms.
- A tapping mode is useful for imaging soft specimens such as polymers or proteins.
- The contact mode is preferred to obtain molecular resolution of a material surface.

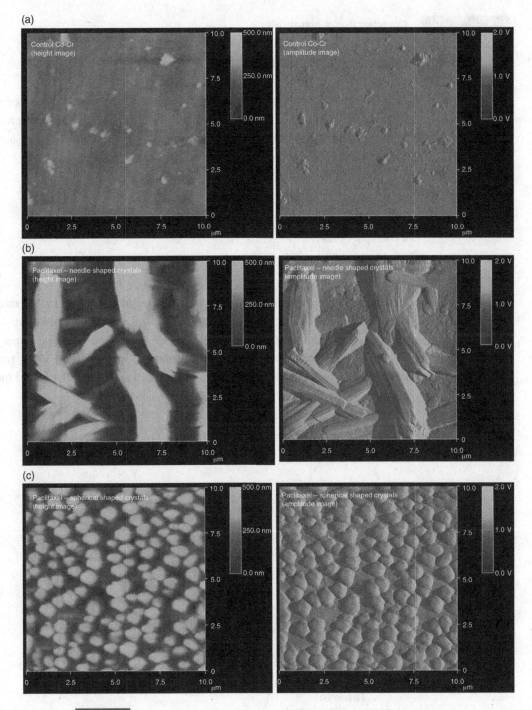

Figure 4.19

AFM tapping mode height (left) and amplitude (right) images of (a) Co–Cr–W–Ni alloy (b) needle shaped paclitaxel crystals coated Co–Cr–W–Ni, and (c) spherical paclitaxel crystals coated Co–Cr–W–Ni. (Reprinted from *Biomaterials*, **31**, Mani, G., Macias, C. E., Feldman, M. D., Marton, D., Oh, S. and Agrawal, C. M., Delivery of paclitaxel from cobalt–chromium alloy surfaces without polymeric carriers, 5372–5384, Elsevier (2010), with permission from Elsevier.)

4.6 Scanning electron microscopy

Scanning electron microscopy (SEM) is commonly used to image the topography of a material surface. Easy to use and maintain, the SEM is capable of producing high resolution 3D-like images with greater depth of field. As shown in Figure 4.20, SEM consists of the following:

- electron gun,
- set of electromagnetic lenses and apertures,
- scanning coils,
- sample chamber,
- detector, and
- computer.

A tungsten filament, typically used as the electron gun, is heated to generate high-energy electrons (30 keV). The resulting electron beam is narrowed down and focused on the sample surface using condenser and objective lenses, respectively. Finally, the electron beam is rastered across the sample surface using scanning coils. The collision of electrons on the sample surface causes elastic and inelastic

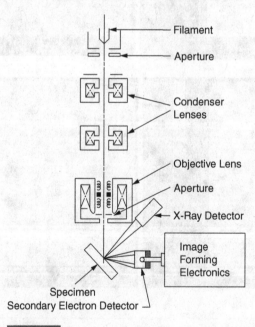

Figure 4.20

Schematic of the structure of scanning electron microscope. (Sibilia, J. P. *A Guide to Materials Characterization and Chemical Analysis*, page no. 143 (1988). Copyright Wiley-VCH Verlag GmbH & Co. KGaA. Reproduced with permission.)

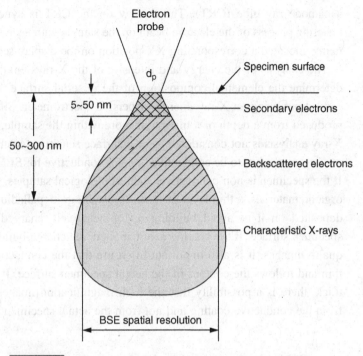

Electron
probe

Specimen surface

d_p

5~50 nm

Secondary electrons

50~300 nm

Backscattered electrons

Characteristic X-rays

BSE spatial resolution

Figure 4.21

Schematic of the generation of secondary electrons, backscattered electrons, and X-rays at different depths of the specimen surface. (Reprinted from *Materials Characterization: Introduction to Microscopic and Spectroscopic Methods*, Leng, Y., page no. 130, John Wiley & Sons (Asia) Pte Ltd (2008), with permission from John Wiley & Sons (Asia) Pte Ltd.)

scattering to generate a variety of signals including secondary electrons (SE), backscattered electrons (BSE), X-rays, auger electrons, and cathodoluminescence radiation (Figure 4.21). These SE, BSE, and X-rays are signals that are commonly detected and analyzed to obtain information regarding topography, atomic number, and chemical composition, respectively. Elastic scattering of electrons results in the transfer of a negligible amount of energy to the colliding atoms, and the electrons bounce off the sample surface as backscattered electrons. Inelastic scattering of electrons results in the transfer of their energy (partly or all) to the colliding atoms and this excess energy is expended by the atoms by releasing secondary electrons or X-rays. Everhart–Thornley (E–T) is the detector system commonly employed in SEM. The secondary electrons generated are attracted to the detector using a positive-charged Faraday cage and are accelerated to a positive-charged scintillator located at the back of the cage. The electrons are converted into photons of light by the scintillator and are transferred to a photo-multiplier tube through a light guide. The photomultiplier tube converts the photons back to electrons and the collected signal is amplified and displayed on a cathode ray tube (CRT). The display on the CRT is synchronous with the rastering process of the electron beam on the sample with each spot on the sample being allocated a corresponding XY position on the display screen.

The analysis of the energy and intensity of the X-rays emitted can be used to determine the elemental composition of the material surface. This is commonly carried out using an X-ray energy dispersive spectrometer. Since the X-rays are produced from a depth of a micron or more within the sample, energy dispersive X-ray analysis is not considered to be a surface sensitive technique.

A specimen has to be conductive or semi-conductive for SEM characterization. If the specimen is non-conductive, such as biological samples, polymers, or most organic materials, a thin layer of metal (typically gold–palladium alloy) has to be deposited on it to avoid building up of negatively charged electrons on the specimen surface. If the coating is not applied, the charge-build may lead to poor quality images. It is also important to ensure that the conductive coating is very thin and follows the contour of the actual specimen surface. If the coating is too thick, there is a possibility that the high magnification images acquired will be from the conductive coating and not from the actual specimen surface.

- SEM is very commonly used to study the topography of any material surface.
- Secondary electron, backscattered electron, and X-ray signals in a SEM are used to study topography, atomic number, and chemical composition, respectively.
- A specimen has to be conductive (or semi-conductive) for characterization using SEM.

Figure 4.22 shows the SEM images of a chemically cleaned Co–Cr alloy and a sandblasted Co–Cr alloy. As seen from the images, a flat topography was observed for the smooth metal surfaces, whereas a roughened surface is obtained after the metal is sandblasted. Thus, SEM is useful for characterizing surface morphology and observing changes therein. It is used extensively to study material surfaces and for visualizing cell–material interactions.

4.7 Transmission electron microscopy

Transmission electron microscopy (TEM) is used to study the microtexture and crystal structure of samples. The principle behind TEM is similar to that for a transmission light microscope. Because the wavelength of electrons is shorter by a

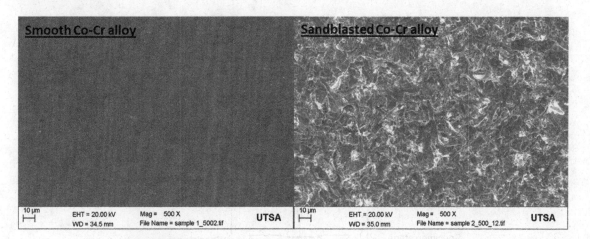

Figure 4.22

SEM images of a smooth Co–Cr alloy (left) and a sandblasted Co–Cr alloy (right).

factor of 10 000 than that of the visible light, the resolution of a TEM can reach up to 0.1 nm. In contrast to SEM, a high degree of skill is needed for sample preparation, instrument operation, and data interpretation of TEM.

As shown in Figure 4.23, the electrons in TEM are generated in a vacuum chamber by an electron gun, and they are converged and focused on a specimen using a series of electromagnetic lenses. When these electrons interact with a specimen, depending on the density of the specimen, some of them undergo scattering, whereas other electrons do not. The non-scattered electrons which pass through the specimens with no interactions are of interest here. These electrons will finally hit the fluorescent screen which is placed at the bottom of the microscope, and an image is generated based on the density of the materials present inside the specimen. If some spots in the specimen are dense, then only a small number of electrons will pass through and they will appear darker on the screen. If the areas are less dense, then a greater number of electrons will pass through and they will appear lighter on the screen. Hence, the differences in the density of the materials give rise to an image with contrasting dark and light areas.

The specimen has to be ultrathin (100 nm) in order to be characterized by using the TEM. This is so that the beam of electrons can transmit through the specimens. The specimen preparation for TEM is laborious. Electrolytic thinning, ion milling and ultramicrotomy are some of the techniques that have been used for making ultrathin specimens. Among these, ultramicrotomy is commonly used for preparing biological and polymeric specimens.

TEM can work in two modes, namely the image mode and the diffraction mode. The microstructure of the material is visualized in the image mode. In the diffraction mode, the crystal structure of the material is determined by recording the

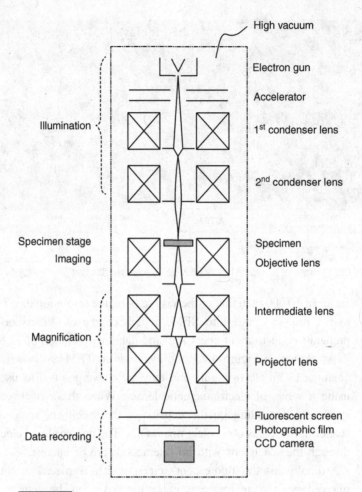

High vacuum

Electron gun

Accelerator

Illumination

1st condenser lens

2nd condenser lens

Specimen stage
Imaging

Specimen

Objective lens

Intermediate lens

Magnification

Projector lens

Fluorescent screen
Photographic film
CCD camera

Data recording

Figure 4.23

Schematic of the structure of transmission electron microscope. (Reprinted from *Materials Characterization: Introduction To Microscopic and Spectroscopic Methods*, Leng, Y., page no. 80, John Wiley & Sons (Asia) Pte Ltd (2008), with permission from John Wiley & Sons (Asia) Pte Ltd.)

electron diffraction pattern on the screen. A single crystal, a polycrystalline material, and an amorphous material will produce a series of spots, rings, and diffuse halos patterns, respectively.

- TEM is used to study microtexture and crystal structure of material surfaces.
- TEM specimen preparation is laborious.

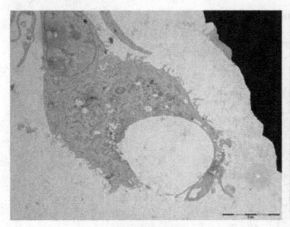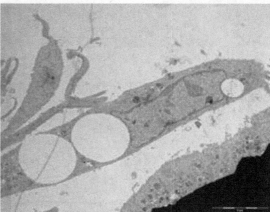

Figure 4.24

TEM images of synoviocytes internalizing microparticles: a synoviocyte undergoing phagocytosis (left) and a synoviocyte that has internalized three microparticles (right). (Reprinted from *Biomaterials*, **30**, Butoescu, N., Seemayer, C. A., Foti, M., Jordon, O. and Doelker, E. Dexamethasone-containing PLGA superparamagnetic microparticles as carriers for the local treatment of arthritis, 1772–1780, Elsevier (2009), with permission from Elsevier.)

As an example, Figure 4.24 shows the TEM images of synovial fibroblasts undergoing phagocytosis.[4] The left image shows that a synoviocyte is in the process of engulfing a microparticle (10 μm). The right image shows the three microparticles that have already been taken up by the synoviocyte. Thus, TEM has been used for studying cellular and subcellular morphology. TEM is also extensively used to characterize nanoparticles and several other nanostructures including dendrimers, carbon nanotubes, and quantum dots.

4.8 X-ray diffraction (XRD)

XRD is a non-destructive analytical technique commonly used to characterize the crystallographic structure and physical properties of materials. An X-ray diffraction pattern of a material is unique and is often referred to as a "finger print" of the material. Shown in Figure 4.25, an X-ray diffractometer consists of an X-ray generator; a diffractometer unit which controls the alignment of X-ray beam and the position and orientation of specimen as well as the detector; a detector; and a computer to collect and process the data. When a beam of monochromatic X-rays is allowed to irradiate on a crystalline material, the X-rays reflect or diffract at different angles with respect to the primary beam. The relationship between the X-ray's wavelength (λ), diffraction angle (θ), and the separation

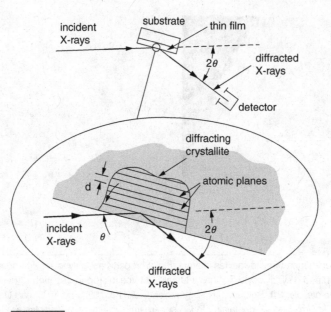

Figure 4.25

Schematic illustrating the principle of X-ray diffraction. (Reprinted from *Encyclopedia of Materials Characterization*, Brundle, C. R., Evans, C. A. and Wilson, S. editors. X-ray Diffraction, Toney, M. F., page no. 200, Elsevier, (1992), with permission from Elsevier.)

distance (d) between the atomic planes of crystal lattice is provided in Eq. (4.9) (Bragg's law):

$$n\lambda = 2d \sin \theta, \tag{4.9}$$

where n is the order of diffraction.

The spacing (d) between the atomic planes of the crystal lattice, calculated from the Eq. (4.9), provides information regarding the dimensions of the unit cell of the crystal, while the intensities of the diffraction peaks provide information regarding the position of atoms in the unit cell of the crystal.

Typically, the XRD data are presented with scattering angle on the x-axis and the intensity of diffraction peaks on the y-axis. An example XRD data set and its interpretation are provided below.

Figure 4.26 shows the XRD pattern for a powder mixture of alumina (Al_2O_3) and silicon (Si).[5] Several peaks observed at different 2θ values represent the finger print region for the crystalline solid, Al_2O_3. Each of the peaks observed in the pattern denotes diffraction from a certain crystallographic plane. The pattern obtained can be readily matched with the standard XRD database to determine crystalline substances in the specimen.

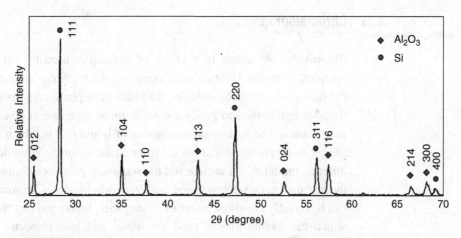

Figure 4.26

XRD spectrum for a powder mixture of Si and Al_2O_3. (Reprinted from *Materials Characterization: Introduction to Microscopic and Spectroscopic Methods*, Leng, Y., page no. 62. John Wiley & Sons (Asia) Pte Ltd (2008), with permission from John Wiley & Sons (Asia) Pte Ltd.)

The XRD diffraction pattern of a fully crystalline polymer is composed of sharp and narrow peaks which arise from the diffractions of several crystallographic planes. The diffraction pattern of an amorphous polymer is composed of broad peaks (halo), which represent the average separation of polymer chains. Thus, the diffraction pattern of a semi-crystalline polymer is the superposition of sharp and narrow peaks on broad peaks.

XRD has tremendous biomedical applications in determining the crystallinity of biomaterials including a variety of metals, polymers, and ceramics. XRD is also used to determine the crystallographic structure of different organic and inorganic materials. It is possible to obtain information regarding more than one compound in a material since the diffraction pattern of one compound does not affect the other. In addition to the crystallographic structure, XRD provides invaluable information regarding phase transformation, crystal size, degree of crystallinity, crystal orientation (texture), and stress or strain measurements. Although XRD does not have the capability to quantify elemental or molecular composition, the XRD data does provide inferred chemical composition once crystalline phases are identified.

- XRD is commonly used to study the crystal structure of metals, polymers, and ceramics.
- The XRD pattern of a material is unique and it represents the finger print of the material.

4.9   Chromatography

Chromatography refers to a group of techniques used to separate and analyze complex mixtures. The two main components of any chromatography system are the stationary and mobile phases. The stationary phase consists of solid or porous particles tightly packed inside a glass or metal tube; this is commonly referred to as a column. The mobile phase consists of liquid or gas which is mixed with the sample to be analyzed and passed through the column. The interactions between different mixtures in a sample and the stationary phase are unique, which facilitate the mixtures to be separated and analyzed individually. Chromatographic systems can be broadly classified under two categories, liquid and gas chromatography, in which the mobile phases used are liquid and gas, respectively. A variety of interactions including hydrogen bonding, van der Waals forces, electrostatic bonding, and hydrophobic interactions occur between the samples and the stationary phase. Based on the separation mechanisms, chromatographic systems can also be classified as follows:

- adsorption chromatography,
- partition chromatography,
- bonded phase chromatography,
- ion exchange chromatography,
- affinity chromatography, and
- size exclusion chromatography.

It is beyond the scope of this chapter to cover all the systems. Hence, this section is confined only to two chromatographic systems that are most commonly used for biomedical applications, namely high performance liquid chromatography (HPLC) and gel permeation chromatography (GPC).

4.9.1   High performance liquid chromatography (HPLC)

The HPLC is used extensively in the biomedical field for the separation and analysis of therapeutic drug molecules, proteins, amino acids, and nucleic acids. The principle of HPLC relies on high pressure to pump the solvent through the column. As with any other liquid chromatography, HPLC can run in either normal phase or reverse phase, depending on the polarity of stationary and mobile phases. Normal-phase HPLC consists of a polar stationary phase and a non-polar mobile phase. In contrast, the reverse-phase HPLC consists of a non-polar stationary phase and a polar mobile phase. In normal-phase HPLC, the stationary phase consists of micron-sized porous silica particles and the mobile phase consists of

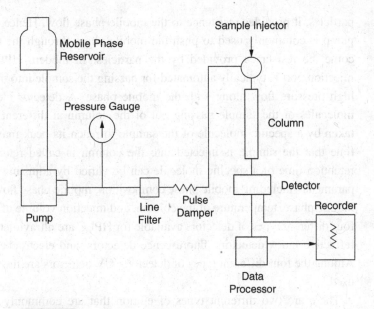

Schematic of a high performance liquid chromatograph. (Sibilia, J. P. *A Guide to Materials Characterization and Chemical Analysis*, page no. 67 (1988). Copyright Wiley-VCH Verlag GmbH & Co. KGaA. Reproduced with permission.)

non-polar solvents such as hexane. When the sample is injected into the column, the polar molecules of the sample interact with polar silica particles and elute at a longer time while the non-polar molecules pass through the column quickly. In the case of reverse-phase HPLC, the silica particles in the column are typically surface modified with long hydrocarbon chain molecules which provide non-polar characteristics to the stationary phase. The sample is mixed with polar solvents (e.g. water and methanol mixture) and allowed to pass through the column. The polar molecules of the sample interact with polar solvents and pass through the column quickly since the interaction with the non-polar stationary phase is minimal. However, the non-polar molecules strongly interact with the non-polar stationary phase and elute at a much slower rate. Reverse-phase HPLC is more commonly used than normal-phase HPLC. As shown in Figure 4.27, a HPLC system consists of a mobile phase reservoir, a high-pressure pump, an injector, a column with the stationary phase, a detector, and a data analysis unit.

The diameter of porous silica particles in the HPLC column typically lies between 3 μm and 10 μm. The significant advantage of using such small size particles is the availability of greater surface area, which increases the interaction between the sample and the stationary phase. This greatly enhances the separation efficiency of the column. When the column is closely packed with numerous particles, it provides resistance to the mobile phase flow. Hence, a high pressure pump is commonly used to push the mobile phase through the column to overcome the resistance provided by the particles. In modern HPLC systems, the injection port is typically automated for passing the sample into the column under high pressure flow along with the mobile phase. A detector identifies different molecules of the sample passing out of the column at different rates. The time taken by a specific molecule of the sample to reach its peak maximum from the time that the sample is injected into the column is called retention time. This retention time of a specific molecule can be varied by adjusting different HPLC parameters including mobile phase composition, mobile phase flow rate, pH of the mobile phase, temperature of the column, and injection volume of the sample. The four different types of detectors available for HPLC are ultraviolet (UV) detectors, refractive index detectors, fluorescence detectors, and electrochemical detectors. Among the four different types of detectors, UV detectors are the most commonly used.

There are two different types of elution that are commonly used in HPLC, namely *isocratic elution* and *gradient elution*. In isocratic elution, the composition of the mobile phase does not change throughout the elution period. However, in gradient elution, the composition of the mobile phase changes in such a way that the solvent strength gradually increases during the elution period. The solvent strength is increased by increasing the percentage of organic solvent (acetonitrile or methanol) in the mobile phase. The two significant advantages of gradient elution are:

- strongly retained compounds move faster, and
- peaks of weakly retained compounds are well resolved.

4.9.2 Gel permeation chromatography (GPC)

Gel permeation chromatography (GPC) is used extensively for characterizing polymers of different molecular weights. Similar to the HPLC, GPC is also a liquid chromatographic system that is used for the separation and analysis of different biological molecules such as proteins, polysaccharides and nucleic acids. Unlike HPLC, which involves chemical interactions between the sample and the stationary phase, there is no such interaction between the sample and the stationary phase in GPC. Also known as *size exclusion chromatography* (SEC), the separation in GPC is mainly based on the size or the hydrodynamic volume of molecules in the sample. A schematic of the GPC is shown in Figure 4.28.

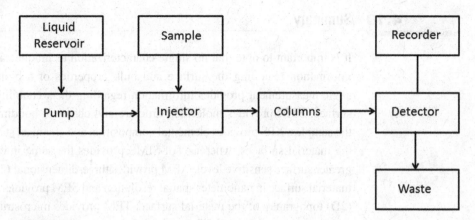

Figure 4.28

Block diagram of a gel permeation chromatograph. (Sibilia, J. P. *A Guide to Materials Characterization and Chemical Analysis*, page no. 82 (1988). Copyright Wiley-VCH Verlag GmbH & Co. KGaA. Reproduced with permission.)

- The interaction (polarity or non-polarity) between the molecules in the samples and the particles in the column form the basis for HPLC separation
- The separation in GPC is mainly based on the size or hydrodynamic volume of molecules in the sample

The column used in GPC is tightly packed with porous beads with pore sizes ranging from 50 Å to 1 000 000 Å. When a polymeric sample containing molecular chains of different molecular weights is allowed to pass through the column, the smaller polymeric molecules traverse through the various narrow pathways available in and between the porous beads and elute at a slow rate. However, the larger polymeric molecules are restricted by size to a reduced number of pathways and pass through fewer beads and elute at a faster rate. Hence, the retention time of polymer molecules is inversely proportional to their size.

The process to separate and analyze complex mixtures requires the initial generation of a calibration graph. This calibration graph is obtained by using polymer molecules of known molecular weight and measuring their retention time. When a polymer sample of unknown molecular weight is injected into the GPC, the retention times for different molecules are obtained and the molecular weights corresponding to the obtained retention times are then determined from the calibration graph.

4.10 Summary

It is important to note that no single characterization technique can provide all the information regarding the surface and bulk properties of a biomaterial. Contact angle measurement provides information regarding the wettability of the material surface. FTIR provides molecular structure and chemical bonding information of the samples. XPS provides elemental composition and chemical state information of the material surfaces, whereas Tof-SIMS provides the same information but at a greater surface sensitive level. AFM provides three-dimensional (3D) images of the material surface in nanometer spatial resolution and SEM provides two-dimensional (2D) topography of the material surface. TEM provides microstructure and crystal structure of samples while XRD provides crystallographic information and physical properties of materials. HPLC separates and analyzes organic and inorganic molecules while GPC does the same operation but for larger molecules such as polymers and proteins. In summary, during the development phase of any medical devices or implants, the success of fully characterizing a material lies in the investigators' ability to access the different analytical instruments, have working knowledge of the analyses, understand the limitations of each technique, adequately prepare the samples, and accurately interpret the data collected.

Suggested reading

- Ratner, B. D. (2004). Surface properties and surface characterization of materials, in Ratner, B. D., Hoffman, A. S., Schoen, F. J. and Lemons, J. E., editors, *Biomaterials Science: An Introduction to Materials in Medicine*. London, Elsevier Academic Press, pp. 40–59.
- Zhang, S., Li, L. and Kumar, A. (2008). *Materials Characterization Techniques*. Boca Raton, CRC Press.
- Pecsok, R. L., Shields, L. D., Cairns, T. and McWilliam, I. G. (1976). *Modern Methods of Chemical Analysis*, New York, John Wiley & Sons.
- Robinson, J. W., Frame, E. M. S. and Frame, G. M. (2005). *Undergraduate Instrumental Analysis*. New York, Marcel Dekker.
- Sibilia, J. P. (1988). *A Guide to Materials Characterization and Chemical Analysis*. Weinheim, Wiley-VCH Verlag GmbH & Co. KGaA.
- O'Connor, D. J., Sexton, B. A. and Smart, R. St. C. (1992). *Surface Analysis Methods in Materials Science*. Berlin Heidelberg, Springer-Verlag.
- Brundle, C. R., Evans, C. A. and Wilson, S. (1992) *Encyclopedia of Materials Characterization*, Stoneham, Butterworth-Heinemann.
- Wachtman, J. B. (1993). *Characterization of Materials*. Manning, Butterworth-Heinemann.

- Vickerman, J. C. (1997). *Surface Analysis – the Principal Techniques*. West Sussex, John Wiley & Sons.
- Brandon, D. and Kaplan, W. D. (1999). *Microstructural Characterization of Materials*. West Sussex, John Wiley & Sons.

References

1. Shabalovskaya, S. A., Rondelli, G. C., Undisz, A. L. *et al.* (2009). The electrochemical characteristics of native nitinol surfaces. *Biomaterials*, **30**(22), 3662–3671.
2. Belu, A. M., Graham, D. J. and Castner, D. G. (2003). Time-of-flight secondary ion mass spectrometry: techniques and applications for the characterization of biomaterial surfaces. *Biomaterials*, **24**(21), 3635–3653.
3. Mani, G., Macias, C. E., Feldman, M. D. *et al.* (2010). Delivery of paclitaxel from cobalt-chromium alloy surfaces without polymeric carriers. *Biomaterials*, **31**(20), 5372–5384.
4. Butoescu, N., Seemayer, C. A., Foti, M., Jordan, O. and Doelker, E. (2009). Dexamethasone-containing PLGA superparamagnetic microparticles as carriers for the local treatment of arthritis. *Biomaterials*, **30**(9), 1772–1780.
5. Leng, Y. (2008). *Materials Characterization: Introduction to Microscopic and Spectroscopic Methods*. Singapore: John Wiley & Sons (Asia) Pte Ltd.

Problems

1. A research group is investigating the interaction of cells with metal specimens coated with a novel molecular coating. The spreading of cells with its characteristic shape was observed only on certain spots of the specimens. The group leader believes that the molecular coating was not homogeneous which led to the poor result. You are asked to determine whether the molecular coating was homogeneous or not in the simplest and quickest way possible. Which instrument will you use? How would you characterize homogeneity/heterogeneity of the coating?

2. The carbonyl (C=O) group is present in different functional groups including amides, ketones and esters. In FTIR characterization, do you expect any changes in the peak positions of C=O groups in amides, ketones and esters? Justify your answer.

3. Match the following (with respect to FTIR characterization).

(a) Attenuated total reflection	(i) Angle of incidence of infrared beam is 85°
(b) Specular reflectance	(ii) Analyze powders, particles and rough surfaces

| (c) Infrared reflectance absorption spectroscopy | (iii) Angle of incidence of infrared beam is 30° |
| (d) Diffuse reflectance infrared Fourier transform infrared spectroscopy | (iv) Information can be obtained at large penetration depth for up to 5 µm |

4. A medical device is coated with four layers of polymers, as shown below. The thickness of each polymer layer is 50 nm. The unique elements present in polymer layers 1, 2, 3 and 4, are sulfur, phosphorous, nitrogen and fluorine. You are provided with XPS to characterize the presence of different layers of polymer coating. As we know that the sampling depth of X-rays in XPS is 1–10 nm, is it possible to determine the presence of unique elements in layers 2, 3 and 4? Justify your answer. (Hint: XPS depth profiling can be used if needed.)

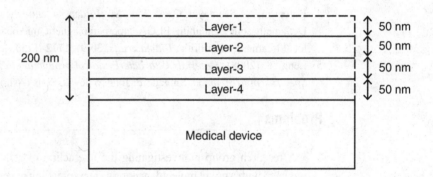

5. Is SIMS a destructive technique? Justify your answer.
6. Which mode (contact or tapping) would you prefer to use when you image proteins using AFM? Justify your answer.
7. Match the following (with respect to SEM characterization).

Signals generated	Information obtained
(a) Secondary electrons	(i) Atomic number
(b) Backscattered electrons	(ii) Chemical composition
(c) X-rays	(iii) Topography

8. What are the two modes in which the TEM can work? Explain your answer.
9. What are the biomedical applications of XRD?
10. What is the primary difference between HPLC and GPC?

5 Metals: structure and properties

Goals

After reading this chapter the student will understand the following.
- Structure and properties of metals commonly used for making biomedical implants and devices.
- Use of different metals as biomaterials.

Metals are extensively used as materials for biomedical implants, devices, and surgical tools. Some of the implants made from metals are shown in Figure 5.1. For example, metals are used for orthopedic reconstructions (implants for artificial hip, knee, shoulder, and elbow joints), fracture fixation (plates, pins, screws, rods, and nails), oral and maxillofacial reconstructions (dental implants and mini-plates), and cardiovascular interventions (stents, heart valves, and pacemakers). In general, metals used for biomedical applications should exhibit the following properties:

- high corrosion resistance,
- biocompatibility,
- high wear resistance,
- excellent mechanical properties.

Most metallic biomaterials have a stable surface oxide layer that enhances their corrosion resistance properties. It is believed that the presence of this stable surface oxide layer is key to the biocompatibility of metals. The mechanical properties of the metal are important and should satisfy the requirements of the specific application in the body. For instance, when a metal is used to augment a bone, the elastic modulus of the metal should be ideally equivalent to that of the bone. If the elastic modulus of the metal is greater than that of bone, then the load experienced by the bone is reduced due to a phenomenon known as stress shielding. This can cause the bone to remodel to adjust to the lower load and

Figure 5.1

Examples of metallic implants showing (left to right) a femoral hip prostheses, an acetabular component of a hip joint, and a fracture fixation plate.

eventually result in the loss of bone quality. In another example, stainless steel is commonly used for making coronary stents due to its well-suited mechanical properties. Stainless steel has good radial strength (due to its high elastic modulus of ~190 GPa), low recoil, good expandability, and sufficient flexibility, which makes it a highly preferred metal for making stents. Several metals such as titanium, stainless steel, cobalt–chromium alloys, nitinol (nickel–titanium alloy), tantalum, and magnesium have been used for a variety of clinical applications, with titanium, stainless steel, and cobalt–chromium alloys being the most commonly used metals. This chapter describes the structure and properties of these metals.

> The three metals most commonly used in biomedical implants and devices are:
> - titanium and its alloys,
> - stainless steel, and
> - cobalt–chromium alloy.

5.1 Titanium and its alloys

Titanium (Ti) is well known for its light weight, excellent corrosion resistance, and enhanced biocompatibility. The density of Ti (4.5 g/cm^3) is significantly lower compared to other metals used for biomaterials such as 316L stainless steel (7.9 g/cm^3) and cobalt–chromium alloy (8.3 g/cm^3). Although it is a light weight material, Ti provides excellent mechanical and chemical properties comparable to 316L stainless steel and cobalt–chromium alloy. Commercially pure Ti (cp-Ti) and Titanium–6Aluminum–4Vanadium (Ti–6Al–4V) alloy are the two titanium-based metals commonly used in the biomedical industry.

Table 5.1a Chemical composition of four grades of cp-Ti and Ti-6Al-4V[a]

| cp-Ti grades and Ti-6Al-4V | Elements | | | | | |
	Carbon (% max)	Oxygen (% max)	Nitrogen (% max)	Hydrogen (% max)	Iron (% max)	Titanium (% max)
Grade-1	0.10	0.18	0.03	0.015	0.20	balance
Grade-2	0.10	0.25	0.03	0.015	0.30	balance
Grade-3	0.10	0.35	0.05	0.015	0.30	balance
Grade-4	0.10	0.40	0.05	0.015	0.50	balance
Ti-6Al-4V	0.08	0.13	0.05	0.0125	0.25	balance[b]

[a] Values were taken from references 1–4.
[b] In addition to 6% Al and 4% V.

Table 5.1b Mechanical properties of four grades of cp-Ti and Ti-6Al-4V[a]

| cp-Ti grades and Ti-6Al-4V | Mechanical properties | | |
	Young's modulus (GPa)	Yield strength (MPa)	Tensile strength (MPa)
Grade-1	103	170	240
Grade-2	103	275	345
Grade-3	103	380	450
Grade-4	104	485	550
Ti-6Al-4V	110	795	860

[a] Values were taken from references 1–4.

The American Society for Testing and Materials (ASTM) International has classified cp-Ti or unalloyed Ti into four different grades, namely Grade-1, Grade-2, Grade-3, and Grade-4 Ti. These classifications are based on the amount of impurities (such as oxygen, nitrogen, and iron) present in the metal. In particular, the amount of oxygen present in Ti has significant influence on its mechanical properties. Tables 5.1a and 5.1b show the chemical composition and mechanical properties (elastic modulus, yield strength and tensile strength) for the four grades of cp-Ti and Ti–6Al–4V, respectively.[1–4] From Table 5.1a, it is evident that when the percentage of oxygen increases from 0.18 (Grade-1) to 0.40 (Grade-4), the tensile strength of cp-Ti increases from 240 MPa to 550 MPa, respectively. Similarly, the increase in oxygen also results in an increase in the yield strength of cp-Ti from 170 MPa (Grade-1) to 485 MPa (Grade-4). In

Ti-based metals commonly used for biomedical implants are as follows.
- Commercially pure Ti (cp-Ti)
  Grade-1
  Grade-2
  Grade-3
  Grade-4
- Ti-6Al-4V alloy

general, an increase in the amount of impurities in cp-Ti leads to increased strength but reduced ductility. The fatigue strength of cp-Ti also increases with increase in the oxygen.

5.1.1   Classification of Ti and its alloys based on crystallographic forms

The cp-Ti is an allotropic material and exists in two crystallographic forms. At temperatures from room temperature up to 882.5 °C, cp-Ti exists as an alpha phase and exhibits a hexagonal close-packed crystal structure. A phase transformation occurs at temperatures above 882.5 °C, with cp-Ti transforming to a beta phase, and exhibits a body-centered cubic crystal structure. The crystal structure of Ti can be stabilized selectively by alloying it with certain elements (see below). Based on the crystal structure, Ti alloys can be broadly classified under three different types: (a) alpha alloys; (b) beta alloys; (c) alpha + beta alloys.

Alpha alloys

By alloying Ti with elements such as aluminum, gallium, or tin, the alloyed Ti retains its alpha-phase structure. The phase diagram of Ti alloy with alpha stabilizers is shown in Figure 5.2a. These alpha Ti alloys are not heat-treatable but are weldable. They exhibit good creep resistance at elevated temperature and do not undergo a ductile–brittle transformation. These alpha Ti alloys are usually used for low-temperature (cryogenic) applications. Alpha Ti alloys which contain very small amounts of beta stabilizers such as Ti–8Al–1Mo–1V or Ti–6Al–2Nb–1Ta–0.8Mo are called near-alpha alloys. Since beta stabilizers are present in very low quantities, these alloys behave more like alpha alloys rather than alpha–beta alloys. Alpha alloys and near-alpha alloys have very limited biomedical applications due to their low strength at ambient temperature when compared to beta and alpha–beta alloys.

> - The various components of a metal alloy form a solid solution.
> - The relative miscibility of these components is a function of temperature and determines the phase and the crystal structure of the metal alloy.
> - The graphical depiction of the phase as a function of temperature is known as the *phase diagram*.

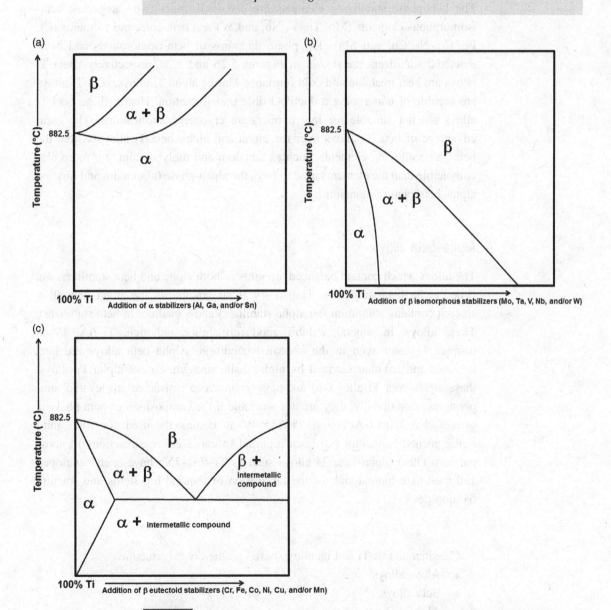

Figure 5.2

(a) Phase diagram of Ti alloy with alpha stabilizers. (b) Phase diagram of Ti alloy with beta isomorphous stabilizers. (c) Phase diagram of Ti alloy with beta eutectoid stabilizers.

Beta alloys

Beta Ti alloys are formed by alloying Ti with elements such as molybdenum (Mo), tantalum (Ta), vanadium (V), niobium (Nb), tungsten (W), chromium (Cr), iron (Fe), cobalt (Co), nickel (Ni), copper (Cu), and/or manganese (Mn). The alloying elements used in this category stabilize the body-centered cubic crystal structure. The beta-phase stabilizing elements are classified under two categories: beta-isomorphous elements (Mo, Ta, V, Nb, and W) and beta-eutectoid elements (Cr, Fe, Co, Ni, Cu, and Mn). The phase diagrams of beta-isomorphous and beta-eutectoid stabilizers are shown in Figures 5.2b and 5.2c, respectively. Beta Ti alloys are heat treatable and cold formable. Unlike alpha Ti alloys, beta Ti alloys are capable of undergoing a ductile–brittle transformation. Hence, these beta Ti alloys are not suitable for low temperature cryogenic applications. The main advantage of beta Ti alloys over the alpha and alpha–beta Ti alloys is that the beta-phase alloying elements, such as tantalum and molybdenum, are more bio-compatible than the elements used to form the alpha-phase (aluminum and tin) and alpha–beta-phase (vanadium).

Alpha–beta alloys

The alloys which contain balanced amounts of both alpha and beta stabilizers are called alpha–beta stabilizers. Ti–6Al–4V is an example of the alpha–beta alloy since it contains aluminum (an alpha stabilizer) and vanadium (a beta stabilizer). These alloys, in general, exhibit good formability, although, Ti–6Al–4V is difficult to form even in the annealed condition. Alpha–beta alloys are heat treatable and are characterized by high tensile strength. Unlike alpha Ti alloys, these alpha–beta Ti alloys do not have good creep resistance at elevated temperatures. Additionally, they are not weldable if the composition of beta phase is greater than 20%. Although Ti–6Al–4V is commonly used for total joint replacements, vanadium has been reported to cause adverse reactions in some patients. Other alpha–beta Ti alloys such as Ti–6Al–7Nb have been developed and used as a biomaterial for the fabrication of femoral hip stems and fracture fixation devices.

> Classification of Ti and its alloys based on the crystal structure.
> - Alpha alloys.
> - Beta alloys.
> - Alpha + beta alloys.

5.1.2 Surface properties

Titanium and its alloys possess a surface oxide layer of thickness 5–20 nm. This passive oxide layer is mainly responsible for its excellent corrosion resistance and biocompatibility. The surface chemistry and topography of Ti can be altered by various methods depending on the application. Mechanical treatments (polishing and sandblasting), chemical treatments (acid etching), and electrochemical treatments (anodization) are commonly used to alter the surface properties of Ti. Details of the different surface modification techniques that can be applied to Ti surfaces are discussed in Chapter 9.

5.1.3 Applications

Titanium and its alloys are commonly used for a variety of biomedical applications including total joint replacements, maxillofacial, craniofacial, and dental implants, pacemaker cases, and housings for ventricular assist devices.

5.2 Stainless steel

Stainless steels are iron-based alloys which contain at least 10.5% chromium. The corrosion resistance of stainless steel is attributed to the formation of chromium oxide (Cr_2O_3) on its surface. The corrosion-resistant properties of stainless steel can be further improved by increasing the chromium content. These properties and other physical and mechanical properties can also be improved by the addition of several other alloying elements. For instance, addition of molybdenum increases pitting corrosion resistance, while the addition of nitrogen increases mechanical strength and pitting corrosion resistance. Based on their microstructure, stainless steel alloys can be classified under four categories:

- martensitic stainless steel,
- ferritic stainless steel,
- austenitic stainless steel, and
- duplex stainless steel.

Classification of stainless steel based on the microstructure.
- Martensitic stainless steel alloys.
- Ferritic stainless steel alloys.
- Austenitic stainless steel alloys.
- Duplex stainless steel alloys.

5.2.1 Martensitic stainless steels

Martensitic stainless steels have a body-centered tetragonal crystal structure. Hardened by heat treatments, these alloys have good mechanical properties and moderate corrosion resistance. The chromium content in martensitic stainless steel varies from 10.5% to 18% and the carbon content can be greater than 1.2%. The amounts of chromium and carbon are adjusted in such a way that a martensitic structure is obtained. Several other elements such as tungsten, niobium, and silicon can be added to alter the toughness of the martensitic stainless steel. The addition of small amounts of nickel enhances the corrosion resistance and toughness, whereas the addition of sulfur improves the machinability of these alloys. Applications of martensitic stainless steels in the medical industry include surgical and dental instruments such as forceps, pliers, scalpels, curettes, dental burs, explorers, and root elevators.

5.2.2 Ferritic stainless steels

Ferritic stainless steels have a body-centered cubic crystal structure, and the chromium content in these alloys can vary from 11% to 30%. Other elements such as niobium, silicon, or molybdenum may be added to obtain specific characteristics in these alloys. Similar to martensitic stainless steel, sulfur or selenium can be added to improve the machinability of these alloys. However, unlike martensitic stainless steel, ferritic stainless steel cannot be strengthened by heat treatment. Additionally, cold working is not commonly performed since it reduces the ductility of these alloys. The few biomedical applications of ferritic stainless steel include handles for instruments and medical guide pins.

5.2.3 Austenitic stainless steels

Austenitic stainless steels have a face-centered cubic crystal structure. The chromium, nickel, and manganese contents in these materials vary from 15% to 20%, 3% to 14%, and 1% to 7.5%, respectively. A variety of other alloying elements such as molybdenum, niobium, silicon, and aluminum can be added to improve the pitting and oxidation resistance. Similar to ferritic stainless steels, these alloys cannot be hardened by heat treatment. However, cold working is commonly performed to harden them. Austenitic stainless steels possess excellent cryogenic properties, high-temperature strength, oxidation resistance, and formability. The amount of nickel directly influences the formability of these alloys, with an increase in nickel resulting in improved formability properties. These steels have been extensively used for medical implants and devices. For example, 316L stainless steel is the most commonly used austenitic stainless steel for implants and devices.

316L stainless steel

316L stainless steel consists of primarily iron (60%–65%), chromium (17%–20%), nickel (12%–14%), and smaller amounts of molybdenum, manganese, copper, carbon, nitrogen, phosphorous, silicon, and sulfur (Table 5.2a). Table 5.2b shows the mechanical properties of annealed and 30% cold worked 316L stainless steel. The letter "L" in 316L stainless steel stands for low carbon content (<0.030%). The low carbon content is highly preferred for excellent corrosion resistance. If the carbon content is greater than 0.030%, then it may precipitate as carbides ($Cr_{23}C_6$) at grain boundaries. The serious drawback of carbide precipitation at grain boundaries is that it reduces the amount of chromium in the regions adjacent to the grain boundaries, thereby reducing the formation of the protective chromium oxide layer. As a result, carbide-precipitated steels are highly prone to corrosion, especially at the grain boundaries. Although chromium provides excellent corrosion resistance, one of its drawbacks is that it stabilizes the ferritic phase, which is weaker when compared to the austenitic phase.

Other commonly employed alloying elements such as molybdenum and silicon also stabilize the ferritic phase. Hence, nickel is commonly added to strengthen the austenitic phase. An exclusive austenitic phase is preferred for 316L stainless steel

Table 5.2a   Chemical composition (% max) of 316L stainless steel[a]

Cr	Ni	Mo	Mn	Si	Cu	N	C	S	P	Fe
17.0–20.0	12.0–14.0	2.0–4.0	2.00	0.75	0.5	0.1	0.03	0.03	0.03	Balance

[a] Values were taken from references 1–3, 5.

Table 5.2b   Mechanical properties of 316L stainless steel[a]

Mechanical properties	Annealed	30% cold worked
Young's modulus (GPa)	190	190
Yield strength (MPa)	172	690
Tensile strength (MPa)	485	860

[a] Values taken from references 1–3.

with no ferritic phase or carbide precipitation. The steel should also be free of sulfide-based impurities which could easily arise from unclean steel manufacturing facilities and cause pitting-type corrosion.

In addition to uniformity in grain size, the ASTM also recommends a grain size of 6 or finer for 316L stainless steel. The grain size is determined by processing conditions such as solidification, cold working, annealing, and recrystallization. 316L stainless steel is typically 30% cold worked to significantly improve mechanical properties such as yield strength, ultimate tensile strength, and fatigue strength (Table 5.2b). Although the ductility of the 316L stainless steel is compromised during cold working, it is not a major concern for the biomedical implant and device industry. Implant grade stainless steels are commonly produced by vacuum melting, vacuum arc re-melting, or electroslag refining. Among these fabrication methods, vacuum melting is the preferred technique used to produce 316L stainless steel because it improves its pitting corrosion resistance and provides a ferrite-free microstructure. The 316L stainless steels are used in many biomedical applications including coronary stents, orthopedic implants, and fracture fixation devices.

5.2.4   Duplex stainless steels

Duplex stainless steels are two-phase alloys that contain equal proportions of ferrite and austenite phases in their microstructure. With a carbon content less than 0.030%, the amount of chromium and nickel content in these steel can vary from 20% to 30% and 5% to 8%, respectively. Minor alloying elements contained in duplex stainless steels include molybdenum, nitrogen, tungsten, and copper. These duplex stainless steel alloys exhibit superior strength when compared to certain grades of austenitic steels. Additionally, these duplex stainless steels have improved toughness and ductility when compared to ferritic stainless steels. However, these alloys are not yet used for medical implants and devices.

5.2.5   Recent developments in stainless steel alloys

Recently, nitrogen-strengthened stainless steels have been developed with improved mechanical properties and corrosion resistance when compared to 316L stainless steel. Nitrogen-strengthened stainless steels are prepared by an electroslag re-melting technique and are commonly classified under ASTM F1314, F1586, and F2229 categories. These alloys are used for fracture fixation devices and other medical implants. The nitrogen contents in ASTM F1314, F1586, and F2229 can vary from 0.20% to 0.40%, 0.25% to 0.50%, and ~1.0%, respectively. They alloys are typically ferrite-free and can be cold worked to show improved tensile strength, fatigue strength, crevice corrosion, and pitting corrosion resistance when compared to the 316L stainless steel. ASTM F2229 is especially attractive to the biomedical industry since it contains no nickel and patient allergies due to nickel can be avoided.

5.3   Cobalt–chromium alloys

The four cobalt–chromium alloys commonly used for biomedical applications are ASTM F75 (Co–28Cr–6Mo casting alloy), ASTM F799 (Co–28Cr–6Mo thermodynamically processed alloy), ASTM F90 (Co–20Cr–15W–10Ni wrought alloy), and ASTM F562 (Co–35Ni–20Cr–10Mo wrought alloy). The chemical composition and mechanical properties of these alloys are listed in Tables 5.3a and 5.3b, respectively. While the ASTM F75 and ASTM F799 alloys possess almost similar compositions, the different processing methods used to make these Co–Cr alloys result in uniquely different mechanical properties (Tables 5.3a and 5.3b). The cobalt content is less in ASTM F90 and ASTM F562 when compared to ASTM F75 and ASTM F799. Additionally, ASTM F562 contains more nickel, whereas ASTM F90 contains more tungsten (Table 5.3a).

Cobalt–chromium alloys commonly used for biomedical implants and devices.
- ASTM F75 (Co–28Cr–6Mo casting alloy).
- ASTM F799 (Co–28Cr–6Mo thermodynamically processed alloy).
- ASTM F90 (Co–20Cr–15W–10Ni wrought alloy).
- ASTM F562 (Co–35Ni–20Cr–10Mo wrought alloy).

5.3.1   ASTM F75

ASTM F75 is a cast Co-28Cr-6Mo alloy. The commercial names for this alloy are Vitallium (Howmedica, Inc.), Haynes Stellite 21 (Cabot Corporation), Protasul-2 (Sulzer AG), and Zimaloy (Zimmer). The significant feature of this alloy is its excellent corrosion resistance even in chloride environments. This is mainly because of its high bulk chromium content and chromium oxide (Cr_2O_3) surface

Table 5.3a   Chemical composition of cobalt–chromium alloys[a]

| Elements | Cobalt–chromium alloys | | | |
	ASTM F75 (% max)	ASTM F799 (% max)	ASTM F90 (% max)	ASTM F562 (% max)
Cobalt	58.9–69.5	58–59	45.5–56.2	29.0–38.8
Chromium	27.0–30.0	26.0–30.0	19.0–21.0	19.0–21.0
Molybdenum	5.0–7.0	5.0–7.0	—	9.0–10.5
Tungsten	0.2	—	14.0–16.0	—
Nickel	2.5	1.0	9.0–11.0	33.0–37.0
Manganese	1.0	1.0	2.00	0.15
Silicon	1.0	1.0	0.40	0.15
Iron	0.75	1.5	3.00	1.0
Carbon	0.35	0.35	0.15	0.025
Nitrogen	0.25	0.25	—	—
Phosphorous	0.02	—	0.04	0.015
Titanium	0.10	—	—	1.0
Sulfur	0.01	—	0.03	0.010

[a] Values were taken from references 1–3, 6.

layer. This alloy is extensively used for biomedical as well as aerospace applications. It is usually shaped into the final device using a procedure called investment casting, which involves the use of wax as an initial mold for the final device. The wax mold is then coated with a ceramic followed by melting the wax which leaves behind a ceramic mold. The alloy is melted at 1350–1450 °C and poured into this ceramic mold. After solidification, the ceramic is broken away, and the metal is further processed to obtain the final device. Three different microstructural features, which tend to adversely affect the mechanical properties of the alloy, have been observed depending on the casting conditions of F75 alloy. (a) Co-rich matrix (alpha-phase) with inter-dendritic and grain boundary carbides such as $Co_{23}C_6$ or $Cr_{23}C_6$ or $Mo_{23}C_6$ is present. Inter-dendritic Co- and Mo-rich sigma inter-metallic and Co-based gamma-phases may also be present. (b) The grain sizes are large. This is a significant limitation since it decreases the tensile strength of the alloy. (c) Casting defects are present. Such defects could cause fatigue fracture of the implant under *in vivo* conditions.

Table 5.3b   Mechanical properties of cobalt–chromium alloys[a]

Properties	ASTM F75		ASTM F799	ASTM F90		ASTM F562	
	As-cast/ annealed	Powder metallurgy/ hot isotactic pressing	Hot forged	Annealed	44% cold worked	Hot forged	Cold worked, aged
Young's modulus (GPa)	210	253	210	210	210	232	232
Yield strength (MPa)	450	841	896– 1200	448–648	1606	965– 1000	1500
Tensile strength (MPa)	655	1277	1399– 1586	951–1220	1896	1206	1795

[a] Values were taken from references 1–3.

5.3.2   ASTM F799

ASTM F799 is known as thermomechanical Co–Cr–Mo alloy because it is processed by hot forging at a temperature of about 800 °C after casting. This thermomechanical processing by hot forging is the main difference between this alloy and the ASTM F75. There are several steps involved in transforming the alloy to the desired shape, and the temperature varies depending on the step. For instance, a high temperature is required for greater deformation, but it results in loss of strengthening. Hence, in most cases, a high temperature is used in initial steps to deform the alloy, and a low temperature is used in later steps to induce cold working. In this way, the shape of the final product can be achieved with greater strengthening. The yield and tensile strengths of ASTM F799 alloys are twice as high as those for as-cast ASTM F75 alloys.

5.3.3   ASTM F90

ASTM F90 is a wrought Co–Cr–W–Ni alloy and its commercial name is Haynes Stellite 25 (Cabot Corporation). The addition of tungsten and nickel to the alloy makes it easy to machine and fabricate. In the annealed state, the yield and tensile strengths of ASTM F90 are equal to those of as-cast ASTM F75 (Table 5.3b). However, in the 44% cold-worked state, the yield and tensile strengths are twice as much as those of ASTM F75 (Table 5.3b).

5.3.4   ASTM F562

ASTM F562 is a wrought Co–Ni–Cr–Mo alloy. Also known as MP-35N, this alloy is well known for its excellent strength, high ductility, and corrosion resistance. It is typically heat treated and cold worked to produce a controlled microstructure with high strength. Under equilibrium conditions, the pure solid cobalt has a face-centered cubic crystal structure above 419 °C and a hexagonal close-packed crystal structure below this temperature. When cobalt is alloyed with other elements, the processing methods involve cold working, which partially transforms the face-centered cubic crystal structure to hexagonal close-packed crystal structure. The hexagonal close-packed phase appears as platelets within face-centered cubic phase grains and this significantly strengthens the alloy. It can also be further strengthened by a heat treatment (430–650 °C), which produces Co_3Mo precipitates on the hexagonal close-packed phase platelets. Hence, the alloy is a multiphase (the "MP" in MP-35N stands for multiphase) material. Its excellent strength is mainly due to the variety of techniques such as cold working, solid solution strengthening, and precipitation hardening involved in its processing. As seen in Table 5.3b, cold worked and aged ASTM F562 has a tensile strength of 1795 MPa, which is among the highest among the metals used for biomedical implants.

5.4   Nitinol

Nitinol (Ni–Ti) is an alloy with near-equiatomic composition (49 atomic%–51 atomic%) of nickel and titanium. Ni–Ti belongs to the class of shape memory alloys which can be plastically deformed at a low temperature but return back to their original pre-deformed shape when exposed to a high temperature. The shape memory capacity of Ni–Ti is the result of its martensitic transformation. At high temperature, the atoms are arranged in an orderly fashion and possess a body-centered cubic crystal structure known as the austenite phase. In this austenitic phase, the alloy is resistant to twisting and bending. At low temperatures, Ni–Ti transforms to a monoclinic distorted crystal structure known as the martensite phase. In the martensitic phase, the alloy can be easily deformed. The temperature at which the austenite phase transforms to martensite phase is called the transformation temperature. The use of the shape memory effect of Ni–Ti in a

Table 5.4 Mechanical properties of Nitinol[a]

Young's modulus (GPa)	83 (austenite phase)
	28–41 (martensite phase)
Yield strength (MPa)	195–690 (austenite phase)
	70–140 (martensite phase)
Tensile strength (MPa)	895

[a] Values were taken from references 1 and 2.

biomedical implant is exemplified by self-expanding vascular stents. These stents have a small diameter at room temperature so that they can be inserted through an artery without causing damage. However, they expand to their preset diameter at body temperature. Ni–Ti is the preferred material for making such self-expanding vascular stents. Initially, the Ni–Ti stent is plastically deformed at room temperature (martensite phase) to crimp it into the delivery system. However, the diameter of the targeted blood vessel is already pre-memorized in the Ni–Ti from when it was in its austenitic phase at an elevated temperature. Hence, when the stent is implanted, it expands to its original shape because the body temperature is above its transformation temperature.

The atomic% composition of nickel in Ni–Ti is crucial in determining the transformation temperature and mechanical properties of the alloy. If the atomic% composition of nickel increases (even by 0.1%), then the transformation temperature decreases accompanied by an increase in yield strength of the alloy. Other elements such as iron and chromium can also be added to decrease the transformation temperature. Common contaminants, such as carbon and oxygen, can change the deformation temperature and affect the mechanical properties. Thus, care should be taken to minimize the concentration of these contaminants. The mechanical properties of Ni–Ti in its austenite and martensite phases are provided in Table 5.4. A range of values is provided since the properties of the alloy depend on the processing conditions.

Ni–Ti is typically prepared by vacuum arc re-melting or vacuum induction melting. Because of the very high reactivity of titanium, these methods have to be performed under a vacuum or in inert atmospheric conditions. In vacuum arc re-melting, the raw materials, nickel and titanium, are used to make a cylindrical ingot, which is used as an electrode. The ingot is placed inside a large crucible (typically copper), which is surrounded by water. An electrical arc generated between the electrode and a small amount of the same alloy at the bottom of the crucible melts the alloy. The molten alloy is then collected at the bottom of the crucible. The water surrounding the crucible controls the cooling rate of the alloy. Since the whole process is carried out under vacuum, this is a high purity alloy with negligible amount of carbon content. In vacuum induction melting, the raw materials, nickel and titanium, are melted in a graphite crucible using electromagnetic induction. The structure of alloy produced by induction melting is more homogeneous than that of arc melting due to the constant stirring of molten alloy by induction. Although the process is carried out in vacuum, the alloy produced by this method has significant amount of carbon due to the use of a graphite crucible.

The corrosion resistance of Ni–Ti is debatable. Although most scientific literature suggests that Ni–Ti is a corrosion-resistant material, some studies have shown that nickel ions leach from the alloy and have toxic effects on the surrounding tissue. Hence, several surface treatments have been developed to form a titanium enriched surface layer, thereby reducing the concentration of nickel at the surface. These treatments include electropolishing, nitric acid treatments, and plasma-immersion ion implantation. Ni–Ti has applications in several medical implants and devices including self-expanding vascular stents, medical staples, blood clot filters, and orthodontic wires.

5.5   Tantalum

Tantalum (Ta) has been used for making biomedical implants and devices, either in its commercially pure (99.9%) state or as an alloying element in titanium alloys. Tantalum is well known for its excellent corrosion resistance and biocompatibility because of a stable surface oxide layer. It has also been used as coatings on other metallic devices, such as 316L stainless steel, to improve the substrate's corrosion resistance and to enhance biocompatibility. The mechanical properties of Ta (annealed) are provided in Table 5.5. The elastic modulus of Ta is close to that of 316L stainless steel; however, the yield and tensile strengths of Ta are lower when compared to those of Ti, 316L stainless steel, and Co–Cr alloys. Although the mechanical properties are slightly inferior to other commonly used metallic biomaterials, Ta has been used for a variety of biomedical applications. Owing to its high density (16.6 g/cm^3), Ta has been used as a radiographic marker for diagnostic applications. Ta has also been used for coronary stents, vascular clips, cerebral covering for cranial defects, fracture fixation, and dental implants.

Table 5.5   Mechanical properties of tantalum (annealed)[a]

Young's modulus (GPa)	185
Yield strength (MPa)	138
Tensile strength (MPa)	207

[a] Values were taken from references 1 and 2.

5.6 Magnesium

The use of magnesium (Mg) for orthopedic applications dates back nearly half a century. Mg is well known for its light weight and biodegradability. Its physical and mechanical properties are listed in Table 5.6, and for comparison, the properties of bone are also provided.

The density, elastic modulus, yield strength, and fracture toughness of Mg are close to that of bone (Table 5.6). In addition, Mg is present naturally in the bone tissue. It is estimated that approximately half of the total physiological Mg is stored in bone. These properties appear to make Mg an attractive material for orthopedic applications. However, the rapid corrosion of pure Mg under physiological conditions seriously limits its use in load bearing applications. An implant used for load bearing applications is expected to maintain its mechanical integrity, typically for a period of 12–18 weeks until it is replaced by natural bone. However, pure Mg degrades within the first few days after *in vivo* implantation. Under physiological conditions, the main degradation products of Mg are magnesium hydroxide ($Mg(OH)_2$) and hydrogen (H_2) gas. Although $Mg(OH)_2$ is not very soluble in water, it reacts immediately with chloride ions in the physiological environment to produce magnesium chloride ($MgCl_2$) which is readily soluble in water. The reactions involved in the corrosion of Mg are as follows:

$$Mg + 2H_2O \rightarrow Mg(OH)_2 + H_2, \tag{5.1}$$

$$Mg(OH)_2 + 2Cl^- \rightarrow MgCl_2. \tag{5.2}$$

The H_2 gas evolved during the degradation of Mg is a significant concern, since it can accumulate as gas bubbles under the skin. Patients with such issues are treated by drawing off the gas using a subcutaneous needle.

In order to improve its corrosion resistance, a variety of elements has been alloyed with Mg. These alloying elements include aluminum, zinc, manganese, zirconium, yttrium, and rare earth elements. The Mg alloys show improved

Table 5.6 Mechanical properties of magnesium and bone[a]

Mechanical properties	Magnesium	Bone
Elastic modulus (GPa)	41–45	7–30
Yield strength (MPa)	65–100	130–180
Fracture toughness (MPa$\sqrt{m}$)	15–40	2–12
Density (g/cm^3)	1.7–2	1.8–2.1

[a] Values were taken from references 1 and 7.

corrosion resistance and mechanical properties when compared to those of pure Mg. However, the degradation products of these alloys should be carefully investigated as the release of different metal ions may have toxic effects in the body.

Mg alloys have been recently used for making biodegradable coronary stents. AE21 (2% aluminum, 1% rare earth metals, and remaining Mg) and WE43 (4% yttrium, 3.4% rare earth metals, 0.6% zirconium, and remaining magnesium) are the two examples of Mg alloys that have been used for making stents. These biodegradable metal stents look promising for growing arteries in children. Although Mg alloys could have potential applications in cardiovascular and orthopedic areas, additional research needs to be conducted to verify its biocompatibility and to optimize its corrosion and mechanical properties for the intended applications.

5.7   Summary

In summary, titanium, stainless steel, and cobalt–chromium alloys are the most commonly used metallic biomaterials. In the titanium family, cp-Ti and Ti–6Al–4V are widely used for a variety of biomedical implants and devices. Based on the percentage composition of impurities, the ASTM classifies cp-Ti into four grades (Grade-1, Grade-2, Grade-3, and Grade-4). The amount of oxygen impurity in cp-Ti has a significant influence on the mechanical properties of the metal. Based on the crystal structure (hexagonal closed-packed crystal structure or body-centered cubic crystal structure, or both), titanium can be classified into alpha alloys, beta alloys, and alpha + beta alloys. Alpha-phase titanium is formed by alloying with elements such as aluminum, gallium, and tin. The biomedical applications of alpha alloys are limited when compared to those of beta alloys and alpha + beta alloys. By alloying titanium with elements such as molybdenum, vanadium, chromium, and iron, the body-centered cubic crystal structure is stabilized, resulting in the formation of beta-phase alloys. Ti alloys with both alpha and beta stabilizers are called alpha + beta alloys. Ti–6Al–4V belongs to this alpha + beta alloy family since it contains alpha (aluminum) and beta (vanadium) stabilizers. The corrosion resistance and biocompatibility of titanium are influenced by its surface oxide layer.

Iron-based alloys with at least 10.5% chromium belong to the stainless steel family. The surface chromium oxide (Cr_2O_3) layer provides corrosion resistance to these alloys. Based on the microstructure, stainless steel is classified into four types: martensitic, ferritic, austenitic, and duplex. Several elements are added to these alloys to further improve their corrosion resistance and mechanical properties. Martensitic and ferritic stainless steels are used for a variety of surgical instruments, while austenitic stainless steel is extensively used for medical implants. 316L stainless steel belongs to austenitic type and is the most commonly used stainless steel grade for implants and devices. With improved mechanical and corrosion resistant properties when compared to the 316L stainless steel, nitrogen-strengthened stainless steels are used in fracture fixation devices.

The four main types of cobalt–chromium alloys commonly used for biomedical applications are the ASTM F75, ASTM F799, ASTM F90 and ASTM F562. While ASTM F75 and ASTM F799 are Co–Cr–Mo alloys, ASTM F90 is a Co–Cr–W–Ni alloy and ASTM F562 is a Co–Ni–Cr–Mo alloy. These alloys possess excellent corrosion resistance and mechanical properties and are commonly used for cardiovascular and orthopedic implants.

Other metals used in the biomedical industry include nitinol, tantalum, and magnesium. Nitinol, a nickel–titanium alloy, belongs to the class of shape memory alloys. At low temperature, these alloys can be plastically deformed and the original undeformed shape can be obtained back by increasing the temperature. These alloys are commonly used for self-expanding vascular stents. Tantalum, on the other hand, is known for its excellent corrosion resistance and biocompatibility properties because of the stable surface oxide layer. However, the mechanical properties (especially yield and tensile strengths) are inferior to other commonly used metallic biomaterials. Similar to tantalum, pure magnesium has limited biomedical applications. It has extremely fast degradation rate. However, magnesium alloys with improved corrosion and mechanical properties have been found applications in cardiovascular and orthopedic areas.

References

1. Davis, J. R. (2003). *Handbook of Materials for Medical Devices*. Materials Park, ASM International.
2. Park, J. B. and Kim, Y. K. (2003). Metallic Biomaterials, in Park, J. B. and Bronzino, J. D., editors, *Biomaterials Principles and Applications*. Boca Raton, CRC Press, pp. 1–20.
3. Brunski, J. B. (2004). Metals, in Ratner, B. D., Hoffman, A. S., Schoen, F. J. and Lemons, J. E., editors, *Biomaterials Science An Introduction to Materials in Medicine*. London, Elsevier Academic Press, pp. 137–153.
4. Freese, H. L., Volas, M. G., Wood, J. R. and Textor M. (2001). Titanium and its alloys in biomedical engineering, in Buschow, K. H. J., Cahn, R., Flemings, M. *et al.*, editors, *Encyclopedia of Materials: Science and Technology*. Headington Hill Hall, Pergamon, pp. 9374–9380.

5. Blair, M. (2001). Stainless steels: cast, in Buschow, K. H. J., Cahn, R., Flemings, M. *et al.*, editors, *Encyclopedia of Materials: Science and Technology*. Headington Hill Hall, Pergamon, pp. 8798–8802.
6. Klarstrom, D. and Crook, P. (2001). Cobalt alloys: alloy designation system, in Buschow, K. H. J., Cahn, R., Flemings, M. *et al.*, editors, *Encyclopedia of Materials: Science and Technology*, Headington Hill Hall, Pergamon, pp. 1279–1280.
7. Staiger, M. P., Pietak, A. M., Huadmani, J. and Dias, G. (2006). Magnesium and its alloys as orthopedic biomaterials: a review. *Biomaterials*, **27**(9), 1728–1734.

Suggested reading

- Brunski, J. B. (2004). Metals in Ratner, B. D., Hoffman, A. S., Schoen, F. J. and Lemons, J. E., editors, *Biomaterials Science: An Introduction to Materials in Medicine*. London, Elsevier Academic Press, pp. 137–153.
- Park, J. B. and Kim, Y. K. (2003). Metallic biomaterials, in Park, J. B. and Bronzino, J. D., editors, *Biomaterials Principles and Applications*. Boca Raton, CRC Press, pp. 1–20.
- Davis, J. R. (2003). *Handbook of Materials for Medical Devices*. Materials Park, ASM International.

Problems

1. You are asked to choose one of the following metals for a bone replacement. Which one will you choose and why?
   (a) Metal-A (Elastic modulus – 240 GPa).
   (b) Metal-B (Elastic modulus – 200 GPa).
   (c) Metal-C (Elastic modulus – 100 GPa).
   (d) Metal-D (Elastic modulus – 25 GPa).
2. Which element of impurity plays a significant role in deciding the mechanical properties of commercially pure titanium?
3. Give an example for alpha–beta alloy with justification.
4. Which surface property is crucial for the corrosion resistance of metallic biomaterials?
5. Which type of stainless steel (martensitic or ferritic or austenitic or duplex) has extensive applications in biomedical implants and devices? Give an example.
6. What is the significant advantage of 316L stainless steel over other grades of stainless steel? Explain your answer.
7. Which technique is commonly used to produce 316L stainless steel? Why?
8. What is the main difference between ASTM F75 and ASTM F799 cobalt–chromium alloys?

9. What is investment casting? Explain the procedures involved in it.
10. What is MP-35N? What are the significant advantages of this alloy?
11. How do self-expanding stents made from Nitinol work?
12. What are the main advantage and disadvantage of tantalum?
13. What is the main limitation of using pure magnesium for biomedical implants and devices?

6 Polymers

Goals

After reading this chapter the student will understand the following.

- Structure of polymers and their common physical states.
- General properties of polymers.
- Common types of polymerization techniques used for production.
- Chemical structure and properties of common biomedical polymers.
- Structure of hydrogels and their general properties.
- Definition and uses of nanopolymers.

In everyday life, we encounter a variety of polymers, some natural and others synthetic. The vast majority of these are carbon-based in nature. They range from synthetic polymers seen in products such as polyethylene grocery bags, polymethylmethacrylate-based window panes, and polystyrene-based eating utensils, to natural polymers such as starch, cellulose and rubber. Polymers used as biomaterials are often similar to these common materials. For example, the polymer most extensively used in total joint prostheses is ultrahigh molecular weight polyethylene – chemically identical to the material used for plastic bags, although having a much higher molecular weight. The same is true for bone cement which is used in conjunction with bone surgery and Plexiglass®, which is used for window panes. Both of these materials are polymethylmethacrylate (PMMA). Of course, any polymer that is used as an implant has to meet strict safety standards as required by governmental and other regulatory agencies and has to be virtually contaminant free.

Polymers have the advantage that they can be easily formed into desired shapes using a variety of techniques such as solution casting, melt molding, or machining. Thus, polymer-based implants are relatively inexpensive to manufacture. Polymers can also be made reactive so that different chemical molecules can be attached to the surface of implants in order to make them more compatible with the surrounding environment in the body. Some polymers are biodegradable in the body. If used to make implants for temporary needs, these polymers offer the advantage that the implant can gradually biodegrade within the body after it has served its function, thus mitigating the potential for any long term complications. If a biodegradable polymer is used as an implant, it can potentially also be designed to release therapeutic drugs or growth factors during the degradation process. On the other hand, despite these favorable properties, polymers are usually not as strong or stiff as metals or ceramics and therefore may not be the correct choice when an implant is required to carry large loads in its function.

The use of polymers as biomaterials has increased significantly over the past 75 years as advances in polymer science have yielded a variety of polymers. In this chapter, we will cover the basic structures of polymers, their general physical and mechanical properties, discuss some of the polymers most commonly used as biomaterials, and then conclude with introductions to hydrogels and nanopolymers.

6.1 Molecular structure of polymers

The Swedish chemist J. J. Berzelius was the first to use the term *polymer* in 1833.[1] Polymers are long-chain molecules or macromolecules that are built by connecting repeat chemical units called "mers." The term is derived from Greek, combining *poly* (many) + *mer* (part). Each molecule of a polymer can consist of hundreds, thousands, or even millions of repeat units. Depending on the type of polymerization, the repeat unit may contain exactly the same atoms as the starting molecule or monomer, or may contain a smaller number of atoms due to the elimination of some during the polymerization reaction. In either case, a single repeat unit is known as the *monomeric* unit or *monomer*. Small chains of up to roughly 10 repeat units are called *oligmers* (from Greek *oligos* meaning few), although there is no hard and fast rule about the number of repeat units needed for the transition from oligmers to polymers. The structural units terminating the ends of a polymer chain molecule are known as *end groups*.

The number of units in a polymeric chain plays a significant role in determining its properties. Let us take the example of polyethylene, which is derived from the basic unit of ethylene gas. As the number of units in the chain increases, the product changes from a gas to a liquid and then to a brittle or waxy solid. As the number increases even more, the polymeric chains become long enough to start entangling with each other and lead to the properties more commonly associated with polymers.

—A—A—A—A—A—A—A—A—A—A—
Homopolymer

Homopolymer with A monomers.

(a)
—A—B—A—B—A—B—A—B—A—B—
Alternating copolymer

(b)
—A—A—B—B—B—B—A—B—A—B—
Random copolymer

(c)
—A—A—A—A—B—B—B—B—A—A—
Block copolymer

Polymerization of A and B monomers to form (a) alternate copolymers, (b) random copolymers, and (c) block copolymers.

The units in a polymer are usually held together by covalent bonds. Adjacent polymeric chains or the different segments of the same chain may bond together by intermolecular forces or van der Waals bonds. In some cases, ionic bonds may also occur in polymers. Covalent bonds are characterized by relatively high energy, fixed angles, and short distances (0.11–0.16 nm), and they determine the mechanical, thermal, chemical and photochemical properties of a polymer. Secondary bonds, on the other hand, govern the physical characteristics of the material such as the dissolution, melting, diffusion, and flow properties – properties which involve the breaking and forming of these bonds and the movement of the molecules relative to each other.

Polymers can be classified in a variety of ways. If only one type of monomer (A) is used to form the polymer, the product, as shown in Figure 6.1, is called a *homopolymer*.

If two types of monomers (A and B) are used in the polymerization reaction, the product is known as a *copolymer*, and there are several possible combinations of copolymers (Figure 6.2). If the monomers alternate, then the product is known as an *alternating copolymer*. If they are attached in a random fashion, then the result is a *random copolymer*. If they attach in blocks, then the product is a *block copolymer*.

Moreover, the placement of the blocks in the backbone of the block copolymer can be alternating or random. The properties of the final product can vary significantly depending on the relative placement of the monomers or blocks and on the starting ratio of the two monomers. The structure of the chains can also significantly influence properties.

As shown in Figure 6.3, polymeric chain structure can also be characterized as *linear*, *branched*, or *networked*. A linear polymer has a backbone analogous to a single string – it has no branches attached to it. However, it may have some pendant groups, such as in the case of polystyrene. Polystyrene has the same structure as linear polyethylene (Figure 6.4). However, in every repeat unit of the polystyrene backbone, there is a closed benzene ring replacing one of the hydrogen atoms attached to a carbon atom. *Branched* polymers arise due to side reactions in the polymerization process and consist of branches attached to the main backbone of the molecule. When the branches connect with adjacent molecules during or after the polymerization process, the product is a *networked* polymer. Other polymeric structures include star, comb, and ladder polymers. The same basic type of polymer can often exist in different forms depending on the conditions during polymerization.

Box 6.1

- Polymers can be classified in a variety of ways.
- Based on the number and arrangement of monomeric units, such as homopolymers, copolymers, random copolymers, block copolymers, etc.
- On the basis of molecular structure such as linear, branched, cross-linked, star, comb, or ladder polymers.
- Polymers can also be classified as thermoplastic or theromset. Thermoplastic polymers soften and flow upon heating and can be remolded or reshaped.
- Thermoset polymers do not flow upon reheating and cannot be reshaped.

Networked polymers are also known as *cross-linked* polymers and may arise if multifunctional monomers are used instead of difunctional monomers. Crosslinking may also take place post-polymerization in an existing polymer. When subjected to a high-energy source, such as electron beam radiation or gamma rays, enough energy may be provided to rupture some bonds within the polymers. These ruptured bonds may then react with adjacent chains to form a network. As the degree of cross-linking increases, the polymer chains lose their ability to slide past each other, and these cross-linked polymers become more rigid and dimensionally stable. Known as *thermosets*, such cross-linked polymers are difficult to melt or dissolve, hence making them difficult to fabricate into products. Usually thermosets are formed into shape by either temporarily disabling the cross-linking, by cross-linking them via heat, or other triggers after they have been put into their final shape.

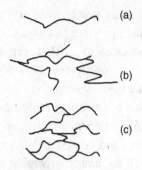

Figure 6.3

Different structures for polymeric molecular chains: (a) linear, (b) branched, and (c) networked (cross-linked).

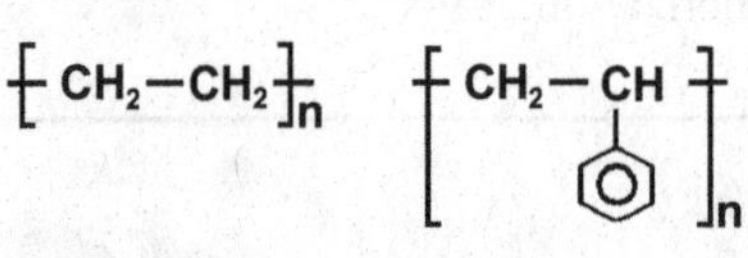

$$\left[\!\!\begin{array}{c} CH_2\!-\!CH_2 \end{array}\!\!\right]_n \qquad \left[\!\!\begin{array}{c} CH_2\!-\!CH \\ \hspace{0.5cm} \bigcirc \end{array}\!\!\right]_n$$

Polyethylene   Polystyrene

Figure 6.4

Structures of polyethylene and polystyrene.

Non-cross-linked polymers are usually linear or branched in structure and can be melted or dissolved. This renders them easy to fabricate into their final product form. Such polymers are categorized as *thermoplastics*. Since the molecules do not form covalent bonds with adjacent chains, the chains can flow past each other and the polymer can behave like a viscous fluid upon heating. These polymers can be repeatedly softened or melted using heat and molded into shape.

Other types of polymer structures could include *star* polymers where multiple polymeric chains emanate from one central point, *comb* polymers where multiple side chains are attached to the same side of a linear backbone, or ladder polymers which are made up of fused ring structures in the backbone.

If a single polymer chain has carbon atoms in the backbone that have the same substituents, then the polymer is symmetric (see Figure 6.5a); if the carbon atoms have different substituents, then the carbon is called asymmetric. Carbon is tetrahedral and thus the asymmetric atoms can exist in two different spatial arrangements. If the transition from one to the other cannot occur without breaking bonds, then each such form is called a *configuration*. In general if the structure can be changed via simple rotation about a bond then it is called a *conformation*.

If the pendant or side groups (such as the methyl group for polypropylene) attached to the carbons in the polymer backbone are all on one side (Figure 6.5b),

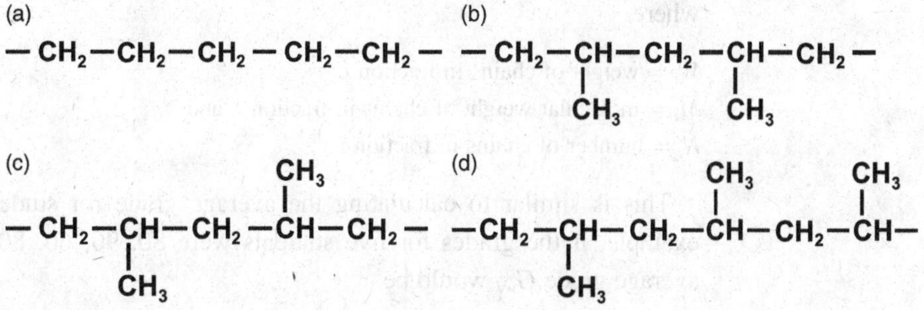

(a)

$$- CH_2 - CH_2 - CH_2 - CH_2 - CH_2 -$$

(b)

$$- CH_2 - CH - CH_2 - CH - CH_2 -$$
$$\qquad\quad |\qquad\qquad |$$
$$\qquad\quad CH_3 \qquad\quad CH_3$$

(c)

$$\qquad\qquad\quad CH_3$$
$$\qquad\qquad\quad |$$
$$- CH_2 - CH - CH_2 - CH - CH_2 -$$
$$\qquad\quad |$$
$$\qquad\quad CH_3$$

(d)

$$\qquad\qquad\quad CH_3 \qquad\qquad CH_3$$
$$\qquad\qquad\quad | \qquad\qquad\quad |$$
$$- CH_2 - CH - CH_2 - CH - CH_2 - CH -$$
$$\qquad\quad |$$
$$\qquad\quad CH_3$$

Figure 6.5

Different forms for a polymeric molecular chain. (a) A polymer with no asymmetry, (b) isotactic – each pendant group is on the same side of the chain, (c) syndiotactic – the pendant groups appear on alternate sides of the chain in a regular fashion, and (d) atactic – the pendant groups appear on opposite sides of the chain in a random fashion.

then it is known as an *isotactic* form. However, if the pendant or side groups are attached alternately on the carbon backbone of the polymers (Figure 6.5c), the form is known as *syndiotactic*. Pendant or side groups can also be randomly attached on the carbon backbone of the polymers, as shown in Figure 6.5d, and these structures are known as *atactic*. Atactic polymers do not usually crystallize and stay in an amorphous form.

6.1.1 Molecular weight

In any polymerization process, it is practically impossible to terminate all molecules at exactly the same chain length with the same number of monomeric units. The lengths of the chains comprising a specimen vary and can be plotted as a distribution. The breadth of the distribution can be controlled by carefully monitoring the reaction parameters and time. However, often in a polymer sample, there is a distribution of chain lengths that could vary from monomers, dimers, and oligmers to chain lengths of several million units. Thus, the molecular weight for the polymer is represented by an average. There are two types of averages that are commonly used, *number average* and *weight average* molecular weight.

The number average molecular weight, M_n, is calculated as follows:

$$M_n = \frac{\sum_{i=1}^{\infty} W_i}{\sum_{i=1}^{\infty} N_i} = \frac{\sum_{i=1}^{\infty} M_i N_i}{\sum_{i=1}^{\infty} N_i}, \tag{6.1}$$

where

W_i = weight of chains in fraction i,
M_i = molecular weight of chains in fraction i, and
N_i = number of chains in fraction i.

This is similar to calculating the average grade for students for a test. For example, if the grades for five students were 80, 90, 60, 80, and 90, then the average grade G_{av} would be

$$G_{av} = \frac{\sum (60 \times 1) + (80 \times 2) + (90 \times 2)}{\sum 1 + 2 + 2} = 80. \tag{6.2}$$

The weight average molecular weight M_w is defined as:

$$M_w = \frac{\sum_{i=1}^{\infty} M_i W_i}{\sum_{i=1}^{\infty} M_i N_i} = \frac{\sum_{i=1}^{\infty} M_i^2 N_i}{\sum_{i=1}^{\infty} M_i N_i}. \tag{6.3}$$

The information provided by each of these two measures of molecular weight is not redundant because M_n is more sensitive to the short chains. However, M_w is more influenced by longer chains because the term for chain molecular weight is squared in Eq. (6.3). The ratio M_w/M_n is known as the *polydispersity index* and is a measure of the width of the molecular weight distribution for a polymer. A higher polydispersity index is indicative of a wider distribution of molecular chain sizes in the sample. Sometimes a polymer may exhibit a bi-modal molecular weight distribution. This may result from the simultaneous use of two different types of catalysts in the polymerization process. Each catalyst would have its own characteristic reaction rate leading to two families of molecules within the polymer.

The molecular weight determines the *viscosity* of a dilute solution of the polymer. The relationship between these two parameters can be described by the Mark–Houwink–Sakurada equation:

$$[\eta] = K M_v^{\alpha} \tag{6.4}$$

where

$[\eta]$ = viscosity at infinite dilution,
K and α = Mark–Houwink constants (available from published tables), and
M_v = viscosity average molecular weight.

The value of M_v usually lies between M_n and M_w. Higher molecular weight polymers are more viscous in solution or melt. This makes ultrahigh molecular weight polyethylene (UHMWPE), which is used as the bearing material in several types of total joint implants, difficult to melt mold.

6.2 Types of polymerization

Polymerization is the process by which the repeat units comprising a polymer are joined together via covalent bonds. There are two main types of polymerization reactions:

- addition, and
- condensation.

Under addition polymerization, monomers with double bonds sequentially attach to the growing end of a forming polymer chain. In this type of polymerization, the atoms in the monomer are directly added to the chain, and thus, the monomer and the repeat unit have the same number of atoms. On the other hand, during condensation polymerization, there is usually the elimination of some atoms as a by-product during the reaction between the monomer and the growing chain. As such, condensation polymerization results in fewer atoms in the reacted repeat unit than in the monomer. In addition, the molecular weight of the chain is not the simple product of the number of repeat units and the molecular weight of the monomer, as it is in the case of addition polymerization.

Polymerization can also be classified as either *step growth* or *chain growth*. In step growth, of which condensation polymerization is an example, reactions take place throughout the monomer matrix, and any two reactive monomer molecules with the correct orientation and energy will react. In fact, during the step growth process, reactions can take place between monomers, dimers, trimers, or oligomers. Since the reactions take place in a multitude of sites, the change in molecular weight occurs slowly, although the monomer is consumed rapidly. Increasing viscosity with reaction time prevents the mobility of molecules and reduces the rate.

On the other hand, in chain growth polymerization, there is a usually an initiator molecule that reacts with a monomer to start the reaction. This initial reaction creates a free radical at the end of the chain which then reacts with the next monomer. In the reaction shown below, an initiator forms a free radical with an unpaired electron which then attacks the double bond of an ethylene monomer. The ethylene is added to the chain, and the free radical is regenerated to react again and continues to add molecules until all the monomers are consumed, or there is a termination step with two chains reacting to quench the free radical.

Initiation: $R\bullet + CH_2{=}CH_2 \Rightarrow RCH_2CH_2\bullet$

Chain growth: $RCH_2CH_2\bullet + CH_2{=}CH_2 \Rightarrow RCH_2CH_2CH_2CH_2\bullet$

In this type of polymerization, the molecular weight of the polymer increases quickly while the consumption of monomers occurs at a slow rate. The reaction continues until virtually all the monomers are consumed. In general, step growth polymerization yields polymers with a narrow range of molecular weight distribution, while the products of chain growth polymerization exhibit a broad distribution.

The *degree of polymerization*, $\overline{DP}$, is defined as the average number of repeat units in the molecules present in a polymer specimen. Thus,

$$M = m\,\overline{DP}, \tag{6.5}$$

where M is the average molecular weight of the polymer, and m is the molecular weight of the repeat or monomeric unit.

If there are N_O molecules initially and the total number of molecules at the end of a reaction period is N, then

$$\overline{DP} = \frac{N_O}{N}. \tag{6.6}$$

The conversion rate, p, for the reaction is

$$p = \frac{N_O - N}{N_O}. \tag{6.7}$$

Thus,

$$\overline{DP} = \frac{1}{1-p}. \tag{6.8}$$

6.3 Physical states of polymers

Polymers can exist in different states depending on the phase, configuration and alignment of their molecular chains. The two main states are amorphous and crystalline. It is common for polymers to exist in a semi-crystalline form wherein both amorphous and crystalline phases coexist.

6.3.1 Amorphous phase

In the amorphous phase, the molecular chains in a polymer are not organized in any particular way, although there may be some orientation of chains due to manufacturing processes such as extrusion, melt molding, or fiber drawing. In general, the chains in an amorphous phase do not follow any pattern, are randomly intermingled, and are best represented by bowl of freshly cooked spaghetti. With low inter-chain interactions, these chains have the flexibility to slide past each other, and the polymer can deform relatively easily.

Segments of atoms in the backbone of a molecular chain can have flexibility to move depending on the bonds holding them in place and the space or free volume surrounding them. This gives the segments the ability to rotate, twist or vibrate in place without the chains having to slide past each other and without breaking any bonds. If this segmental mobility is limited to only vibrations, the polymer behaves like a glass and it cannot undergo more than a few percent strain deformation without failing. If the polymer is heated, the energy thus supplied results in increased kinetic energy for the segments and larger amplitude of motion. However, the motion is still limited to short range. As the temperature is increased, there is a critical point at which a very definite change occurs. At the critical point, the polymer loses its rigid or *glassy* structure and becomes more rubber-like, soft, and malleable. The temperature at this critical point is known as the glass transition temperature, T_g.

In this rubber-like phase, the molecular segments have sufficient energy to vibrate, twist and turn. Segments of up to 20–50 atoms can be involved. Movement of such large segments requires more space between the molecules or a larger free volume. This causes an increase in specific volume, which can be measured using a variety of techniques. The rubber-like polymer is not a liquid because although there is some segmental motion, the ability of chains to slide or flow past each other is still limited by several factors such as cross-linking, hydrogen bonds between adjacent chains, branching, pendant groups, and chain entanglements. Chain entanglements are similar to knots that may form in a mass of yarn. When an external force is applied, the chains may initially slide past each other but then reach a point where the entanglements prevent any further movement. Thus, a polymer in this rubber-like amorphous phase does not flow like a viscous fluid. Instead it softens, deforms easily, and may not retain its original shape. This is reflected in a change in the elastic modulus or stiffness of the polymer after it reaches its T_g.

Examples of polymers in a glassy state at room temperature include polycarbonate ($T_g \sim 150\ °C$), which is often used as bulletproof "glass" and for canopies in fighter jets. Other examples are polymethylmethacrylate (Plexiglass™, $T_g \sim 105\ °C$) and polystyrene ($T_g \sim 100\ °C$). On the other hand, polymers that exist in a flexible phase at room temperature are already above their T_g at this temperature. An example is LDPE or low density polyethylene, which is used for making plastic bags and has a $T_g = -20\ °C$.

Several factors influence the glass transition temperature. Since this transition point is dependent on segmental rotations, any property that affects these rotations will have an influence on T_g. For example, chemical structure plays a large role. Double bonds in the backbone, bulky pendant groups, branching, and polarity all tend to restrict rotation, thus necessitating a higher energy level for segmental movements and ultimately a higher T_g. Molecular weight also plays a role because longer chains have a higher probability of entanglement and lower rotational ability. Additionally, a higher molecular weight implies a lower number of chain ends and thus a lower free volume. Thus, the T_g of a polymer can change with increasing molecular weight. An example is polystyrene, which at a low molecular weight ($M_n = 3000$) has a T_g of 40 °C, but at a higher molecular weight ($M_n = 300\,000$) has a T_g of 100 °C. It is also instructional to compare polyethylene ($T_g = -20$ °C) to polystyrene ($T_g = 100$ °C) where the main difference in structure is the bulky benzene pendant group present in the repeat units of polystyrene (see Figure 6.4).

The transition between the glassy and rubber-like phases is important for biomaterials. Materials used for biodegradable sutures, for instance, should have a T_g below room temperature so that the sutures stay flexible. Conversely, load carrying devices such as biodegradable plates and screws for fracture fixation should stay rigid in the body and should have a T_g significantly above body temperature (37 °C).

6.3.2   Crystalline phase

Polymers do not crystallize in the way small molecules do, where the crystals may contain hundreds or thousands of molecules and may grow to macroscopic sizes. The crystallites in a polymer form when portions of a molecular chain or several chains organize themselves in closely packed linear patterns. Segments of the same molecular chain can contribute to several different crystallites and the portions of the chains in between these crystallites are known as *tie molecules* or *tie chains*. Shown schematically in Figure 6.6, the phase in between the crystallites remains unorganized and amorphous.

- Above the glass transition temperature, T_g, amorphous polymers exist in a rubbery state.
- Rotation and movement of segments of the molecular chain with respect to each other is responsible for the soft, flexible state of the polymer.
- Double bonds in the backbone of the polymer, large pendant side groups, polarity, and cross-linking all interfere with segmental movement and result in higher T_g.

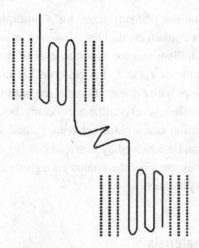

Figure 6.6

Schematic showing a molecular chain folding to form a local aligned and closely packed structure and a crystalline area. Dotted lines are indicative of other molecular chains participating in the crystal. Non-aligned areas in the polymer stay amorphous.

There are different kinds of crystallite structures that can form in polymers. Spherulites are aggregates of fiber or hair-like strands arranged as clusters. These usually form when there is nucleation in a polymer melt or solution and there is no externally applied force. A drawn fibrillar morphology occurs when there are drawing forces applied. Crystallites with an epitaxial morphology are the result of a number of lamellar growths off long chains and have a shish-kebab morphology. These usually form when the polymer melt or solution is stirred during solidification.

Any structural or chemical properties that enable close packing of molecules assist in crystallization. Thus, flexible backbones can facilitate crystallization and so can a strong polarity. Conversely, large and bulky pendant groups and branching of molecular chains deter crystallite formation.

Since the crystallites represent more tightly packed chains, crystalline or semi-crystalline polymers are usually stiffer, tougher, more resistant to solvents, and opaque. The higher stiffness and lower solubility of the crystalline phase compared to the amorphous phase are caused by strong intermolecular forces, which result from the close packing. The higher toughness is a reflection of the intermixing of the amorphous and crystalline regions while the opacity is a result of light scattering by the crystallites.

Crystalline polymers have a melting temperature T_m. However, usually the melting takes place over a narrow range of temperatures instead of one specific temperature because each specimen contains polymeric chains of various molecular weights and crystallites of many sizes. Fully amorphous polymers do not have a T_m as they do not contain crystallites. However, most non-amorphous polymers are only semi-crystalline, and so, they may exhibit both a T_g and a T_m. As a general guideline, the ratio of T_m to T_g lies between 1.4 and 2.0 (T_g is lower than T_m). Crystallization in a polymer does not take place until the temperature is lowered below T_m, and no effective crystallization occurs below T_g. Thus, the maximum rate of crystallization takes place between T_g and T_m. On the other hand, an amorphous phase can be achieved by the rapid cooling of a polymer from a melt. This fast cooling does not give the chains enough time to organize and orient themselves to form crystallites.

6.4 Common polymeric biomaterials

6.4.1 Polyethylene

The most common form of polyethylene is low density polyethylene (LDPE) which was first made by reacting ethylene gas under high pressure (100–300 MPa) in the presence of catalysts. This technique yields a mostly branched type of molecular chain structure, less compact packing, and hence the low density. Starting in the 1950s, the Ziegler–Natta catalyst was developed, and this led to the polymerization of high density polyethylene under lower pressure (10 MPa). The resultant high density polyethylene (HDPE) is linear in structure, more crystalline, and has a higher melting point. Extending the polymerization time leads to even higher molecular weight polyethylene with molecular weights in the range 3×10^6 to 10×10^6 daltons. This latter material is known as ultrahigh molecular weight polyethylene (UHMWPE) and is used extensively in orthopedic total joint implants.

UHMWPE has superior wear properties compared to other polymers. Additionally, it is tough and ductile and so it is used as the bearing material in total hip, knee, and shoulder prostheses. However, the very high molecular weight of the material also poses several difficulties because it makes UHMWPE difficult to process into the final form compared to other polymers. This material is difficult to injection mold and has to be machined, extruded, or compression molded. It is also not easy to dissolve in solvents.

Various grades of UHMWPE are commercially available for surgical implants. The properties of these grades can differ from each other and may also vary among production lots. Standards formulating agencies such as the American Society for Testing and Materials (ASTM) stipulate and specify minimum acceptable properties for medical grade UHMWPE.

Although UHMWPE has good wear characteristics in terms of degree of wear, it produces submicron and nano-sized wear debris (Figure 6.7) in large quantities which can overwhelm the body's ability to successfully remove the material. This problem is aggravated by the oxidation of UHMWPE if it is sterilized in air using gamma radiation. Although evidence was accumulating for years, issues linking the wear debris to osteolysis or loss of bone in the vicinity of the implant came to the forefront only in the 1990s. Various remedies have since been introduced including cross-linking the polymer using electron beam or gamma radiation, and sterilizing the components under vacuum. These process changes have resulted in the reduction of the overall wear of the polymer and a decrease in the generation of polymeric wear particles.

Box 6.2

- Osteolysis or bone loss in the vicinity of total joint prostheses was first thought to be due to bone cement or PMMA. Hence it was called cement disease.
- It was later discovered that this osteolysis, which often results in the loosening of implants, is caused by the wear particles shed by the UHMWPE components of the prostheses.
- Initially only micron-sized UHMWPE wear particles were detected. It was estimated that billions of such particles were released into the surrounding tissue every year.
- In later years, with more research and better microscopes, it was discovered that nano-sized particles were also present in large numbers.
- In recent years, cross-linking of the UHMWPE in a non-oxygen environment has led to significantly reduced wear and osteolysis.

6.4.2   Polymethylmethacrylate (PMMA)

The chemical structure of polymethylmethacrylate (PMMA) is shown in Figure 6.8. PMMA is used as bone cement in the field of orthopedics to stabilize total joint prostheses as well as a bone substitute in pathologic vertebral and other fractures. The basic material used for orthopedic applications is chemically the same as Plexiglass™, which is sold commercially in hardware stores. All commercially available bone cements are based on the same monomer: methylmethacrylate (MMA). This material is an ester of methacrylic acid

(a)   (b)

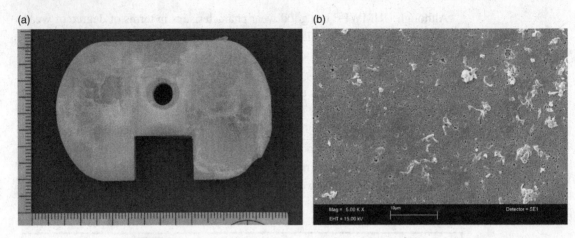

Figure 6.7

Picture of (a) a damaged UHMWPE meniscus replacement plate retrieved from a patient that had undergone total knee replacement (courtesy of Melinda Harman, PhD, and Estefania Alvarez, PhD, from the Department of Bioengineering, Clemson University); and (b) UHMWPE wear particles from a hip prosthesis.

and has to be polymerized *in situ* for clinical applications. However, the straightforward *in situ* polymerization is not performed for the following three main reasons:

- the polymerization reaction would be very slow,
- the MMA undergoes approximately 20%–21% shrinkage upon polymerization, and
- the polymerization reaction is exothermic and can cause tissue damage due to an undesirable increase in temperature.

Box 6.3

- The inclusion of antibiotics with bone cement or PMMA has led to a decrease in infections.
- However, the inclusion of antibiotics also leads to a lower strength and worse fracture properties for the PMMA.

Thus, the bone cement used clinically is available as a kit that contains a dry component and a liquid component. The dry powder component consists of pre-polymerized PMMA beads, barium sulfate (opacifier), and dibenzoyl peroxide (initiator). The initiator is the source of free radicals that starts the reaction.

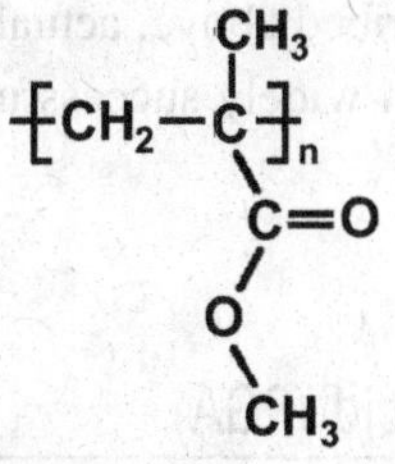

$$\left[CH_2 - \underset{\underset{\underset{CH_3}{O}}{\overset{C=O}{|}}}{\overset{CH_3}{\underset{|}{C}}} \right]_n$$

Figure 6.8

Chemical structure of polymethylmethacrylate (PMMA).

The opacifier is added to make the bone cement visible on radiographs. Sometimes, antibiotics can be mixed with the PMMA to fight infection.

The liquid component contains the MMA monomer and N,N-dimethyl-*p*-toluidine, which is an accelerator. The latter is included to increase the rate of breakdown of the dibenzoyl peroxide and the production of free radicals. Once the solid and liquid components are mixed, the initiator, aided by the accelerator, produces free radicals which drive the polymerization of the monomer. The polymerization reaction is exothermic in nature. An energy of approximately 52 kJ per mole of MMA is released and causes an increase in temperature which may rise to far above 100 °C. In most commercially available bone cement kits, the mixing ratio is two to three parts dry powder component to one part liquid component. This reduces the amount of shrinkage and the generation of heat.

The polymerization reaction is characterized by different time periods. The *dough time* typically lasts for 2–4 min and is the time elapsed from the point of the initial mixing of the solid–liquid components to the time when the mixture has reached enough viscosity that it can be handled as a mass. The time period between the end of the dough time and the point where the polymer is too hard to mold is known as the *working time*.

Bone cement does not adhere well to either metal or adjacent bone. It is not an adhesive and functions mostly as a space-filler or grout. Its adhesion to metal is improved if the metal is chemically pre-coated with PMMA. Increased surface roughness of the metal and higher porosity of adjacent bone both result in better infiltration of the polymer and better interlocking. On the other hand, the presence of air bubbles at the metal interface leads to decreased adhesion and possible failure. Other potential problems with the use of PMMA in medical applications include the release of monomer into the blood stream, leading to toxic effects such as a drop in blood pressure and death in extreme cases.

Notwithstanding the potential issues described above, actual problems are rare. The use of PMMA as bone cement has been widely successful and is a common practice in the surgical field.

6.4.3 Polylactic acid (PLA) and polyglycolic acid (PGA)

The chemical structure of polylactic acid (PLA) and polyglycolic acid (PGA) are shown in Figure 6.9. Both PLA and PGA, and their family of materials, are biodegradable polymers that are used extensively as biomaterials for implants in a variety of medical applications. These polymers are poly α-hydroxy acids and are linear polyesters. Their properties have been investigated since the 1950s, but the interest in their use as medical implants grew after the pioneering work of Kulkarni and colleagues in the 1960s.[2] These polymers and their copolymers are now used extensively in the field of orthopedics as fixation devices for bone and soft tissue in the form of biodegradable plates, screws, and anchors (Figure 6.10). They are also very popular as the scaffolding material for tissue engineering applications and are commercially available as synthetic bone grafts. Additionally, they are used for a variety of dental and controlled drug-delivery applications.

Both the PLA and PGA can be produced using polycondensation techniques to polymerize lactic and glycolic acids, but the yield is usually low molecular weight

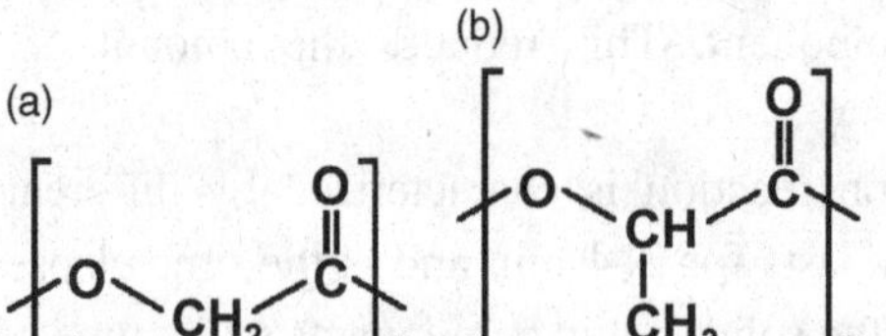

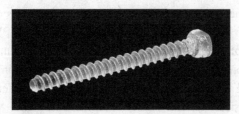

Figure 6.9

Chemical structure of (a) polyglycolic acid (PGA); and (b) polylactic acid (PLA).

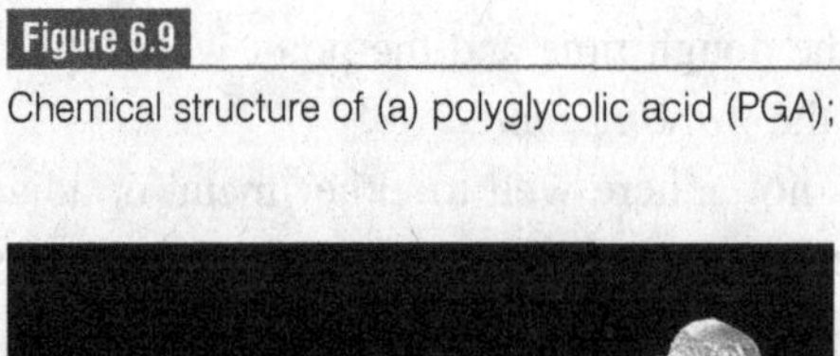

Figure 6.10

Example of a typical commercially available medical device made from polylactic acid and polyglycolic acid.

materials. Higher molecular weights can be obtained using ring-opening melt condensation polymerization of lactide and glycolide dimers with catalysts based on antimony, tin, titanium, aluminum, or zinc among others. Stannous octoate is commonly used as a catalyst.

A system or molecule is chiral when it not identical to its mirror image, that is, the mirror images are not superimposable. Chiral molecules are known as enantiomers. PLA is a chiral compound and can exist in different enantiomeric states: L and D. Thus, PLA can exist in D or L forms or as their racemic (equal amounts of each enantiomer) mixture DL. The DL-PLA has a lower crystallinity than D-PLA or L-PLA. In contrast, PGA exists in only one form.

Copolymers of PLA and PGA are relatively easy to synthesize and exist in various ratios. For example, Vicryl®, which is widely used for absorbable sutures, is a 90%PGA–10%PLA copolymer. The 50%PGA–50%PLA copolymer biodegrades relatively faster than the other copolymers and is often used as a carrier for drug delivery and for fabricating scaffolds for tissue engineering.

The degradation of these polyesters takes place primarily through non-specific hydrolytic scission of their ester bonds. Upon hydrolysis, PGA is converted to glycolic acid, which then reacts to form glycine. The glycine formed ultimately enters the body's natural tricarboxylic acid cycle and is reduced to water and carbon dioxide. These waste products are excreted in urine or through the respiratory process. Monomeric units of PGA can be directly excreted through urine. Also, there is evidence that PGA can be degraded by certain enzymes, especially those with esterase activity. In the case of PLA, it is first converted to lactic acid upon reaction with water. Lactic acid is a chemical that naturally occurs in the body and is processed by the body to yield water and carbon dioxide.

The physical and degradation properties of the PLA and PGA families of polymers and copolymers are affected by the chemical structure, molecular weight, molecular packing, and copolymer ratios of the polymers. For example, L-PLA is more crystalline and degrades slower than DL-PLA (a mostly amorphous polymer). This is because of the tightly packed L-PLA crystalline structure, which limits the access of water molecules to the ester bonds and thus reduces the rate of hydrolytic scission. The same degradation phenomenon is true when comparing the degradation rates of PLA and PGA. As shown in Figure 6.9, the PLA molecule has a methyl group as a pendant group, which offers stearic hindrance and makes it more hydrophobic. This methyl group provides protection to the backbone, thus yielding a lower degradation rate compared to PGA which has a hydrogen atom in place of the methyl group. On the other hand, PGA molecules without the pendant group can be packed more tightly and thus yield a more crystalline polymer. PLA polymers, because of their higher hydrophobicity and lower crystallinity, are more soluble in organic solvents than PGA.

The molecular weight and the crystallinity both affect the mechanical properties of the poly alpha-hydroxy acids. Both higher molecular weight and higher crystallinity result in improved mechanical properties of the polymers up to a limit. Biodegradation results in reduced molecular weight and altered crystallinity and thus can result in significant changes in properties.

The hydrolytic degradation reaction of these polyesters is catalyzed by a low pH environment. As the first direct products of degradation for both PLA and PGA are acids, this sets up the potential for autocatalysis. In autocatalysis, the degradation products make the reaction go faster and accelerate the degradation rate. This can be especially true in the interior of large solid implants made of PLA or PGA materials, where there are chances of accumulation of acidic by-products. When these polymers are used as implants, another issue related to the acidic degradation products is the detrimental effect of the acids on the surrounding cells. This can be an issue when PLA or PGA materials are used as implants in areas of the body that have low vascularity and where the body is unable to clear the degradation products in an expedient manner. When used for tissue engineering scaffolds, the relative hydrophobic nature of these polymers can prevent good cell adhesion. This problem can be addressed by surface treatment of the polymers by gas plasma in an oxygen environment or other chemical means to attach more hydrophilic moieties to their surface while still retaining the advantages of their bulk properties and biodegradability. The principle of plasma treatment is discussed in Chapter 9.

6.4.4 Polycaprolactone (PCL)

The chemical structure of polycaprolactone (PCL) is shown in Figure 6.11. It is a biodegradable polymer that is often used for tissue engineering applications as well as for drug delivery devices. It is made by the ring-opening polymerization of ε-caprolactone in the presence of a catalyst such as stannous octoate.

This polymer has a low melting point (59–64 °C) and a glass transition temperature of approximately −60 °C. Compared to PLA, PCL is more rubbery in nature at room temperature and biodegrades at a slower pace. The slower

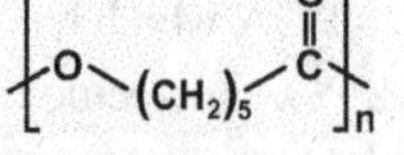

Figure 6.11

Chemical structure of polycaprolactone (PCL).

biodegradation is because of its relative high crystallinity and hydrophobicity. It is also easily soluble in organic solvents and has the ability to form blends. PCL degrades through the hydrolytic cleavage of its ester bonds and can be used to release drugs over a prolonged period. For example, it has been used to deliver the contraceptive levonorgestrel (a synthetic version of progestogen) for more than a year through an implantable device. PCL has also been studied to deliver anti-cancer drugs through nanoparticles. In some case these particles can be delivered via an injection.

6.4.5   Other biodegradable polymers

Polyanhydrides possess high hydrolytic instability and are used primarily for drug delivery. These are highly reactive materials which degrade by surface erosion and have been investigated for the delivery of a variety of drugs, proteins, and growth factors including chemotherapeutic agents, insulin, heparin, growth factors, and alkaline phosphatase.

Other biodegradable polymers used as a biomaterial include polyorthoesters, polycarbonates, and poly(p-dioxanone).

6.4.6   Polyurethanes

Polyurethanes have been sold commercially since the 1930s for a variety of applications including insulation, seat covers, and others. They were investigated for biomedical applications starting in the 1960s. Polyurethanes can be fabricated from isocyanates, which can be either aliphatic or aromatic. The isocyanates react with organic hydroxyls to form the urethane (—O—CO—NH—) linkages. The chemical structure of polyurethane is shown in Figure 6.12. Polyurethanes can be synthesized to have both hard and soft segments. This versatility can result in a wide variety of mechanical and physical properties. Pre-polymerized diisocya-nates, such as 4,4-metyl-diphenyl-diisocyanate with a diol as a chain extender, are commonly used to create hard segments. The soft segments can be polyether glycols such as polyethylene glycol. The relative ratios of the hard and soft segments and their respective molecular weights can yield stiffer or softer materials and determine their hydrophilic nature.

Because of their good mechanical properties and blood biocompatibility, poly-urethanes have been used for a number of biomedical applications including leads for pacemakers, catheters, heart valves, and ligament reconstruction.

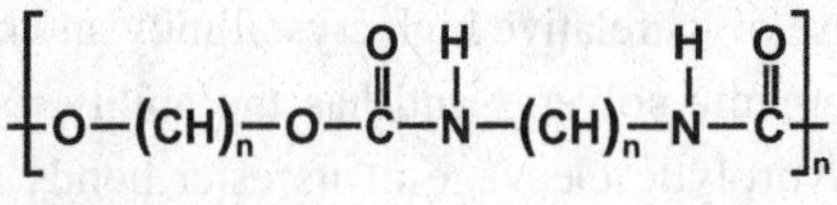

$$\left[O-(CH)_n-O-\overset{O}{\underset{}{C}}-\overset{H}{\underset{}{N}}-(CH)_n-\overset{H}{\underset{}{N}}-\overset{O}{\underset{}{C}} \right]_n$$

Figure 6.12

Chemical structure of polyurethane (PU).

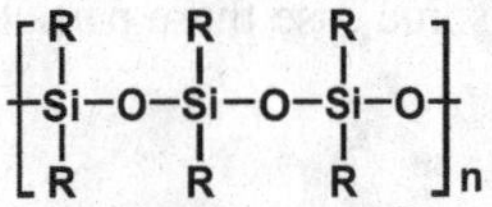

$$\left[\overset{R}{\underset{R}{Si}}-O-\overset{R}{\underset{R}{Si}}-O-\overset{R}{\underset{R}{Si}}-O \right]_n$$

Figure 6.13

Chemical structure of silicone.

6.4.7 Silicones

The chemical structure of silicone is shown in Figure 6.13. These polymers are characterized by alternating silicon and oxygen atoms in their backbone. Organic groups are attached to the silicon atoms to complete the repeating unit which is known as siloxane. The most common organic group is the methyl group, and the attachment of the methyl group on silicones forms polydimethylsiloxane or PDMS. Other groups such as vinyl or phenyl can be substituted for the methyl group. The combination of organic groups attached to a non-organic backbone gives silicone polymers a range of unique properties. As a result, they are used widely in the construction, aerospace, and electronics industries, in addition to the biomedical industry.

Silicones can be cross-linked to form elastomeric 3D networks. The cross-linking can be achieved using different techniques including radicals, or condensation and addition type reactions. In most cases, fillers are used to enhance the mechanical properties and hardness of these silcone elastomers. For most medical applications, fused silica is used as the filler. Particles (10 nm) of this amorphous material fuse to form aggregates, which in turn interact with other aggregates to form agglomerates with high surface area. The silica is added to the elastomer prior to cross-linking. The polymer forms hydrogen bonds with the silica and attaches to the large surface area of the nanoparticles. This gives the polymer good tensile strength, high elongation ability, and higher viscosity. The silicones used for biomedical applications usually have a glass transition temperature below room temperature.

Compared to polymers with a carbon backbone, siloxane molecular chains are highly flexible and can have a variety of configurations due to low resistance to rotation, low inter-chain interactions, and larger bond lengths and angles. The chains orient themselves to present the maximum number of methyl or other groups to the outside, which makes them hydrophobic. These materials can easily form films and have a high permeability for gases such as nitrogen, oxygen, and water vapor.

Silicone elastomers are used extensively in the biomedical industry. They have good blood biocompatibility and thus are used for many cardiovascular applications including catheters. In the area of orthopedics, they are used for prostheses to replace finger joints, carpal bones, and toes. They also have large-scale use in breast reconstruction and augmentation as these procedures often use silicone-based implants. Silicone-based implants are also used as for jaw augmentation, chin augmentation, and nasal supports. In general, implants of this material are successful, although there have been reports that the gradual deterioration of the elastomer can lead to the failure of prostheses such as finger joints.

6.5 Hydrogels

Gels are solid, jelly-like materials, which exhibit no flow when in a steady state. In general, hydrogels are 3D structures in which hydrophilic, water-insoluble, polymeric chains are dispersed in water and maintain their shape due to the presence of cross-linking and strong water retention. The cross-linking in hydrogels can be physical (chain entanglements) or chemical (van der Waals, covalent, ionic, or hydrogen bonds). Hydrogels can be colloidal in nature with water as the dispersion medium. Because of their physicochemical properties and high water content (they can contain more than 99.9% water), hydrogels have found numerous applications in the pharmaceutical and biomedical fields.

Hydrogels have polymeric structures that are dispersed in water where their molecules are tightly bound to water molecules, leading to the formation of a swollen semi-solid structure. Shown in Figure 6.14 is a representation of the interaction of a polymeric chain with water molecules. The water-holding capacity of the hydrogel is an intrinsic property of the polymer and depends on the chemical nature of the polymer's backbone and more importantly on the chemistry of its functional groups. However, the shape and strength of the hydrogel depend

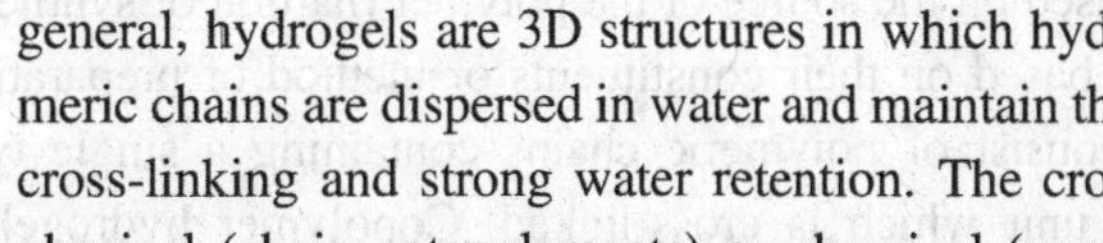

Figure 6.14

Schematic representation of interaction (absorption) of water molecules (circles) on a polymeric chain.

(a)   (b)

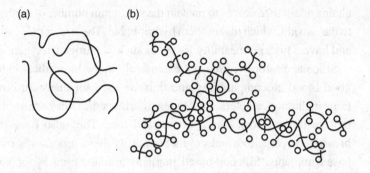

Figure 6.15

Schematic representation of (a) network of polymeric chains in dry form, and (b) water molecules absorbed onto polymeric chains in a hydrogel leading to water-swollen 3D structure.

on the type and degree of cross-linking. Figure 6.15a illustrates a network of polymeric chains in dry form. When mixed with water, the polymeric chain interacts with water molecules and subsequently causes swelling (Figure 6.15b).

Hydrogels can be classified in a variety of ways. For example, they can be classified based on the source of the polymer (natural or synthetic), or they can be categorized based on their constituents or method of preparation. Homopolymer hydrogels consist of polymeric chains containing a single type of hydrophilic monomeric unit which is cross-linked. Copolymer hydrogels use cross-linked polymers that have two types of monomer, at least one of which is hydrophilic. Multipolymer hydrogels have polymers with more than two types of monomer. Interpenetrating network hydrogels are prepared by first producing a cross-linked network. This network is then infiltrated with another monomer solution, which is then reacted to form its own network enmeshing the first network. Other classification systems are based on the hydrogels' ionic charge, structure, type of cross-links, or application (Table 6.1).[3]

Hydrogels based on natural polymers like chitosan, gelatin, and alginate can have variations in their composition and thereby in their properties. Synthetic polymers such as acrylate-based polymers, on the other hand, can be produced with high fidelity in their molecular weight and composition; hence their physicochemical properties are more consistent and uniform than natural polymer-based hydrogels.

Depending upon the type or overall charge of the hydrogel needed, a variety of functional groups can be utilized for synthesis of the polymer. Table 6.2 shows examples of different functional groups and the type of hydrogel charge obtained.

In recent years, there has been a surge in the synthesis of smart polymers, which can be used to formulate stimuli-responsive hydrogels. Smart polymers can exhibit a rapid change in water affinity when an environmental factor (pH,

Table 6.1 Classification systems for hydrogels

Classification	Property/type
Source	Natural
	Synthetic
Component or method of preparation	Homopolymer
	Copolymer
	Multipolymer
	Interpenetrating polymer
Electric charge	Anionic
	Cationic
	Neutral
	Zwitterion
Physical structure	Amorphous
	Semi-crystalline
	Hydrogen bonded
Cross-link	Physical entanglement
	Covalent bond
	van der Waals interaction
	Hydrogen bond
Function	Stimuli responsive
	Superabsorbent
	Biodegradable

temperature, light) is changed. Stimuli responsive hydrogels are usually reversible in their nature and show transition from solution (sol) to gel form depending on the stimulus presented. Poly(N-isopropyl acrylamide) (PNIPAm) is a polymer highly studied for its temperature-regulated sol–gel transition. This polymer has both hydrophilic and hydrophobic moieties in its structure. When dispersed in water at a temperature lower than 32 °C, hydrophilic interactions with water molecules dominate and the polymer chains stay in extended form, giving it a solution form. When the temperature is raised above 32 °C, the hydrogen bonds between the polymer and the water molecules become less stable, and hydrophobic interactions between isopropyl groups on the polymer become thermodynamically favored leading to the formation of compact polymer structures and subsequent precipitation of the polymer in the solvent. The temperature at which this transition begins is known as the *critical temperature* for that solvent–polymer pair. The schematic in Figure 6.16 depicts the sol–gel transition for temperature-responsive polymers.

Table 6.2   Functional groups for polymers used for hydrogels

Functional group	Charge	Example
Carboxyl, sulfonic	Anionic	Acrylates
Quaternary ammonium	Cationic	Chitosan
Non-ionic groups	Neutral	Poly-NIPAm
Carboxylic and quaternary ammonium	Zwitterion	Polypeptides

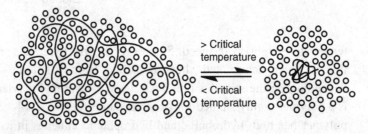

Temperature-dependent sol–gel transition. Below the critical temperature, the polymer is in expanded form due to hydrophilic interactions with water molecules. Above the critical temperature intra- and inter-chain hydrophobic interactions lead to the formation of compact polymer aggregates surrounded by water molecules.

The stimulus for sol–gel transition can also be provided by pH. Hydrogels that are pH sensitive are formed by polyelectrolyte polymers which have weak acids or bases in their structure and can accept or release protons depending on the pH of the environment. The transition of such polymers from compact to expanded state

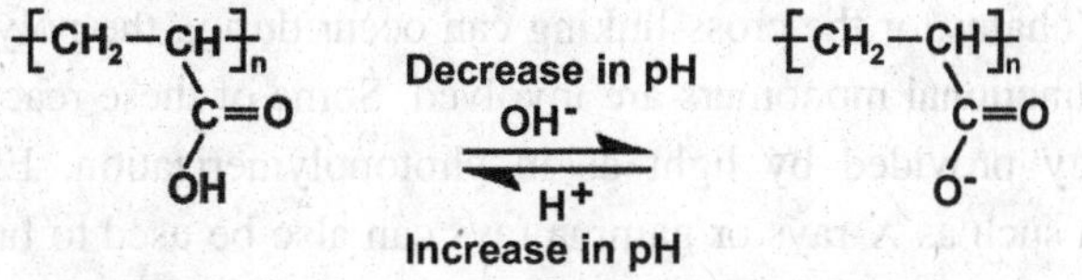

Figure 6.17

The pH-dependent ionization of acidic polyelectrolytes, e.g. poly(acrylic acid).

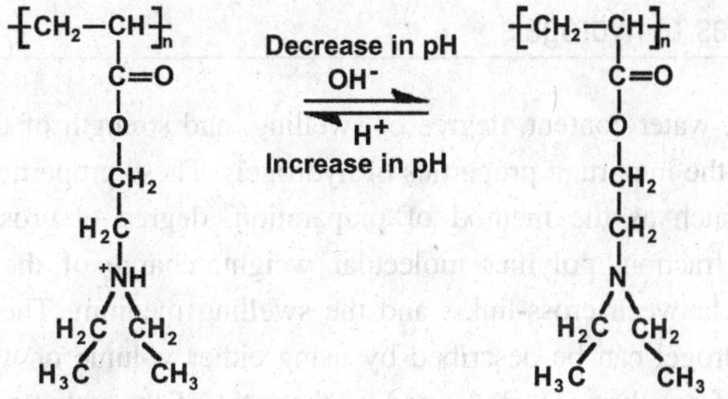

Figure 6.18

The pH-dependent ionization of basic polyelectrolytes, e.g. poly(N,N'-diethyl aminoethyl methacrylate).

can be orchestrated by changing the pH. Polyacidic polymers are non-swollen at low pH and expand at high pH because ionization at high pH values leads to solubilization. On the other hand, polybasic polymers show the opposite behavior compared to the polyacidic polymers. Ionization of polyelectrolytes with change in the pH is illustrated in Figures 6.17 and 6.18.

Similarly, other types of stimuli-responsive hydrogels also utilize the effect of stimulus on the water affinity of the polymer chains to achieve sol–gel transitions.

6.5.1 Synthesis of hydrogels

Hydrogels are prepared by swelling cross-linked polymer networks in water. The fabrication of polymers with good water absorption capacity is usually the first step. To this end, polymers with desirable functional groups have to be synthesized or an existing polymer first modified. The synthesis of application-specific copolymers or block copolymers is also common. The next step is the fabrication of cross-linked networks and these can be generated using a variety of methods. For example, short di- or multifunctional linkers can be used to react with long polymer chains, or the cross-linking can occur during the polymerization process if multifunctional monomers are involved. Some of these reactions can be driven by energy provided by light as in photopolymerization. Electron beams and radiation such as X-rays or gamma rays can also be used to break existing bonds in polymeric chains to create free radicals that react to create a cross-linked structure.

6.5.2 Properties of hydrogels

Porosity, water content, degree of swelling, and strength of the 3D network are some of the important properties of hydrogels. These properties are dependent on factors such as the method of preparation, degree of cross-linking, polymer volume fraction, polymer molecular weight, charge of the polymer, average distance between cross-links, and the swelling medium. The swelling behavior of a hydrogel can be described by using either volume or weight. The volume degree of swelling, Q, is defined as the ratio of the volume of the swollen gel divided by the volume of the dry sample:

$$Q = V_{gel}/V_p = \text{volume of swollen gel/volume of dry polymer.} \qquad (6.9)$$

Similarly, the weight degree of swelling, q, is defined as the ratio of the weight of the swollen gel divided by the weight of the dry sample. The degree of swelling of a hydrogel is important because it influences the gel's mechanical properties, the diffusion of solutes through the hydrogel, and the surface and optical properties of the gel.

6.5.3 Applications

Hydrogels have been used extensively for soft contact lenses since the 1960s. The reason for their popularity is because they provide comfort for the user and also have better oxygen permeability in comparison to the hard lenses made from PMMA. The first soft contacts were made of poly(2-hydroxyethyl methacrylate) or poly-HEMA, which was developed by Otto Wichterle, a Czech chemist and inventor. This material has several advantages including high flexibility and dimensional stablility, resistance to changes in pH and temperature, ease of fabrication into contact lenses, and is relatively inexpensive to produce. A major disadvantage of polyHEMA is that it has limited oxygen permeability because it depends upon water to transport oxygen across the material, and water has a limited ability to dissolve and carry oxygen. To overcome this problem, more hydrophilic monomers such as N-vinyl pyrrolidone (NVP) or methacrylic acid (MA) have been added to the methyl methacrylate (MMA) in a variety of products.

Silicone-based contact lenses provide excellent oxygen permeability and durability but do not permit water to pass through. In addition, the hydrophobic nature of silicone-based contact lenses leads to lipid deposition. In response to these issues, silicone–hydrogel contacts have been developed which combine silicone with hydrophilic monomers that form hydrogels. This combination leads to high oxygen permeability while adding wettability and fluid transport properties to the material.

Hydrogels are also used for drug delivery based on a variety of delivery platforms. For example, the drug can be evenly distributed in the matrix of a dry hydrogel. Once in contact with water, the hydrogel swells and gradually releases the drug through diffusion. Alternatively, the drug can be stored in a reservoir surrounded by a layer of hydrogel, and as water penetrates the hydrogel and reaches the inner core, the drug is released gradually. The properties of the hydrogel can be designed to release the drug over a desired period of time following pre-determined release profiles.

In recent years, hydrogels have been used in the field of tissue engineering where cells can be imbedded within the gels and implanted in the body. Other uses include various applications in the area of wound healing, where sheets of hydrogels can be used to cover wounds and keep them moist under aseptic conditions.

6.6 Nanopolymers

Nanopolymers are polymers that are available in the form of particles, fibers, tubes, or other shapes that have at least one dimension of less than 100 nm. Although the chemical structure of these polymers is the same as the bulk polymer, the particles usually have very high surface area to volume ratio. This high surface area leads to a significant increase in their ability to interact with their environment and their neighboring particles, leading to increased chemical reactivity, strength, hardness, heat resistance, and other properties.

Electrospun nanofibers, which have a diameter of less than 100 nm, are used to develop scaffolds for tissue engineering applications. Cells are introduced into these porous scaffolds so that they can proliferate, form extracellular matrix, and generate new tissue. The high surface area of these fibers, compared to microfibers, provides additional surface for cells to attach.

Nanopolymers in the form of particles are being studied extensively as drug delivery vehicles, especially in the area of cancer therapy. These particles are small enough to pass through the leaky vasculature found in many tumors and penetrate the walls of malignant cells where they can deliver their payload of drug in a very efficient manner leading to a high efficacy. The particles can have an inner core or reservoir containing the drug with an outer shell made of the polymer. Nanotubes can carry the drug in their hollow centers, thus protecting it from degradation in the blood stream. The polymer can be designed to release the drug in a controlled fashion. In addition, targeting vectors can be attached to the nanotubes or nanoparticles, which help them to locate and congregate at the target before releasing the drug. These vectors could consist of biological moieties such as ligands, peptides or proteins, or magnetic materials such as ferric oxide. In the latter case, the particles can be driven to the target site by using external magnetic stimuli.

Recent research is revealing that nanopolymers can have direct beneficial therapeutic effects. It has been determined that in animal models, an injection of polyethylene-oxide (PEO) increases tissue perfusion and decreases mean blood pressure. PEO nanoparticles interact with endothelial cells and increase the production of endothelial nitric oxide (eNO) and its attendant enzyme (eNOS). It is known that eNOS over-expression and eNO induction can provide cardioprotection. Thus, it may be possible to reduce permanent damage to the cardiac muscle after a myocardial infarction or heart attack by using PEO nanoparticles.

6.7 Summary

Polymers are large molecules with repeating chemical units or monomers. They can be produced by addition or condensation polymerization. While addition polymerization involves monomers being directly added to the growing polymeric chain, the growth of the repeat units during condensation polymerization involves the elimination of some atoms via a chemical reaction and the formation of a by-product. Generally, polymers produced by common techniques contain chains with different molecular lengths, and so, the molecular weight is determined using different ways of calculating averages. In addition to the chemical elements of a polymer, its properties are also dependent on the form of its polymeric chains such as linear, branched, or cross-linked structures. Also, whether the polymeric molecules are arranged in an amorphous or crystalline state plays a significant role in the polymer's properties. The polymers most often used as biomaterials are similar to those widely used in everyday life. These polymers include polyethylene, polymethylmethacrylate, silcones, and others. In recent years, hydrogels have been explored for a variety of applications in biomedicine, and the area of nanopolymers shows great promise.

References

1. Berzelius, J. J. (1833). *Jahresberitche*, **12**, 63.
2. Kulkarni, R. K., Pani, K. C., Neuman, C. and Leonard, F. (1966). Polylactic acid for surgical implants, *Arch. Surg.*, **93**, 839.
3. Akio, K. and Yoshito, I. (2001). *Hydrogels for Biomedical and Pharmaceutical Applications. Polymeric Biomaterials*, revised and expanded. CRC Press.

Suggested reading

Rodriquez, F. (1996). *Principles of Polymer Systems*. Washington D.C., Taylor & Francis, ISBN 1–56032–325–6.

Stevens, M. P. (1998). *Polymer Chemistry: An Introduction*. Oxford University Press, ISBN-10: 0195124448.

Ratner, B. D., Hoffman, A. S., Schoen, F. J. and Lemons, J. E., editors (2004). *Biomaterials Science: An Introduction to Materials in Medicine*, London, Elsevier Academic Press.

Peppas, N. A. (1997). Hydrogels and drug delivery, *Crit. Opin. Colloid Interface Science*, **2**, 531–537.

Peppas, N. A. (2004). Hydrogels, in *Biomaterials Science: An Introduction to Materials in Medicine*, Ratner, B. D., Hoffman, A. S., Schoen, F. J. and Lemons, J. E., editors. London, Elsevier Academic Press, pp. 100–106.

Problems

1. Ethylene ($CH_2{=}CH_2$) and propylene ($CH_2{=}CHCH_3$) are each separately polymerized to yield polymer A and B respectively. Which of the polymers would have a higher glass transition temperature? Which would more likely be crystalline? Explain your answers.

2. Explain the difference between branched and cross-linked polymers. Which would be expected to be more stiff? Why?

3. What is the difference between isotactic, syndiotacytic, and atactic forms of polymers? Which are more likely to be crystalline in nature?

4. You are given two samples of the same polymer. One is amorphous and the other is mostly crystalline. Which will be more difficult to dissolve in a solvent? Which is more likely to be opaque?

5. The average molecular weight of a sample of PMMA (Figure 6.8) is 75 650. What is the degree of polymerization for the specimen?

6. Three polymer samples with the following properties are mixed together to get 1 kg of polymer mixture. What is overall M_n?

Sample	M_n	Weight% in mixture
A	2.3×10^5	40
B	3.2×10^5	35
C	5.1×10^5	25

7. Use the chemical structures of PLA and PGA to explain the differences in their rates of hydrolytic degradation.

8. Describe the different systems used to classify hydrogels.

9. What are smart hydrogels? Give an example and explain how it works.

10. How do polyesters such as PLA and PGA degrade? What are some properties that affect their rate of degradation? Explain how autocatalysis can influence their degradation.

11. As part of an investigation into a failed implant, a polymeric impurity has been detected as an inclusion. Owing to the small size of the sample you have been provided with a very dilute solution of this polymer. An analysis provides data as shown in the table below. Given this information, calculate the number average molecular weight (M_n) of the polymer. If the polydispersity index for the sample is 1.5, what is its weight average molecular weight (M_w)?

Size (Da)	5000	10 000	20 000	30 000	40 000	45 000	50 000	52 000
Number of chains	100	150	500	1000	5000	8000	7000	6000

7 Ceramics

Goals

After reading this chapter, students will understand the following.
- The general definition of a ceramic.
- Common properties of ceramics.
- Different classifications used for ceramics.
- Properties of different bioceramics.
- Different technologies used for fabricating nanoceramics.

The use of ceramics in medicine dates back many centuries, with reports of artificial teeth found in Egyptian mummies.[1] Besides ceramics developed for medical applications, other engineering ceramics include semiconductors, dielectrics, high temperature superconductors, magnets, and piezoelectrics. However, what is a ceramic? In general, a ceramic is defined as an inorganic, non-metallic material that consists of two or more metallic and non-metallic elements. Unlike metals and polymers, which comprise mainly of metallic and covalent bonding, respectively, ceramics are made up of ionic and covalent bonding.

Depending on the atomic arrangements, ceramics can either exist as amorphous or crystalline structures. An example of an amorphous ceramic is glass, whereas an example of a crystalline ceramic is porcelain. In an amorphous structure, the atoms are arranged randomly or with high degree of short-range order and absence of long-range order. A short-range order refers to the tendency for an ordered atomic arrangement within one or two atom spacings, whereas a long-range order refers to an ordered atomic arrangement over a larger distance. Figure 7.1a shows a schematic drawing of a non-crystalline (glass) silicon dioxide, with random, short-range order atomic arrangement. As an example of the long-range order observed in crystalline ceramics, Figure 7.1b shows a schematic drawing of a crystalline silicon dioxide, with atoms arranged in an ordered pattern.

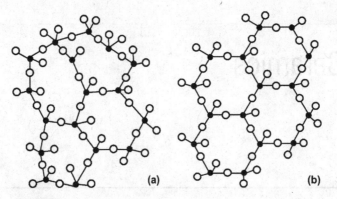

Schematic renditions of atomic arrangement of (a) an irregular arrangement of silicon dioxide (SiO_2) ions indicating amorphous structure, and (b) a regular arrangement of SiO_2 ions indicating crystalline structure. Symbols: ● = Si^{4+}; ○ = O^{2-}.

7.1   General properties

The atomic bonds in a ceramic crystal have both covalent and ionic characteristics. These strong bonds are responsible for the great stability of ceramics and impart very useful properties such as hardness, high modulus of elasticity, and resistance to heat and chemical attack. Ceramics are also strong in compression but weak in tension. As shown in Figure 7.2, this unique mechanical property has allowed stones and bricks to be laid on top on each other in compression, especially in buildings. However, ceramics are brittle which causes them to fracture rapidly or shatter instead of undergoing plastic deformation. Ceramics can either be opaque or transparent. Figure 7.3 shows different ceramic dental implants that exhibit such opaque or transparent properties. Another unique property of ceramics is that they are good insulators. Metals such as copper and iron that are good conductors of electricity, have electrical resistivity in the range of 10^{-6} Ω cm. In contrast, the ceramics such as alumina and silica have electrical resistivity in the range of 10^{14} Ω cm.

- Ionic and covalent bonds are two of the common chemical bonds for ceramic materials.
- Ceramics are brittle as a result of their strong bonds.
- Ceramics are dimensionally stable, in addition to their high resistance to abrasion and their chemical inertness.

Figure 7.2

Superior compressive loading ability of ceramics exhibited by the ability to stack stones on top of each other without failure.

7.2   Classifications

There are numerous ways to classify ceramics; they can be classified based on their general form, composition, or reactivity.

7.2.1   Classification based on form

Depending on the specific function, ceramics can be classified into the following three general forms:

- powders,
- coatings, and
- bulk shapes.

Figure 7.3

Dental implants made of alumina. Different optical properties of the implants are exhibited, with single crystal alumina implants being transparent and polycrystalline alumina implants being opaque.

Powders are dry, solid particles of various sizes that are not cemented together. These particles flow freely when shaken or tilted. Fine grain powders have a tendency to form clumps, whereas coarser grain particles (or granular particles) do not form clumps except when wetted. Coatings, in general, refer to films or deposits on substrates, whereas bulk shapes refer to the densified form of the ceramics.

7.2.2 Classification based on composition

Ceramics can be classified as oxides, non-oxides, and composites. Both oxide and non-oxide ceramics are chemically inert. However, oxide ceramics such as alumina and zirconia are oxidation resistant, electrically insulating, and have generally low thermal conductivity. Other examples of oxide ceramics include magnesium oxide (MgO), aluminum titanate ($Al_2O_3 \cdot TiO_2$), and lead zirconate titanate (Pb $[Zr_xTi_{1-x}]O_3$). In contrast, non-oxide ceramics possess low oxidation resistance, are electrically conducting, and have high thermal conductivity. There are many different non-oxide ceramics, such as carbides, borides, nitrides, and silicates. Composite ceramics are made of different combinations of oxides and non-oxides and thus, in general, have variable thermal and electrical conductivity.

7.2.3 Classification based on reactivity

In addition to general forms and compositions, ceramics can be classified according to their reactivity. In this classification, ceramics are grouped as one of the following:

- inert ceramics,
- degradable or resorbable ceramics, and
- surface reactive ceramics.

An inert ceramic is chemically stable, that is, these ceramics do not corrode, wear, or react to the host environment. Little or no chemical change occurs during the long-term exposure of inert ceramics to the physiological or host environment. An example of inert ceramics is the alumina that is used in the fabrication of total hip replacement prostheses as well as dental crowns. Other examples of inert ceramics include glass, zirconia, and diverse forms of carbon. Resorbable and surface reactive ceramics react to the host or physiological environment resulting in surface or bulk chemical changes. Bioactive glass is an example of surface reactive ceramics. When exposed to fluid in a biological environment, bioactive glass forms a bioactive carbonate apatite layer on its surface. Hydroxyapatite and tricalcium phosphate are examples of resorbable ceramics and these ceramics are capable of degrading in the presence of a biological environment.

7.3 Bioceramics

The name "bioceramics" is given to ceramics that are used in medical applications. Today, a wide range of ceramic and glass materials are used for biomedical applications, ranging from bone implants to biomedical pumps. These ceramics are used in structural functions as joints or tissue replacements as well as being used as coatings to improve the biocompatibility of metal implants. Chapter 13 also provides a discussion on ceramics that can be fabricated to function as resorbable lattices, thereby providing temporary structures or scaffolds for tissue regeneration. Ceramics have also been used as carriers for growth factors, antibiotics, and drugs.

Box 7.1

- The first generation bioceramics are inert ceramics used in medicine, whereas the second generation bioceramics include degradable and surface reactive ceramics.
- Ceramics used as temporary structures or scaffolds in regenerative medicine are known as the third generation bioceramics.

Important characteristics of bioceramics include their mechanical integrity and their physical and chemical compatibility in the presence of host or biological environments. Implantable bioceramics should have the following properties:

- non-toxic,
- non-carcinogenic,
- do not induce allergic reactions,
- do not induce inflammatory response,
- induce tissue regeneration if needed, and
- induce tissue integration if needed.

Some of the bioceramics that are commonly used today in medicine and dentistry are discussed below.

7.3.1   Silicate glass

Silicate glass is commonly used for windows, drinking glasses, and bottles. Other applications of glass include ocular prostheses, sensors, fiber lasers, and optical fibers. The composition of glass is primarily silica (SiO_2), also known as silicon dioxide. The SiO_2 acts as the glass former or the basic building block of glass. As one of the most abundant oxides in the earth's crust, SiO_2 commonly occurs as sandstone, silica sand, or quartzite and is formed by strong, directional covalent bonds. It has a well-defined local structure with four oxygen atoms at the corners of a SiO_4 tetrahedron surrounding a central, network-forming cation, which in this case is the silicon atom (Figure 7.4). The ideal bond angle of a tetrahedral such as methane (CH_4) is 109.5°, whereas the bond angle of O–Si–O in a SiO_4 tetrahedron varies slightly from this ideal bond angle.

Silica is formed by interconnecting multiple SiO_4 tetrahedral units into a three-dimensional (3D) network structure. Two SiO_4 tetrahedrals can be connected by an oxygen atom known as the bridging oxygen. The angle of Si–O–Si bond between the two SiO_4 tetrahedrals is 144°, and this bond angle can vary from 120° to 180°. The flexibility of the oxygen bridge between the two silicon atoms permits SiO_2 to exist as a polymorphic material, exhibiting both amorphous and crystalline structures. Its properties permit it to be used for abrasives, refractory materials, fillers, and optical components as well as in high temperature and corrosive environments.

The non-bridging oxygen in silicate glass is a negatively charged oxygen atom that is bonded to only one silicon atom. The presence of the non-bridging oxygen in silicate glass decreases the overall structural bonding within this ceramic and is critical for lowering of the melting temperature as well as decreasing the viscosity

(a)   (b)

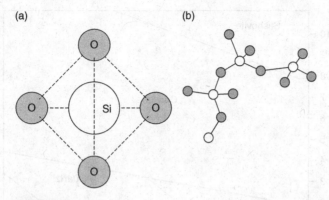

Figure 7.4

A schematic showing (a) the tetrahedron structure of silica (SiO_2), with a central silicon atom surrounded by four oxygen atoms, and (b) the bridging oxygen connecting the two tetrahedra. Symbols: $\circ = Si^{4+}$; $\bullet = O^{2-}$.

of the ceramic during melting. Stability of the non-bridging oxygen atom is accomplished by bonding to network modifying cations such as oxides or alkaline earth ions. Examples of the network modifying oxides include oxides of boron, germanium, phosphorus, sodium, potassium, rubidium and cesium, whereas alkaline earth ions that are used for network modification include magnesium, calcium, strontium, and barium.

Fused or vitreous silica is an amorphous phase made of high-purity SiO_2. It is used in a variety of applications including gas transport systems, laser optics, fiber optics, waveguides, electronics, vacuum systems, and furnace windows. During its application, glass may experience changes in pressure and temperature, which may result in alteration of its properties. As shown in Figure 7.5, SiO_2 exhibits different crystalline forms under different temperatures and pressures. The three crystalline forms of SiO_2 are quartz, cristobalite, and tridymite, with each of the crystalline forms capable of exhibiting allotrophic transformation.

Box 7.2

- Physical properties of vitreous silica include excellent resistance to chemicals, minimal thermal expansion, and high refractoriness.
- Vitreous silica can either be transparent or opaque. Opacity is attributed to bubbles within the glass that scatter light thus giving it a milky appearance.

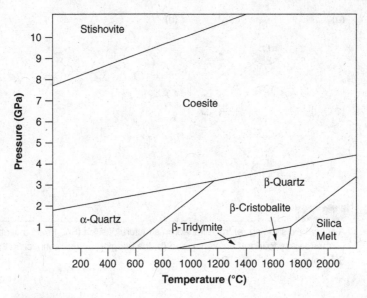

The stability of the SiO_2 crystalline forms is highly dependent on the energy of the crystal structure. At any given temperature and pressure, the most stable crystalline form is the structure with the lowest energy. Changes in pressure and temperature can result in two types of transformations, namely displacive and reconstructive. Differences between these two transformations are as follows.

• Displacive transformation occurs when atoms are displaced, but no interatomic bonds are broken during a temperature change. However, there may be slight changes in angles between the atoms.

• Reconstructive transformation occurs when interatomic bonds are broken and there is remaking or reformation of Si–O bonds.

The following are three characteristics of a displacive transformation in silica.

• Depending on the pressure, displacive transformation occurs at specific temperatures and results in distortions of the tetrahedral framework. Under normal pressure, the transformation from low or α-quartz (rhombohedral or trigonal close-packed structure) to high β-quartz (hexagonal close-packed structure) occurs at 573 °C. The associated phase diagram (Figure 7.5) also shows that the transition temperature rises rapidly with increasing pressure. Low or α-tridymite (orthorhombic close-packed structure) is transformed to high or β-crystobalite (hexagonal close-packed structure) at 117 °C. Similarly, low or α-crystobalite (tetragonal close-packed structure) is transformed to high or β-crystobalite (cubic close-packed structure) at 270 °C. Regardless of the temperature, coesite and stishovite exist at high pressures.

- Transformations are rapid or instantaneous. During the transformations from the α to β phase of each polymorphic form, atoms in the crystal lattice are slight displaced simply through the alterations of the bond angles and the length of the chemical bonds. For example, the Si–O–Si bond angle is changed from 144° to 153° when α-quartz is transformed to β-quartz. Similarly, the Si–O–Si bond angle for tridymite is changed from 140° to 180° when α-tridymite is transformed to β-tridymite. The transformation from the α to β phase of critobalite results in a change in Si–O–Si bond angle from 147° to 151°.

- Linear expansion occurs during displacive transformation, thus resulting in the high temperature form of SiO_2 having a greater specific volume. As shown in Figure 7.6, both cristobalite and quartz expand more than tridymite. Since dental investment materials contain SiO_2, the linear expansion is an important property. The expansion ability of the dental investment material helps to compensate for the shrinkage on cooling of metal alloys during casting. Fused quartz does not undergo displacive transformation and thus exhibits little thermal expansion.

Unlike displacive transformation, the breaking of bonds during reconstructive transformation requires more energy, thereby resulting in reassembly of atoms and extensive rearrangement of the crystal structure into a different structure. Complete reconstructive transformations between polymorphs also usually occur more slowly. At 872 °C, β-quartz undergoes reconstructive transformation to form β-tridymite. As the temperature continues to be raised to 1470 °C, β-tridymite undergoes reconstructive transformation to form β-cristobalite. When the temperature is raised above 1700 °C, fused quartz is formed.

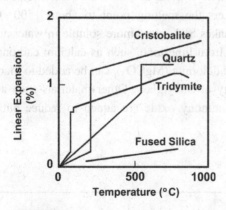

Figure 7.6

Thermal expansion of the allotropes of silica.

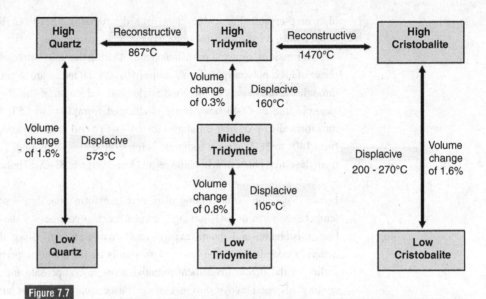

Summary chart showing the temperatures and changes in volume of each polymorph during displacive and reconstructive transformations of silica.

A summary of reconstructive and displacive transformations of silica poly-morphs is shown in Figure 7.7.

At high temperatures, SiO_2 fuses together. Rapid cooling of this fused SiO_2 locks the atoms in a rigid structure and does not give them enough time for the formation of an ordered or crystalline structure. Depending on the application and the properties needed, the characteristics of silicate glass can be varied by changing the composition and cooling rate. SiO_2 softens at temperatures above 2300 °C, but the addition of soda ash, such as sodium carbonate (Na_2CO_3) to silica-based glass lowers the melting point to about 1500 °C. However, the addition of Na_2CO_3 makes SiO_2 glass more soluble in water and thus is usually less desirable. Oxides from limestone, such as calcium carbonate ($CaCO_3$), cal-cium oxide (CaO), and dolomite ($MgCO_3$), can be added to increase the hardness and chemical durability of SiO_2 glass. Other materials such as sodium sulfate, sodium chloride, or antimony oxide are added to reduce bubbles in the glass during processing.

7.3.2   Alumina (Al_2O_3)

Aluminum oxide, also known as alumina (Al_2O_3), is commonly used in orthoped-ics and dentistry. First used as an implant biomaterial in the 1960s, today's Al_2O_3

devices are made from high purity Al_2O_3; because of its inertness, biocompatibility and excellent wear resistance, these devices are used in hip and knee prostheses. In dentistry, aluminous porcelain is a major ingredient in the fabrication of crowns. Shown in Figure 7.8 is a fabricated alumina crown. The addition of Al_2O_3 in porcelain crowns also serves as an intermediate to produce ceramics with high viscosity as well as low firing temperatures.

High purity Al_2O_3 is a fine white material that is similar in appearance to common salt. Besides being known as alumina, Al_2O_3 is also called corundum, sapphire, ruby, or aloxite. Al_2O_3 is insoluble in water and organic liquids and is used in the abrasive, ceramics, and refractory industries. However, Al_2O_3 is slightly soluble in strong acids and alkalies. Alumina is found in nature as native corundum or in ores such as bauxite and cryolite. Bauxite usually consists of two forms of Al_2O_3 – a monohydrate form called boehmite ($Al_2O_3.H_2O$) and a trihydrate form called gibbsite ($Al_2O_3.3H_2O$). The alumina content in bauxite ranges from 30% to 54%, with the remainder consisting of silica, iron oxides, and titanium dioxide. The commonly available α-Al_2O_3 can be prepared by calcining alumina trihydrate.

Using the Bayer process (discovered and patented in 1887 by Karl Bayer, an Austrian chemist), the extraction process begins by grinding bauxite ore to a size that will maximize solid–liquid contact. Sodium hydroxide is then added to the

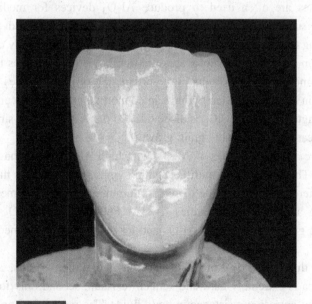

Figure 7.8

Alumina crown after casting. Surface finishing is performed to achieve the natural appearing gloss of a natural tooth. (Courtesy of Dr. Namsik Oh, Inha University Hospital, South Korea.)

ground bauxite ore to produce a slurry. Depending on the content, boehmite dissolves in 10% sodium hydroxide at temperatures above 220 °C under high pressure, whereas gibbsite dissolves in 10% sodium hydroxide at temperatures below 150 °C. During this process and in the presence of steam, the silicate component of the bauxite is chemically attacked by the sodium hydroxide to form sodium aluminate solution. This reaction is rapid and can be represented by Eq. (7.1) and Eq. (7.2):

$$2NaOH \ + \ Al_2O_3 \cdot 3H_2O \rightarrow \ 2NaAlO_2 \ + \ 4H_2O \qquad (7.1)$$

$$2NaOH \ + \ Al_2O_3 \cdot H_2O \rightarrow \ 2NaAlO_2 \ + \ 2H_2O \qquad (7.2)$$

By passing the contents through other vessels at reduced temperature and pressure, impurities such as titanium oxide, iron, and silica compounds are removed. The sodium aluminate solution is then left to precipitate as aluminum trihydrate ($Al_2O_3 \cdot 3H_2O$), followed by heat treatment above 1100 °C to evaporate any moisture present. The resulting end product after heat treatment is commercially pure alumina. Depending on the heat treatment conditions, Al_2O_3 exists in several allotropic forms (α, δ, γ, $\acute{\eta}$, θ, ρ, or χ phases). The most thermodynamically stable form is the α-Al_2O_3, and thus this is used as a biomaterial. With a melting temperature of about 2040 °C, pressing and sintering of the polycrystalline α-Al_2O_3 powder can occur at temperatures between 1600 °C and 1800 °C, and these processes are often used to produce Al_2O_3 devices for medical applications. Fluxes such as CaO and magnesia (MgO) are often added to Al_2O_3 during sintering to decrease the sintering temperature. Doping of fully dense Al_2O_3 with less than 0.5% MgO inhibits abnormal grain growth, whereas the presence of a small amount of CaO results in the production of liquid phases or segregation at the grain boundaries, thereby inducing abnormal grain growth and a possible loss of strength over time. SiO_2 is also added to the Al_2O_3 during sintering to impede densification and promote grain growth.

There are many grades of Al_2O_3, with high Al_2O_3 grades having at least 99% purity. The amount of impurities and alloying agents (such as fluxes) account for the differences between the different alumina grades. The American Society for Testing and Materials specifies (ASTM F603–78) that Al_2O_3 implants should contain greater than 99.5% Al_2O_3 and less than 0.1% combined SiO_2 and alkali oxides.

The different forms of Al_2O_3 are all dense, non-porous, and nearly inert. Alumina is extremely hard and scratch resistant, second only to diamond. It has excellent corrosion resistance in body fluids. The mechanical properties of Al_2O_3 such as strength, fatigue resistance, and the fracture toughness of polycrystalline α-Al_2O_3 are a function of grain size, porosity, and purity. The finer the grain size or the higher the specific surface area of α-Al_2O_3, the higher will be its opacity. Similarly, the finer the grain size or higher the specific surface area of α-Al_2O_3, the higher will be its strength. As indicated in the ASTM F603–83, an average grain size of less than 4 μm and greater than 99.7% purity will exhibit excellent compressive and flexural strength. An increase in grain size above 7 μm has been reported to result in a 20% decrease in its mechanical properties.[3] High density Al_2O_3 with a grain size of less than 4 μm is also known to exhibit a very low surface roughness (<0.02 μm) and high surface wettability or surface energy (2.64 J/m^2), thereby leading to a very low coefficient of friction when sliding against itself or polyethylene.[4] This property of high wettability is attributed to the non-saturation of oxygen ions on its outer surface. When implanted in the body, the non-saturation of oxygen ions allows the formation of an absorbed layer containing water and biological molecules and thus limits direct contact with the articulating solid surface.

7.3.3   Zirconia (ZrO_2)

The evaluation of zirconia (ZrO_2) ceramics for medical applications began in the late 1960s. Also known as zirconium oxide, ZrO_2 is a white crystalline oxide of zirconium. It is a well-known polymorph that occurs in three polymorphic forms. At room temperature it exists as a monoclinic crystal structure. At temperatures between 1170 °C and 2370 °C, ZrO_2 exists in a tetragonal phase. A tetragonal–monoclinic transformation occurs when ZrO_2 ceramics are cooled below 1170 °C. This transformation results in a volume expansion of approximately 3%–5%. Cracks are formed as a product of the stresses generated by such a volume expansion. At temperatures above 2670 °C, a cubic phase is formed. Similar to the tetragonal–monoclinic transformation, the cubic–tetragonal transformation also induces volume expansion, large stresses, and cracks upon cooling from high temperatures. As a result of these volumetric expansions due to temperature-sensitive phase transformations, ZrO_2 ceramics are not suitable for use as a biomaterial.

To prevent volume expansion during phase transformations, several different stabilizing oxides, such as yttria or yttrium oxide (Y_2O_3) and magnesium oxide (MgO) are added to ZrO_2 to stabilize the tetragonal and/or cubic phases by producing multiphase materials known as partially stabilized zirconia (PSZ). ZrO_2 is very useful as a biomaterial in its "stabilized" state. Depending on the amount of the stabilizer added, cubic ZrO_2 is generally the major phase in the microstructure of PSZ at room temperature, and monoclinic and meta-stable tetragonal ZrO_2 make up the minor phases. If sufficient quantity of the

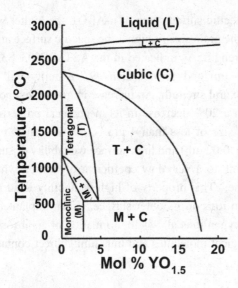

Figure 7.9

Zirconia–yttria phase diagram (with kind permission from Springer Science + Business Media).[5]

metastable tetragonal phase is present, then an applied stress, magnified by the stress concentration at a crack tip, can cause the tetragonal phase to convert to monoclinic, with the associated volume expansion. This phase transformation can then put the crack into compression, retarding its growth, and enhancing the fracture toughness. This mechanism is known as transformation toughening, and significantly extends the reliability and lifetime of products made with stabilized ZrO_2. All PSZ degrade rapidly when exposed to water vapor at elevated temperatures, typically in the range of 200 °C to 300 °C. Additionally, all toughened ZrO_2 ceramics show a degradation of properties with increasing temperature and are thus limited to applications at temperatures below 800 °C.

With reference to the ZrO_2–Y_2O_3 system in Figure 7.9, PSZ can be obtained depending on the concentration of Y_2O_3. For example, the addition of Y_2O_3 to ZrO_2 replaces some of the Zr^{4+} ions in the ZrO_2 lattice with Y^{3+} ions. This replacement of the Zr^{4+} ions produces oxygen vacancies, as three O^{2-} ions replace four O^{2-} ions. It also permits Y_2O_3-stabilized ZrO_2 to conduct O^{2-} ions, provided there is sufficient vacancy site mobility, a property that increases with temperature. A 3%–8% mol Y_2O_3 doping results in a partially Y_2O_3-stabilized ZrO_2 ceramic consisting of a cubic matrix with tetragonal inclusions. However, in this PSZ system, it is also possible to obtain ceramics formed at room temperature with a metastable tetragonal phase only, called tetragonal ZrO_2 polycrystal (TZP), using smaller concentrations of stabilizer. TZP materials, containing approximately 2%–3% mol Y_2O_3, are completely made up of tetragonal grains with sizes of the order of hundreds of nanometers. The fraction of tetragonal phase retained at room temperature is dependent on the size of grains, Y_2O_3 content, and degree of constraint exerted on them by the matrix. However, at 8% mol Y_2O_3, a fully Y_2O_3-stabilized ZrO_2 ceramic is obtained consisting of only a cubic crystal structure. During thermal cycling, the stabilized cubic phase is not as stable as the monoclinic phase and reverting back to the monoclinic phase results in cracking of the ZrO_2 ceramics.

Proposed in the mid 1980s as an alternative ceramic material to Al_2O_3 for femoral ball heads, PSZ ball heads possess higher wear resistance, higher bending strength and fracture toughness when compared to Al_2O_3 ball heads. The PSZ ball heads are also more superior than the metallic ball heads since they possess properties such as higher wear resistance and higher corrosion resistance. In the early 1990s, PSZ was also used as a dental implant. Today, the PSZ ceramics are commonly used in the fabrication of subframes for the construction of dental restorations and ball heads in artificial hip implants. Clinical reports on the implantation of more than 300 000 TZP ball heads indicate benefits such as the lower incidence of revision surgery.[6]

7.3.4 Carbon

Carbon is a common element which exists in many allotropic forms such as crystalline diamond, graphite, non-crystalline carbon, or quasicrystalline pyrolitic carbon. It can be fabricated as powders, fibers, sheets, blocks, and thin films. Properties that make carbon desirable for a number of applications include:

- excellent electrical and thermal conductivity,
- low density,
- sufficient corrosion resistance,
- low elasticity, and
- low thermal expansion.

Carbon was one of the first materials used for filaments in incandescent lamps, whereby cotton and bamboo fibers were carbonized to form carbon fibers. Carbon fibers can also be produced by carbonization of polyacrylonitrile, resulting in very high tensile strength and elastic modulus. In the field of biomaterials, pyrolitic carbon is widely used for implant fabrication and surface coatings. It is often used as an artificial heart valve material due to its excellent strength, wear resistance and durability, and thromboresistance. Pyrolytic carbon is also used as a biomaterial for the fabrication of small joint implants such as fingers and spinal inserts.

The pyrolytic carbon used in implants exists in two forms, namely low temperature isotropic (LTI) carbon and the ultralow temperature isotropic (ULTI) carbon. LTI carbon exhibits good biocompatibility with blood and soft tissues and is highly thromboresistant. Popular in the mid 1960s, LTI carbon also possesses the following properties:

- excellent durability,
- high strength, and
- excellent wear and fatigue resistance.

In addition to being used as a bulk biomaterial for medical devices, pyrolytic carbon is also used as a coating. Owing to the relatively low density and brittleness of carbon, LTI and ULTI carbon coatings have been deposited on heart valve prostheses. Typically produced using a chemical vapor deposition process (see Chapter 9 for description of the deposition process), the coatings are also doped with up to 20 wt% silicon to improve stiffness, hardness, and wear resistance.

7.3.5 Calcium phosphates (CaP)

Calcium phosphates (CaP) ceramics are of special interest to the biomaterials community because of their occurrence in normal (bone) and pathological calcification (arteries) in the body and their association with the formation, progression and arrest of enamel and dentin caries. Several different forms of CaP ceramics can be found in the human body, and these include:

- amorphous calcium phosphate (ACP),
- brushite or dicalcium phosphate dehydrate (DCPD) – $CaHPO_4 \cdot 2H_2O$,
- monetite or dicalcium phosphate anhydrous (DCP) – $CaHPO_4$,
- octacalcium phosphate (OCP) – $Ca_8H_2(PO_4)_6 \cdot 5H_2O$,
- whitlockite or tricalcium phosphate (β-TCP) – $Ca_3(PO_4)_2$, and
- apatite, calcium-OH-apatite or commonly known as hydroxyapatite (HA) – $Ca_{10}(PO_4)_6(OH)_2$.

In addition to the naturally occurring forms of CaP ceramics, these ceramics can be synthetically produced in the laboratory. The synthetically produced CaP ceramics are similar in composition, biodegradation, bioactivity, and osteoconductivity to the biological apatites. The most commonly CaP phases used in biomedical applications are HA, TCP, biphasic CaP (combination of HA and TCP), and OCP (precursor to HA).

In general, the degradation rate of CaP ceramics is dependent on their crystallite size, degree of crystallinity, and chemical composition. Fine crystals or smaller crystallite-sized CaP ceramics tend to degrade more rapidly when compared to larger crystals. It is also known that the amorphous phase dissolves more rapidly when compared to more crystalline CaP ceramics. The degradation rate of CaP ceramics is also affected by substitutions in the apatite. Substitution of CO_3, Sr^{2+}, or Mg causes greater solubility, while substitution of fluoride causes lower solubility of synthetic and biological apatites. The incorporation of fluoride increases the growth rate of bone mineral apatite crystal. The stability of fluorine containing HA is due to the effect of fluorine substitution for hydroxyl (OH) molecules in the apatite which results in a reduced reaction surface.

In addition to the chemical and physical properties of CaP ceramics, the degradation of each CaP phase is unique and is dependent on the phase and the biological environmental conditions. The availability of phase diagrams is very useful in predicting which CaP phases will be stable under the specific solution conditions dictated by:

- pH, and
- degree of solution saturation with respect to calcium and phosphate ($H_2PO_4^-$, HPO_4^{2-}, PO_4^{3-}) ions.

A 2-dimension solubility isotherm of the different CaP ceramics, as shown in Figure 7.10a, can be used to predict net deposition or dissolution of a CaP phase. Solution conditions above and to the right of the curve of a specific CaP phase will allow a net deposition of ions, whereas conditions below and to the left of the curve will allow the CaP ceramics to experience a net dissolution. For example, it is predicted that, at 25 °C, a pH of 6, and calcium ion (Ca^{2+}) concentration of 10^{-2} M, DCPD or brushite ($CaHPO_4 \cdot 2H_2O$) would be the most stable CaP phase. Similar to the 2-dimension solubility isotherm, a 3-dimension solubility isotherm can also be produced indicating the stability of CaP phases under different conditions of pH, calcium concentrations, and inorganic phosphate concentrations (Figure 7.10b). Solubility isotherms are usually accurate in predicting experimental outcomes and have been used to determine the conditions under which a specific CaP phase will form.

The solubility isotherm is also useful for clearly showing solution conditions at which the ion products are equal to the solubility products for a given phase. In general, a salt dissolves into a solution until saturation or equilibrium is reached, resulting in an equilibrium constant. This reaction of an AB salt with water can be written as in Eq. (7.3):

$$A_a B_b \rightarrow aA^+ + bB^-. \tag{7.3}$$

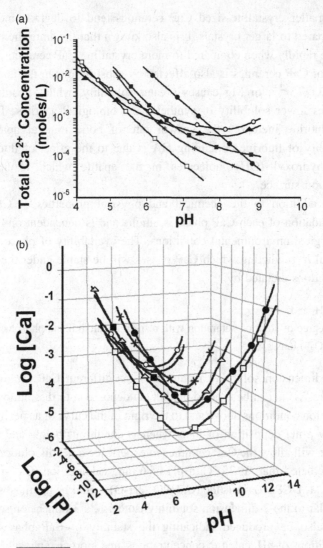

Figure 7.10

(a) The 2-dimension solubility isotherms[7] and (b) 3-dimension solubility isotherms[8] of different calcium phosphate ceramics at 25 °C. Figure (a) is reprinted with permission from SAGE Publications. Figure (b) is reprinted with permission from the Japanese Society for Dental Materials and Devices. Shading; □ = HA; ▲ = DCPA; o =DCPD; ■ = OCP; △ = β-TCP; × = α-TCP; ● = TTCP.

The general equilibrium constant (K) for the above reaction can be written as in Eq. (7.4):

$$K = [A^+]^a[B^-]^b, \tag{7.4}$$

where K refers to the product of the concentration of the ions that are present in a saturated solution of an inorganic compound. At equilibrium, K is also known as

Table 7.1 Solubility product constant, K_{sp} of different calcium phosphates at 25 °C

Calcium phosphates	K_{sp} (mol/l)
Dicalcium phosphate dehydrate, brushite (DCPD)	$10^{-6.59}$
Octacalcium phosphate (OCP)	$10^{-96.6}$
Whitlockite, β-tricalcium phosphate (β-TCP)	$10^{-28.9}$
Hydroxyapatite (HA)	$10^{-116.8}$

the *solubility product constant* (K_{sp}). The less soluble a substance, the lower will be its K_{sp}. In reference to Table 7.1, the K_{sp} of HA ($10^{-116.8}$ mol/l) is lower than the K_{sp} of DCPD ($10^{-6.59}$ mol/l). As a result, HA is more stable or less soluble when compared to DCPD.

When the ionic product that is present in a solution exceeds the K_{sp}, there will be a net deposition on the crystal surface. When this ionic product is less than the K_{sp}, there will be a net loss of ions from the crystal surface. Note that K_{sp} is pH dependent and a decrease in pH results in a higher K_{sp}. This means that a higher concentration of component ions is required to maintain the dissolution–deposition equilibrium. The solubility isotherms in Figure 7.10 show the concept that a rise in K_{sp} is indicated as pH falls, meaning that higher concentrations of ions or ionic products in a solution are required to prevent dissolution of CaP ceramics.

In addition to the pH and degree of solution saturation by calcium and phosphate ions, the stability of CaP ceramics is also influenced by their protonation state of phosphate ions when immersed in solution. The bonding of calcium ions to monovalent phosphate ions ($H_2PO_4^-$) and trivalent phosphate ions (PO_4^{3-}) is stronger compared to the bonding of calcium ions to divalent phosphate ions (HPO_4^{2-}). In general, the weaker the bonding of calcium to the phosphate ions, the lower the stability of CaP ceramics.

7.3.6 Hydroxyapatite (HA)

The most commonly known crystalline CaP biomaterial is hydroxyapatite (HA). Also known as calcium hydroxide phosphate, pure HA has the chemical formula $Ca_{10}(PO_4)_6(OH)_2$. Biological apatites are found naturally in bone and teeth minerals. Unlike pure HA, biological apatites contain chemically substituted ions found in the body. Ions such as HPO_4^{2-}, CO_3^{2-} and F^- can partially substitute for PO_4^{3-} and OH^-, whereas Mg^{2+} and Sr^{2+} substitute for Ca^{2+}. Thus biological apatites are not pure HA.

Synthetic HA can be produced using a range of techniques including the following:

- hydrothermal conversion of natural coral,
- sol–gel synthesis,
- co-precipitation,
- solid-state reactions,
- microemulsion synthesis, and
- mechanochemical synthesis.

Box 7.3

- Bone is made of 33 wt% organic matrix and 67 wt% minerals.
- The mineral portion of bone is made of poorly crystallized carbonate-containing apatite phase and an amorphous calcium phosphate phase.
- Owing to the presence of other constituents, the Ca/P ratio in bone and teeth can be different from synthetic HA.

Stoichiometric HA is known to have a calcium to phosphorus (Ca/P) mole ratio of 1.67. Structurally, pure HA has a hexagonal close-packed arrangement with cell dimensions of 9.42 Å for the a- and b-lattice spacings, and a cell dimension of 6.88 Å for the c-lattice spacing (Figure 7.11). Fluorine, chlorine and hydroxyl groups can be mutually substituted. Thus, HA, fluorapatite, and chlorapatite can conform to the empirical formula $Ca_5(PO_4)_3(OH, F, Cl)$. Carbon dioxide (CO_2) is sometimes found in small amounts as an essential constituent, giving rise to carbonation varieties. The carbonation involves the substitution of carbonate (CO_3) groups for the phosphate (PO_4) groups within the apatite structure. It is common for carbon to substitute for phosphorus or

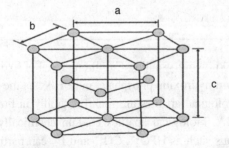

Figure 7.11

The hexagonal close-packed arrangement of pure HA, showing the a-, b-, and c-lattice spacings.

calcium in apatites at high temperatures. In many instances, chemical substitutions are often accompanied by a change in the cell dimensions of the crystallographic HA structure.

Box 7.4

- The high affinity of proteins for HA surfaces allows it to be used as adsorbent in chromatography to separate proteins.
- HA is an osteoconductive material.
- As a biocompatible material, HA is used to enhance bone healing and to establish a high interfacial bone–implant strength.
- Owing to its brittle nature, HA is used as coatings on metallic implant surfaces rather than as solid ceramic implants for dental and orthopaedic applications.

Like other ceramics, HA is known to be very brittle but is strong in compression. Its strength is dependent on grain size, with finer grain sizes associated with higher strength. The compressive strength of HA is also dependent on its density and porosity, which in turn are dependent on the sintering temperature and time.

Hydroxyapatite (HA) is one of the many crystalline forms of CaP ceramics and is classified as a resorbable bioceramic. The solubility behavior of HA is highly dependent on the various factors discussed earlier, such as pH and the concentration of the ions that are present in a saturated solution in which HA is immersed. Physical properties of HA affecting solubility behavior include:

- porosity,
- grain or crystal size,
- crystal perfection, and
- inclusion of chemical impurities.

The chemical reactivity of HA varies inversely to its crystalline perfection, crystal size, and crystal dimension. Larger HA crystals have a lower solubility rate when compared to finer or smaller HA crystals. Additionally, inclusions of chemical impurities in HA result in changes in the solubility behavior of the ceramic. Incorporation of fluorine for hydroxyl ions lowers the solubility rate of HA, whereas the incorporation of sodium and carbonate ions enhances its solubility.

Figure 7.12

Different forms of HA synthesized in the laboratory. (a) HA pellets of different sizes. (b) Block porous HA scaffolds of different shapes and dimensions.

Aside from the physical properties of HA, its solubility is also dependent on the biological environment, such as the pH and the ionic concentration of the solutions, as well as the protonation state of the phosphate ions when immersed in a solution. Calcium ions in the HA lattice are held in place by trivalent phosphate ions (PO_4^{3-}). A fall in the solution pH can cause a change in the protonation state of some trivalent phosphate ions (PO_4^{3-}), converting them to divalent phosphate ions (HPO_4^{2-}). This conversion weakens the calcium–phosphate bonding and thus allows some calcium to be released into the solution. Divalent phosphate (HPO_4^{2-}) released into the solution binds with hydrogen ions to form monovalent phosphate ions ($H_2PO_4^-$). This binding of HPO_4^{2-} with hydrogen ions in solution acts as a buffer, altering the pH of the solution and slowing down the HA degradation.

Although HA is commonly produced as a powder, it can also be fabricated into many shapes or forms. Shown in Figure 7.12 are examples of granular HA beads of different sizes as well as HA fabricated into scaffolds of different shapes and sizes.

7.3.7 Tricalcium phosphate (TCP)

Another popular crystalline CaP ceramic used in the fabrication of medical devices is tricalcium phosphate (TCP). Having the chemical formula, $Ca_3(PO_4)_2$, TCP can exists in two forms, namely α-TCP and β-TCP. The Ca/P molar ratio in both forms is 1.5. However, with the K_{sp} of α-TCP ($10^{-25.5}$ mol/l) being higher than the K_{sp} of β-TCP ($10^{-28.9}$ mol/l), the rate of solubility for α-TCP is more rapid compared to β-TCP. Therefore α-TCP degrades and resorbs more quickly in the body. Knowing the rate of degradation for each form of TCP helps allow the biomaterial to be tailored for various purposes, depending on the desired outcome. Additionally, TCP has a higher solubility rate when compared to HA since the K_{sp} for TCP is higher than the K_{sp} for HA ($10^{-116.8}$ mol/l).

To take advantage of the benefits of both HA and TCP, biphasic CaP ceramics have also been introduced as biomaterials. These composites benefit from the osteoconductivity of HA and the absorbability of the TCP. The TCP, as it dissolves, supplies the localized environment with a concentrated source of calcium and phosphorus. The HA slows the degradation of the structure, giving more support to the healing tissue and acts as a template for the cells to use and grow. The resorbability of biphasic calcium phosphates increases as the TCP to HA ratio increases.

7.3.8   Calcium sulfate (CaSO$_4$·H$_2$O)

Calcium sulfate (CaSO$_4$·H$_2$O), a mineral that is obtained through various mining techniques, must be processed prior to any medical use. Calcination treatment converts calcium sulfate into calcium sulfate hemihydrates, better known as Plaster of Paris. There are two forms of hemihydrates, namely α-hemihydrate and β-hemihydrate. These forms differ in crystal size, surface area, and lattice imperfections. Although both forms have similar chemical structures, the α-hemihydrate is much harder and more insoluble than the β-hemihydrate. Hydration of the hemihydrates results in an exothermic reaction, resulting in the dissolution of the hemihydrates and precipitation of a hydrated calcium sulfate. Impurities in the precipitation process can alter the rate of precipitation and the final hydrated calcium sulfate. The presence of proteins and inorganic salts found in a wound site are just a few components that can alter the setting of the calcium sulfate.

Calcium sulfate possesses many attributes that make it a potential biomaterial for bone regeneration. These attributes include:

- complete and rapid resorption, and
- biocompatibility.

Despite these attributes, there are disadvantages to the use of calcium sulfate, including its rapid rate of resorption which does not provide sufficient time for the formation of new tissue. Additionally, like any other ceramic, calcium sulfate cannot be used in weight bearing applications without additional load-bearing fixation. Improvements in calcium sulfates have included the addition of other ceramic materials or polymers to slow down the degradation rate and to increase its mechanical properties.

7.3.9 Bioactive glass

Glasses that are designed to induce specific biological activity are known as bioactive glasses. Some of these biological activities result in the bonding of the bioactive glass to bone. Originally suggested by Larry Hench in the early 1970s, bioactive glass involves a silicate glass-based system. Since then, several formulations of bioactive glass have been developed. For example, the 45S5 bioactive glass is known to contain 45% SiO_2, 6% P_2O_5, 24.5% CaO, and 24.5% Na_2O, whereas the 52S4.6 bioactive glass is known to contain 52% SiO_2, 6% P_2O_5, 21% CaO, and 21% Na_2O.

There are generally two methods for synthesizing bioactive glass. The first synthesis method involves cooling a mixture of raw materials from the liquid state and is very much dependent on the cooling rate. Typical raw materials used for the synthesis include high-purity silica, phosphorus oxide, and calcium and sodium carbonate. Other salts and oxides have also been added to introduce other elements to the glass composition. While slow cooling rates allow crystallization of the material to occur as a result of atomic arrangement, a vitreous phase is formed when silicate-based melted mixtures are cooled at normal cooling rates. Depending on the composition, melting the silicate glasses is usually done in the range of 1300 °C to 1450 °C. Bioactive glass powders or granules can then be produced by quenching the melt in acetone. Specific shapes and forms can also be produced by casting the melt in appropriate molds. Weighing, mixing, melting, homogenizing, and forming of the bioactive glass must be done without introducing impurities or losing volatile constituents such as Na_2O or P_2O_5. The second synthesis method involves a sol–gel process, whereby controlled hydrolysis and condensation of metal alkoxides are performed to form a suspension of colloidal particles. Polycondensation of the colloidal suspension results in the formation of an interconnected gel-like network structure. Heat treatment of the gel-like network structure is required to form the bioactive glass.

In the absence of contact with a solution, bioactive glasses are inert. In the presence of a physiological solution, rapid surface reaction of the bioactive glass with the solution results in the formation of a silica-rich gel layer within an hour. There is a rapid ionic exchange of Na^+ with the H^+ or H_3O^+ from the solution. Loss of soluble silica in the form of $Si(OH)_4$ to the solution occurs, resulting from

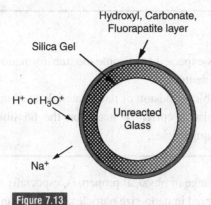

Hydroxyl, Carbonate,
Fluorapatite layer

Silica Gel

H^+ or H_3O^+

Unreacted
Glass

Na^+

Figure 7.13

Reactions of bioglass in physiologic solution.

breaking of Si–O–Si bonds and the formation of Si–OH (silanols) at the bioactive glass–solution interface. Condensation and re-polymerization of a silica-rich layer occurs on the bioactive glass surface. This silica-rich gel layer is favorable for the rapid nucleaction, formation, and growth of the apatite layer. Calcium ions and phosphate groups migrate to the bioactive glass surface through the silica-rich layer, resulting in the formation of an amorphous $CaO–P_2O_5$-rich film. Crystallization of the $CaO–P_2O_5$-rich film occurs by incorporation of OH^-, CO_3^{2-}, or F^- anions from the solution to form a mixed hydroxyl, carbonate, and fluorapatite layer. Figure 7.13 shows that final by-product of bioactive glass as a result of its presence in solution.

7.4 Nanoceramics

Rapid advances in materials science and engineering have resulted in the development of nano-size ceramics for applications in imaging and diagnoses, implantology, anti-cancer therapy, drug delivery, and gene therapy. In general, the transition from micron-size particles to nano-size particles typically results in alteration of physical properties including:

- greater surface area,
- greater surface area to volume ratio,
- increase in grain boundaries,
- greater porosity,
- increased surface roughness,
- increased hydrophilicity, and
- increased surface reactivity.

Box 7.5

- Nanotechnology is expected to become the transformational technology of the twenty-first century.
- Despite the favorable explosion of nanomaterials used in today's applications, there is also a current concern on the possible toxic health effects of these nanoparticles.

In many instances, the change in physical properties, especially the greater surface area to volume ratio observed in nano-size particles, results in an increasing role of atoms on the surface of the particle relative to that of those in the interior in determining the behavior of the material. As a result of this increased dominance, the interaction of the particles with other materials is affected. For example, in nanocomposites, the increased surface area of the nanoparticles results in strong interactions between the constituent materials thereby altering properties such as increasing the strength and/or increasing chemical or heat resistance. Additionally, due to the increase in total surface area, which permits faster dissolution of the nanoparticles in blood and accelerated absorption in the human body, nanoparticles may also be used to improve the delivery of poorly water-soluble drugs.

Using a variety of processes, nano-size particles of 100 nm diameter or less can be produced either using a "top-down" process or a "bottom-up" process. Also known as a solid-state process, Figure 7.14 shows that the "top-down" process uses bulk materials as starting materials. Nanomaterials are fabricated utilizing the method of attrition or mechanical synthesis, that is, the grinding of macro- or micron-scale particles using a ball mill, a planetary ball mill, or other mechanical size reducing mechanism. The properties of nanoparticles formed from grinding or milling are affected by the milling material, milling time, and atmospheric medium. Additionally, contaminations can also be introduced depending on the milling material.

As shown in Figure 7.14, the "bottom-up" process involves the building or assembly of nanomaterials from an atomic scale. Such a process utilizes the method of vapor condensation or chemical synthesis to produce nanophase ceramics. The condensation approach requires pyrolysis to evaporate the ceramics. The vapors formed are then air-cooled by rapid condensation to form the nano-size particles. Particles formed are often in aggregates and agglomerates. An example of the condensation approach is the physical vapor synthesis process. As shown in Figure 7.15, the physical vapor synthesis process involves the feeding of a solid ceramic into the reactor where plasma energy is then used to generate a vapor at high temperature of the order of 10 000 K. Plasma energy is achieved with thermal

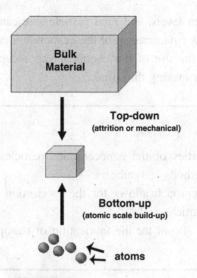

Figure 7.14

The "top-down" and "bottom-up" approaches to fabricating nanoparticles.

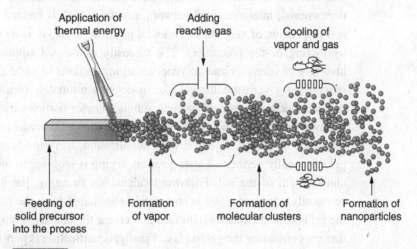

Figure 7.15

Vapor phase plasma synthesis of nanoparticles.

plasma torches such as DC plasma jet, DC arc plasma, and radiofrequency induction plasmas. Ceramic nanoparticles are then formed by cooling the vapors in air at a controlled rate. The vapor can also be rapidly cooled by quenching in the presence of a reactant gas such as oxygen to form nano-size ceramic particles. The nano-size particles synthesized using the condensation approach are easy to disperse in liquid media, and the resulting dispersions are highly stable, low viscosity, and easy to handle. Additionally, the use of the condensation approach results in low contamination levels, and final particle size can be controlled by varying the temperature, gas environment, or the evaporation rate. Other condensation approaches include the use of the chemical vapor deposition technique (discussed in Chapter 9) for making thin films.

Box 7.6

- Structure and properties of the nanoceramics particles produced are dependent on the methods of synthesis.
- The condensation approach allows for the fabrication of non-porous nano-size ceramic particles.
- The sol–gel process allows for the fabrication of nanoporous ceramic particles

---

Other "bottom-up" processes include the use of the sol–gel approach which is a wet-chemical synthesis that comprises a sol solution. In general, a three-dimensional, interconnected network, known as gel, is created from a sol which is a suspension of very small, colloidal particles. Low viscosity sol can be formed by mixing of the precursors. The viscosity of the sol solution increases as it undergoes hydrolysis and polycondensation reactions to yield a gel-like colloidal suspension. Over time, the gel-like suspension ultimately forms an interconnecting network composed of discrete submicrometer particles dispersed to various degrees in a host fluid. Gelation occurs with the completion of the network formation of the sol. Aging, drying, stabilization, and densification of the formed gel are usually performed after gelation. Aging is required to increase the density and strength of the gel. Following calcination or aging, the gel is dried as the physically adsorbed water is completely eliminated from the pores. After drying, the gel is then chemically stabilized to increase the density, strength, and hardness, thereby converting the gel to glass. Finally, densification is performed to eliminate pores within the formed sol–glass or ceramics.

The ultimate size of the nanoparticles synthesized using the sol–gel approach is controlled by stopping the reaction or by choosing chemicals to form stable particles that stop growing at a certain size. Additionally, the rate at which the liquid phase is removed determines the porosity of the nanoparticles formed. Although the sol–gel approach is generally favorable due to low cost and high yield compared to the other approaches, purity of the nanoparticles may be an issue as a result of contamination from the precursor chemicals. The sol–gel approach has also been used in combination with other processes to form nano-ceramic powders. For example, nanophase alumina powders have been synthesized by employing the Bayer process followed by sol–gel and sintering. The size of the particle and its morphology is dependent on the calcination or aging temperature and processing time.

Uses of nanoceramics include structural applications (nozzles and filters, and coatings), electronic and optical applications (nanotube arrays and optical filters), energy storage (fuel and solar cells), and biomedical applications (drug delivery). As a result of their high surface area, nanophase alumina (Al_2O_3) particles have been used in applications such as nanoprobes. The quality and reproducibility of alumina fine coatings can be improved using nanophase alumina, which also improves the scratch resistance of the coating. As stated previously, the strength and toughness of Al_2O_3 are highly dependent on the grain size. As such, the nano-size Al_2O_3 generally exhibits higher strength and toughness compared to micron-size Al_2O_3. Similarly, advances in technology have also allowed the fabrication of nano-size stabilized zirconia, exhibiting several favorable properties including:

- significant reduction in sintering temperature,
- ability to deform superplastically under applied stress,
- higher diffusivity, and
- higher ionic conductivity.

The above properties contribute to the production of sprayed coatings with superior mechanical attributes.

Al_2O_3–zirconia composite materials fabricated by using powder mixture processing have been investiagted as an alternative to monolithic materials. Micro-nanocomposites can be obtained with the matrix containing crystals in the micron-size range and the second phase consisting of nanoparticles located in intragranular and intergranular positions. The large surface area of the nano-particles in the second phase allows for the increased interaction between the two phases, thereby leading to the following properties:

- increased strength,
- increased chemical resistance, and/or
- increased heat resistance.

In comparison to the monolithic Al_2O_3, mechanical properties of the zirconia–Al_2O_3 nanocomposites are also improved as a result of the presence of dislocations around the particles which allow the release of residual stress in the matrix. Similarly, a new class of dense composites with a microstructured Al_2O_3 matrix and nanozirconia particles located at the grain boundaries and intergranular positions present unique properties. The presence of a compressive residual stress around the zirconia nanoparticles is capable of retarding the growth of cracks when an external stress is applied. In addition, the high surface area, porosity, and reactivity of nano-size Al_2O_3 allow for it to be used in conjunction with a polymer matrix for the delivery of poorly water-soluble drugs.

Advances in technology have also allowed the fabrication of nanostructured carbon such as carbon nanotubes. Currently produced by using methods such as arc discharge, laser ablation, or chemical vapor deposition, carbon nanotubes were first reported in the mid 1970s with increasing attention to their development in the early 1990s. Defect-free carbon nanotubes are classified as either single-walled nanotubes (SWNTs) or as multi-walled nanotubes (MWNTs). One difference between SWNTs and MWNTs is that the SWNT is a hollow cylinder of a graphite sheet with hexagonal structural features, whereas the MWNT is a group of coaxial SWNTs.

Box 7.7

- Carbon nanotubes possess extraordinary strength, have unique electrical properties, and are efficient conductors of heat.
- Chemical bonding of nanotubes comprises entirely of sp^2 bonds and is similar to graphite. This bonding structure is stronger than the sp^3 bonds found in diamond and alkanes.

Applications for carbon nanotubes include a wide range of industries such as biomedical, electronics, optics, and nanotechnology. In many instances, carbon nanotubes are used as reinforcements or as additives to improve a material's properties. Their properties can be further improved by covalently or non-covalently adding functional groups to the side walls and tips of the carbon nanotubes. Current investigations of carbon nanotubes for medical applications include its use as a scaffold material for tissue engineering. Used as fillers in collagen matrices, collagen–carbon nanotube composite matrices have been found to be useful for tissue regeneration.[9] The ability to integrate carbon nanotubes with biomolecules such as drugs and proteins has also allowed for the generation of complex nanostructures with controlled functions and properties. For example, functionalized carbon nanotubes have been shown to penetrate target cells and facilitate the delivery of plasmid DNA.[10]

Similar to nanophase Al_2O_3 and zirconia, nanophase calcium phosphates possess properties such as reduced grain size, reduced pore size, and improved wettability. When CaP is used to assist in bone regeneration, these changes in properties can be modulated to control protein interactions and subsequently enhance osteoblast adhesion and long-term functionality. For example, NanOss® bone void filler (Angstrom Medica) approved by the FDA in 2005, is an engineered synthetic bone developed from nanocrystalline calcium phosphate. It is prepared by compressing a nanopowder and heating it to form a dense, transparent, and nanocrystalline material. Being highly osteoconductive and because it remodels over time into human bone, the NanOss® bone void filler is used in various sports medicine, trauma, spine, and general orthopedics applications. Other nano-sized HA products include Ostim®, an injectable bone matrix in the form of a paste. Sold within the European Union after receiving its CE marking in 2002, Ostim® is indicated for metaphyseal fractures and cysts, acetabulum reconstruction, and periprosthetic fractures during hip prosthesis exchange operations, osteotomies, and filling cages in spinal column surgery.

Box 7.8

- Biological apatite found in the body consists of poorly crystalline non-stoichiometric nano-sized, needle-like crystals.
- In a colloidal form, nano-sized calcium phosphate in milk is involved in stabilizing casein micelles.
- Used in the food industry, nano-sized calcium phosphates are used as a nutritional supplement and a raising agent.

Other uses of nano-sized HA particles include their use for vaccine adjuvant and for protein and drug delivery. Since the transport of nanoparticles across the cell membrane is largely dependent on the particle size, a 20 nm particle or smaller can easily pass through the cell membrane or the membrane channels. However, the speed at which nanoparticles are transported across the cell membrane is not dependent on the particle size. In general, optimal delivery of drugs, protein, or gene requires a different optimal particle size, depending on the application.

7.5 Summary

This chapter provides the definition of a general ceramic as well as information on some of the ways to classify ceramics. Some of the common bioceramics, including nanoceramics that are used today in the fabrication of medical devices and their properties, are also discussed. In general, ceramics are either bioinert, biodegradable, or surface reactive. Additionally, in many instances, ceramics are often used in combination with other biomaterials in the medical industry in order to optimize the overall implant properties. An example of such a combination includes the coating of ceramic nanoparticles with polymers to prevent agglomeration. Overall, the parameters to be considered prior to the use of ceramics in biomedical applications should include their physical and chemical behavior, and their biological response *in vivo*. Additional factors that need to be considered should include the fabrication ease, cost, and the site of application.

References

1. Anjard, R. (1981). Mayan dental wonders. *J. Oral Implant.*, **9**, 423–426.
2. Wenk, H.-R. and Bulakh, A. (2003). *Minerals – Their Constitution and Origin*. Cambridge University Press.
3. Hench, L. L. (1996). Ceramics, glasses, and glass-ceramics, in *Biomaterials Science, An Introduction to Materials in Medicine*, Ratner, B. D., Hoffman, A. S., Schoen, F. J. and Lemons, J. E., editors. San Diego, Academic Press, pp. 73–84.
4. McHale, J. M., Auroux, A., Perrotta, A. J. and Navrotsky, A. (1997). Surface energies and thermodynamic phase stability in nanocrystalline aluminas. *Science*, **277**, 788–791.
5. Scott, H. G. (1975). Phase relations in the zirconia–yttria system. *J. Mater. Sci.*, **10**, 1527–1535.
6. Clarke, I. C. (1992). Role of ceramic implants. *Clin. Orthop.*, **28**, 19–30.
7. Chow, L. C. (1988). Calcium phosphate materials: reactor response. *Adv. Dent. Res.*, **2**, 181–184.
8. Chow, L. C. (2009). Next generation calcium phosphate-based biomaterials. *Dent. Mater. J.*, **28**, 1–10.
9. MacDonald, R. A., Laurenzi, B. F., Viswanathan, G., Ajayan, P. M. and Stegemann, J. P. (2005). Carbon–carbon nanotube composite materials as scaffolds in tissue engineering. *J. Biomed. Mater. Res.,* Part A, **74A**, 489–496.
10. Cai, D., Mataraza, J. M., Qin, Z. H. *et al.* (2005). High efficient molecular delivery into mammalian cells using carbon nanotube spearing. *Nat. Methods*, **2**, 449–454.

Suggested reading

• Park, J. B. (2008). *Bioceramics: Properties, Characterizations, and Applications*. Springer, ISBN 0387095446.
• Kokubo, T. (2008). *Bioceramics and Their Applications*. CRC Press, ISBN 1420072072.

Problems

1. In general, what are ceramics composed of?
2. How can ceramics be classified?

3. What are the general properties of ceramics?

4. What are the differences between bioinert, biodegradable, and bioactive ceramics?

5. Name the two types of phase transformations that can occur in silicate glass. What are the differences between these two different phase transformations?

6. What are the properties of alumina?

7. What is the advantage of using alumina as a biomaterial?

8. What are the problems associated with the use of zirconia ceramics as implants, and how are these problems resolved?

9. Which allotropic form of carbon is commonly used as a biomaterial?

10. Are biological apatites similar to pure hydroxyapatite in terms of composition and structure?

11. How many phases have been reported for the mineralized portion of bone?

12. What factors affect the solubility rate of calcium phosphate ceramics?

13. Describe the reactions leading to the formation of an apatite layer on bioglass ceramics.

14. What are the differences between micro-sized ceramics and nano-sized ceramics, and how are these nano-sized ceramics fabricated?

15. How does the increased surface area of nanoparticles affect the overall properties of these materials? What properties can be affected?

8 Natural biomaterials

> **Goals**
>
> After reading this chapter, students will understand the following.
> - Properties that qualify natural biomaterials for biomedical applications.
> - Different major classifications of natural biomaterials.
> - Properties of various natural biomaterials.

What makes natural materials unique and piques the interest of biomaterials scientists, engineers and clinicians? There is a belief that all aspects of materials created naturally have a useful purpose or function. Utilization of such materials thus allows these materials to perform a combination of diverse functions such as intracellular communications and storage. In general, the properties of natural materials are dependent on their composition. For example, the physical–chemical properties of monomers and their sequences determine the properties of polymeric natural biomaterials. Like synthetic materials used for biomedical applications, it is expected that the natural biomaterials should satisfy requirements such as

- being non-toxic,
- being non-inflammatory,
- being non-allergenic,
- having satisfactory mechanical properties,
- being capable of inducing cell attachment and differentiation if needed, and
- having low cost.

Natural biomaterials possess most of the above properties because they are found in biological systems and work well within their respective environments. Favorable characteristics of natural materials include facilitating cell attachment, enhancing the mechanical properties of synthetic biomaterials, and their ability to bind and deliver macromolecules. These desirable characteristics allow natural materials to be used in various biomedical applications including tissue engineering and regenerative medicine. With the exception of corals, which are deposits of calcium carbonates, most of the common natural materials are polymeric in nature and are either protein-based or polysaccharide-based materials. Examples of protein-based natural polymers include collagen, gelatin, silk fibroin, fibrin, and elastin, whereas examples of polysaccharide-based natural polymers include chitosan, starch, alginate, hyaluronan, chondroitin sulfate, and dextran. In this chapter, we will discuss some of the natural biomaterials that are commonly used today in the fabrication of medical devices.

8.1   Collagen

Collagen belongs to a naturally occurring family of proteins that is found exclusively in animals. Derived from the Greek word for glue, the word collagen was originally used to describe the components of connective tissues that yield gelatin on boiling. Being a protein-based polymeric material, collagen offers unique structural properties and has the advantage of mimicking many features of extracellular matrix. It can potentially direct the migration, growth, and organization of cells during tissue regeneration and wound healing. Additionally, collagen also offers the potential for stabilizing encapsulated and transplanted cells.

Box 8.1

- The most abundant protein in the human body, collagen is part of the extracellular matrix that provides the foundation for all connective tissues.
- The tensile strength and structural integrity of the different tissues are attributed to the presence of collagen.

The collagen used for medical applications is often derived from bovine, equine or porcine sources. The bovines used for deriving medical collagen are closely monitored and certified free from bovine spongiform encephalopathy. It is not uncommon to find collagen being used in combination with silicones, glycosaminoglycans, fibroblasts, and growth factors. Collagen is also available commercially as non-prescription amino acid supplements to enhance joint mobility. In order to minimize the possibility of immune reactions, collagen from human sources is also available. The human-derived collagen is usually from donor cadavers and placentas. Biomedical applications utilizing collagen include bioprosthetic implants and tissue engineering of a variety of organs. Additionally, it is also sold as gels or mats to the medical community for wound dressing applications. One of its many applications in surgery includes being used in the construction of artificial skin substitutes for burn patients. An example of such collagen products include Apligraff®, a dermal matrix for artificial skin sold by Organogenesis Inc. (Massachusetts, USA). Other applications include the use of collagen as a drug delivery platform and tissue regeneration. Medtronic Sofamor Danek (Minnesota, USA) uses collagen in their InFUSE® Bone Graft product to deliver bone morphogenetic protein-2 for spinal fusion, trauma, and oral-facial applications, whereas Sulzer Calcitek, (California, USA) commercializes Bio-mend®, a pure collagen type-I membrane product for periodontal tissue regeneration. Collagen has also been used for the fabrication of tube allografts for guiding peripheral nerve regeneration and for vascular prostheses. Other applications of collagen include its use as the main ingredient for some cosmetic makeup.

Box 8.2

- There are various subtypes of collagen barrier membranes, depending on the source (porcine or bovine) and the sites (tendon, intestine, or dermis) where the collagen is derived.
- Ideally, the membranes should be biocompatible, have the ability to exclude tissues or cells, provide space maintenance and have acceptable handling properties.

Making up about 30% of the total proteins, collagen is the primary protein of the skin and connective tissue such as bones, tendon, and ligaments. Collagen also makes up 1% to 2% of muscle tissue. Used as a biomaterial, collagen possesses the following desirable properties:

- it can be resorbed into the body,
- it is non-toxic, and
- it produces minimal immune response, even between different species.

Like many polymeric biomaterials, collagen can be processed into porous sponges or scaffolds, gels, and sheets. Although collagen is naturally derived and possesses favorable biocompatibility properties, it does not possess adequate mechanical properties to prevent contraction or deformation of the collagen-fabricated device as a result of cells pulling or reorganizing the collagen fibers. Additionally, the degradation rate of collagen can be unpredictable. These deficiencies can be overcome through modifications such as cross-linking or the combined use of collagen with other biomaterials to improve its mechanical properties as well as to alter its degradation properties.

There are at least 28 different molecular types of collagen that have been identified, with each collagen type having its unique composition, structure, function, and tissue specificity. Of these 28 different types, types I to V are the major collagen types in existence, making up about 90% of the collagen in the body. Possessing fibrous collagen domains and non-collagenous domains, all collagens are first synthesized as precursor molecules called procollagens that are secreted into the extracellular matrix. Structurally, the procollagen is flanked by a trimeric globular C-propeptide domain on the C terminal and by a trimeric N-propeptide domain on the N terminal. Collagen molecules known as tropocollagens are formed when the C and N terminal propeptide domains are cleaved by the procollagen C-metalloproteinase and the procollagen N-metalloproteinase, respectively. The resultant tropocollagens are flanked by short extra-helical telopeptides which are critical for fibril formation.

Box 8.3

- Telopeptides do not possess the triple-helical structure.
- Telopeptides are not made up of repeating glycine–X–Y structure.
- Telopeptides make up 2% of the collagen molecules.

Figure 8.1 shows the hierarchy for collagen formation. Being the basic unit for the collagen domain, each tropocollagen is approximately 300 nm long, 1.5 nm in diameter and has a molecular weight of approximately 283 kDa. The unique feature of each molecule is that it comprises three long left-handed helical polypeptides or protein chains with a sequence of glycine–X–Y as the regular arrangement of amino acids and with approximately three amino acid residues per turn. The X and Y in the sequence can be any amino acids. In most instances, the X is proline and Y is hydroxyproline for about 1/6 of the total sequence. The sequence of glycine–X–Y is common for all collagen types, with disruption of the sequence at certain locations within the triple helix of the collagen domain.

Each of the three long left-handed helical polypeptide chains, referred to as an alpha-chain (α-chain), is generally more than 1000 amino acid residues long. Of the three α-chains in each collagen molecule, two of the chains have 1056 amino acid residues and are called the α1-chains. The sequence of the third chain, which is known as α2-chain, is different from the α1-chains in that it has only 1029 amino acid residues. By twisting or intertwining the three α-chains together around a central molecular axis to form a triple-helical structure called a collagen monomer

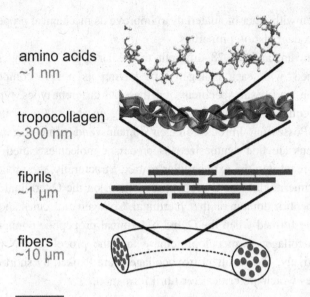

amino acids
~1 nm

tropocollagen
~300 nm

fibrils
~1 µm

fibers
~10 µm

Figure 8.1

Schematic showing the hierarchy of collagen, with structural features ranging from the amino acid sequence, tropocollagen molecules, collagen fibrils to collagen fibers. This is reproduced with permission from the Proceedings of the National Academy of Sciences (Copyright 2006 National Academy of Sciences, USA), and Professor M. J. Buehler.[1]

or a gamma-structure (γ-structure), the collagen molecule provides the overall strength of the collagen protein. Since the interior of the helical structure has no space for large molecules, glycine, being the smallest of the 22 known amino acids, is found in the interior of the helical structure. Proline and hydroxyproline are larger and more bulky amino acids when compared to glycine. As such, in the sequence of the α-chains, proline and hydroxyproline are pointed outward of the helical structure. Such a configuration allows the polypeptide chain to adopt a less constraining conformation. In general, the function of the proline and hydroxyproline is to aid in the formation of the helical structure. Additional factors contributing to the stabilization of the triple-helix structure include:

- tight fit of the amino acids within the triple helix,
- formation of inter-chain hydrogen bond between the carbonyl oxygen atom of the proline (X residues) in one chain and amide hydrogen atom of glycine in the adjacent chain,
- formation of inter-chain hydrogen bond between the hydroxyl group of the hydroxyproline (Y residues) in one chain and amide hydrogen atom of glycine in the adjacent chain, and
- presence of water molecules that contribute to the formation of the inter-chain hydrogen bonds.

- The difference between the polypeptide chains within the collagen molecule is the positions of the X and Y amino acids in the triple helix.
- In general, the X and Y amino acids in the collagen molecule are proline and hydroxyproline, respectively.
- The left-handed triplet helices are held together by inter-chain hydrogen bonds

Schematically shown in Figure 8.1, the tropocollagens are assembled into fibrils, which in turn are then assembled to form collagen fibers. Upon cleaving of the trimeric globular C-propeptide domain and the trimeric N-propeptide domain from the procollagen, the resulting tropocollagens spontaneously self-assemble into cross-striated fibrils that have minimal surface area to volume ratio. Additionally, oxidative deamination of specific lysine and hydroxylsine residues in the collagen molecule by lysyl oxidase allows the fibrils to be stabilized by covalent cross-linking. Although collagen fibril formation is a self-assembly process that is entropy driven, the assembly of collagen into fibrils is dependent on the type of collagen. For example, fibril formation in type-II collagen has a slower assembly rate when compared to type-I collagen.

Type-I collagen is the most abundant collagen in the human body and is also the predominant form of collagen used in biomaterials application. This collagen type is found in skin, dentin, cornea, blood vessels, bone, tendon, ligament, and fibrocartilage. It is also found in the scar tissue as tissue heals. Type-I collagen aggregates into fibrils. Depending on age and tissue, the diameter of the type-I collagen fibril can vary from 50 nm to 500 nm. These fibrils are important structural building units, forming fibers and fiber bundles that are aligned nearly parallel to the direction of load. In the dermis, fibrils of type-I collagen form loose networks, whereas in bone, type-I collagen forms the fibrillar phase of the mineralized cortical and trabecular composites.

Type-II collagen is found mostly in articular cartilage, forming a dense network array of individual thin fibrils, with an interfibrillar matrix typically made of proteoglycans, glycoproteins, non-collagenous proteins, and water. The main function of type-II collagen in cartilaginous tissues is to provide tensile integrity to the tissue, whereas the interfibrillar matrix is responsible for the swollen state. Together, the collagen and the matrix act as a shock absorber in our joints and vertebrae.

Type-III collagen is a homotrimer composed of three $\alpha 1$(III) chains and resembles other fibrillar collagens in its structure and function. The elastic properties of type-III collagen may be due to disulfide bonds. Additionally, there is no lysyl oxidase-dependent cross-link in the C-terminal end. Similar to type-I collagen, type-III collagen is synthesized as procollagen. The difference between type-I collagen and type-III collagen is that the N-terminal propeptide is more often attached in the mature, fibrillar type-III collagen than in type-I collagen. Being the second most abundant collagen in humans, type-III collagen accounts for 20% of the collagen in adult skin, while the remaining is type-I collagen. It is more abundant in the skin of newborn, with the elastic properties of type-III collagen contributing to the suppleness of the skin in newborns. Additionally, this elastic property also accounts for the flexibility of blood vessels. Besides the skin and blood vessels, type-III collagen is also found in ligaments as well as in internal organs.

Type-IV collagen is found in basement membranes of the eye lens, walls of blood vessels, kidneys, and basal lamina structures in the skin. Within these membranes and basal lamina structures, type-IV collagen interacts with non-collagenous components to form meshes or networks, thereby acting as a filtration system for cells, molecules, and light. Light filtration is performed by the basement membrane in the lens capsule of the eye, whereas the filtration of waste products from the blood is performed by the glomerulus basement membrane of the kidney. Similarly, the transport and movement of oxygen and nutrients out of the circulatory system and into the surrounding tissues are governed by the basement membrane in the walls of blood vessels. In the skin, the basal lamina not only delineates the dermis from the epidermis but also governs the transport of materials in and out of the dermis.

8.2   Elastin

Elastin-based biomaterials have been used in several medical applications such as vascular stents, and for repairing skin, bladder, intestine, fallopian tubes, esophagus, stomach, lung, colon, and the heart. They have also been used for cosmetic implantation as in breast implants. Found naturally, elastin is an insoluble extracellular matrix protein and is synthesized as a soluble precursor of about 70 kDa. It is found abundantly in locations such as the skin, bladder, lungs, arteries, and intervertebral discs, where elasticity, flexibility, and strength are of major importance. Elastic ligaments and large arteries are made of 70% and 50% elastin, respectively. The lung is made of 30% elastin, whereas skin is made up of 2%–4% elastin. The presence of elastin allows these organs to retain their shape after stretching or contracting. Flexibility allows the organs to undergo repetitive strain and high deformation without rupture, to store the energy involved in the deformation, and to recover to their original state when the stress is removed. The driving force for the spontaneous recoil of the stretched elastin is entropic in origin. In fact, the elasticity of elastic fibers in these organs is dependent on the entropy of relaxation of elastin. Other properties of elastin include resistance to the action of acids, alkalis, and preteolytic enzymes, except elastase which specifically degrades elastin. Elastin is also an important load bearing tissue when mechanical energy is required to be stored. As such, elastin is becoming an increasingly popular biomaterial due to the following properties:

- ability to stretch and relax, and
- high stability with a half-life of 70 years.

Box 8.4

- Elastin-based biomaterials are increasingly popular for applications in tissue engineering.
- Produced as an extracellular matrix, elastin provides tissues and organs with elastic properties.
- Unlike collagen, whose molecules are closely packed to form the triple-helix structures, elastin is relatively loose, with its unstructured polypeptide chains covalently cross-linked to form a rubber-like elastic meshwork, thereby allowing tissues to stretch or contract.

Composed of fibrillin and about 786 amino acids such as glycine, valine, alanine, and proline, elastin possesses an irregular or random coil conformation. It is highly insoluble due to inter-chain cross-linking. 75% of the amino acids present in elastin are hydrophobic residues. In humans, elastin is synthesized as a soluble precursor known as tropoelastin. Cells that produce tropoelastin include smooth muscle cells, endothelial cells, fibrobalsts, and chondrocytes. The tropoelastin is capable of self-assembly under physiological conditions. Catalyzed by lysyl oxidases, elastin is covalently stabilized by inter-chain cross-linking the side chains of the lysine residues in the tropoelastin, resulting in a highly insoluble polymer that is resistant to enzymatic, chemical, and physical degradation.

Incorporation of elastin in biomaterials becomes important for applications where elasticity and biological effects of such biomaterials can be exploited. Biomaterials derived from natural elastin have been used as autografts, allografts, and xenografts for applications in burn wounds, coronary artery bypass, and aortic heart values. However, calcification of elastin, elastin-based biomaterials, or tropoelastin materials becomes a problem when implanted *in vivo*. This phenomenon is attributed to elastin serving as a nucleation site for mineralization. Current technologies to overcome this limitation include the pre-treatment of the elastin with irradiation, aluminum oxide treatment, or acyl azide cross-linking.

Since it is difficult to obtain pure, tissue-derived elastin that is free of globular protein contamination, synthetic polymeric elastin can be synthesized based on the elastin pentapeptide sequence using solution chemistry, solid-phase approaches, and recombinant genetic engineering. By using chemical, enzymatic, and gamma-irradiation mediated cross-linking, various forms of elastin biomaterials can be processed, including sheets, fibers, and tubular constructs. Incorporation of cell-binding sites and drugs in elastin polymers can be used to expand the functionality of the biomaterials and promote cell adhesion and growth. Inclusion of specific cross-links and amino acid sequences in the elastin polymers enable the control of the degradation rate and thus the drug released. Additionally, the incorporation of motifs in elastin polymers also allows for gel formation, stimuli-responsive characteristics, biodegradation, and biorecognition.[2]

Through the use of recombinant genetic engineering, synthetic elastin-like polypeptides are capable of exhibiting a thermally reversible property. The recombinant engineered polypeptides are made of repeated pentapeptide chains of amino acids, having the sequence of valine–proline–glycine–X–glycine. The amino acid, X, can be any amino acid except proline. Different types of elastin-like polypeptides can thus be synthesized by altering the amino acid, X, within the polypeptide sequence. These synthesized polypeptides are soluble in aqueous solution at low temperatures and become insoluble and aggregate or become more ordered at a critical high temperature. Such a critical temperature, termed as the "inverse transition temperature," (T_t), is similar to the lower critical solution temperature exhibited by thermally responsive polymers. Resolubilization of the insoluble and aggregated recombinant engineered elastin-like polypeptides occurs as the solution is cooled below T_t. Additionally, elastin-like polypeptides can also be engineered to swell and contract by controlling the polymer's T_t. Altering the amino acid, X, within the polypeptide sequence yields different combinations of elastin-like polypeptides with different T_t. This unique thermally reversible property, coupled with the ability of recombinant genetic engineering to synthesize polypeptides with a precise molecular weight and low polydispersity, allows these engineered elastin biomaterials to sense alterations in their chemical microenvironment, to release biomolecular cues, or to elicit a mechanical response.

Box 8.5

- Lower critical solution temperature refers to a critical temperature below which components of a mixture are miscible.
- Precipitation as a result of a change in phase occurs when the mixture is heated above the lower critical solution temperature.

8.3 Silk

Like collagen and elastin, silk is a structural protein that is naturally spun into polymeric fibers by some silkworms, spiders, scorpions, mites, and flies. Shown in Figure 8.2 is a representative photograph of raw silk fibers that can be used as a starting material for the fabrication of medical and non-medical products. Long used as a biomaterial for medical sutures, silk is known for its biocompatibility, slow degradability, and excellent mechanical properties.

As the larva or caterpillar of the domesticated silkmoth, *Bombyx mori*, the silkworm produces silk fiber to weave its cocoon. The silk fiber is primarily made of a glue-like non-filamentous protein, sericin, that surrounds the fibrous protein known as fibroin. In general, sericin in *Bombyx mori* silk makes up about 25% of the fiber, whereas fibroin makes up the majority of the other 75% of the fiber. The amount of sericin and fibroin content is also dependent on the location of silk, with more sericin found in the outer layer. Impurities in *Bombyx mori* silk include carbohydrates (1.2%–1.5%), waxes (0.4%–0.8%), inorganic salts (0.7%), and pigment (0.2%). Acting as a gum binder for maintenance of structural integrity, sericin is more water soluble than fibroin. Properties of sericin include

- being insoluble in cold water,
- easily hydrolyzed,
- soluble in hot water,
- resists oxidation,

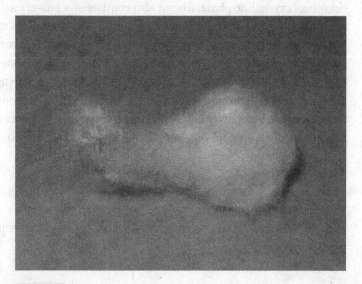

Figure 8.2

Photograph of raw silk fibers spun from the silkworm *Bombyx Mori*.

- antimicrobial,
- UV resistant, and
- absorbs and release moisture easily.

Fibroin is made up of three primary peptides, a 325 kDa–350 kDa Heavy (H)-chain, a 25 kDa–26 kDa Light (L)-chain, and a 30 kDa fibrohexamerin (P 25) which is basically a glycoprotein linked by disulfide bonds. With a molar ratio of 6:6:1 for H-chain, L-chain, and P 25, respectively, these three fibroin components are common among the different silk fibers produced by the Lepidoptera insects. Additionally, the primary structure of the fibroin is made of four regions. Three of these regions are made up of antiparallel beta sheets and have a recurrent amino acid sequence (alanine–glycine–serine–glycine–alanine–glycine)$_n$ as their primary structure, with glycine (G), alanine (A), and serine (S) making up 45.9 mol%, 30.3 mol%, and 12.1 mol% of the overall fibroin composition, respectively. The fibroin is also made up of 5.3 mol% tyrosine (Y), 1.8 mol% valine (V), and 4.7 mol% other amino acids. Being short side-chain amino acids, they form into stacks of beta sheets and thereby result in organized crystalline regions that are responsible for the tensile strength and toughness of the silk fiber. Of these three crystalline regions, region 1 is highly organized and is made up of a highly repetitive GAGAGS sequence. Region 2 is relatively less repetitive compared to region 1. Being semi-crystalline, region 2 contains the GAGAGY and/or GAGAGVGY sequences. Region 3 is very similar to region 1, except that it also contains the additional alanine–alanine–serine (AAS) sequence. In addition to the beta-sheet crystalline phase, fibroin also consists of a non-crystalline fourth region made of microvoids and amorphous structures. The amorphous region contains the negatively charged, polar, bulky hydrophobic and aromatic residues. Unlike the crystalline regions which fold into beta sheets, the amorphous regions exhibit helical, spring-like properties that allow fibers to be elastic or extended. As a result, the silk fibroin protein can be described as a block copolymer containing crystalline domains of predominantly highly repetitive amino acid sequences, interrupted with amorphous domains consisting of the bulkier side-chain amino acids.

- Sericin is primarily amorphous and acts as a gum binder to maintain structural integrity.
- Fibroin is a semi-crystalline polymer and is less water soluble compared to sericin.

At different regions of the gland on the *Bombyx mori* silkworm, fibroin and sericin are separately synthesized and stored. Mixing of the fibroin and sericin occurs within the anterior section of the gland, ultimately resulting in drawing or spinning of the soluble silk at the spinneret. At a spin rate of 1 cm per second and a high shear rate of 2–400 cm per second, the *Bombyx mori* silk is made up of two fibroin protein threads glued together with a sericin protein gum, thereby resulting in a single thread of 10–25 μm diameter. As the silk fibers are spun at ambient temperatures in aqueous solution, the spin rate and high shear rates result in the silk fibers undergoing conformational changes and loss of water. In general, the spun silk fiber possesses the following the unique properties:

- insoluble in water, dilute acids, alkali, and most organic solvents,
- resistant to degradation by most proteolytic enzymes,
- hydrolyzable in concentrated sulfuric acid,
- hygroscopic,
- tensile strength of 650 MPa and modulus of about 15 GPa,
- elastic, with elongation to break at about 15%–20%, and
- poor conductor of electricity.

Unlike silk produced from *Bombyx mori*, it is not possible to maintain domesticated spiders to produce massive amounts of silk. However, spider silks are also intriguing since they are remarkably strong and elastic, but are extremely lightweight. These silk polymers exhibit superior mechanical properties, with a nominal fracture strength of about 1100 MPa. In general, these unique properties of spider silk are the result of the composition of the protein fiber. Although different species of spider have different silk composition, the protein fibers, in general, are made up of a sequence of alternating glycine and alanine or alanine-rich blocks, separated by segments of amino acids with bulky side-groups. Like silk produced from *Bombyx mori*, these sequences of alternating glycine and alanine or alanine-rich blocks in spider silk self-assemble into beta sheet structures, with stacks of beta sheets forming the crystalline portion of the silk fiber, whereas segments of amino acids with bulky side-groups form the amorphous or semi-amorphous portion, contributing to the elasticity of the fiber. The presence of crystalline regions separated by amorphous portions within the fiber governs the properties of the silk. Additionally, spider silk is environmentally non-toxic and degradable since it is spun using water as the solvent at near ambient temperatures and pressure. Although there is commercial interest in duplicating spider silk artificially, it is currently difficult to find a commercially viable process to mass-produce spider silk.

As a medical device, silk is a popular biomaterial used for sutures in operating rooms. Examples of commercially available silk sutures include DemeTech's Silk sutures (DemeTech Corporation, Florida, USA), TRUSILK (Sutures India, Bangalore, India), and PERMA-HAND® Silk Suture (Ethicon, New Jersey,

USA). These sutures are produced from silk derived from the silkworm larva's cocoon and are generally indicated for use in soft tissue approximation and ligation, including use in cardiovascular, ophthalmic, and neurological surgery. These sutures are non-resorbable since they degrade very slowly and exhibit a gradual loss of tensile strength. Other biomedical applications of silk include drug delivery and tissue engineering.

8.4 Chitosan

Chitosan is a polysaccharide-based polymer. Like any polysaccharide, the monomers of chitosan are monosaccharides. Polymerization occurs by linking the monosaccharides together through the formation of O-glycosidic bonds between any of the hydroxyl groups of a monosaccharide, thereby allowing the polysaccharides to form both linear and branched polymers. In general, the physical properties of chitosan, including solubility, gelation and surface properties, are governed by differences in the monosaccharide composition, chain shapes, and molecular weight.

Being a cationic polymer, chitosan is derived from chitin, a polysaccharide found in cell walls of some fungi, cuticles of insects, as well as in the exoskeletons of shellfish such as shrimp and crabs. Chitin is the major tensile element of the exoskeleton. As shown in Figure 8.3, chitin is made up of a long linear polymeric chain of N-acetyl-D-glucosamine.

With the deacetylation (or removal of the acetyl functional group, $COCH_3$) of more than 50% of the repeat units of chitin, chitosan is a linear polysaccharide composed of randomly distributed β-1,4-D-glucosamine copolymer (deacetylated unit) and N-acetyl-D-glucosamine (acetylated unit). The removal of the acetyl functional groups from the glucosamine repeat units allows for tighter packing and an increase in the degree of crystallinity. Shown in Figure 8.4 is a chemical structure of chitosan with 100% of its acetyl functional groups removed.

The degree of deacetylation (DDA), which is the measure of molar fraction of glucosamines to N-acetyl glucosamines, generally ranges from about 60% to 100% for chitosan. With a DDA between 70% and 95%, typical commercially available chitosan has a molecular weight between 10 kDa and 1000 kDa. Many of the physical and chemical properties of chitosan are influenced by the DDA and the molecular weight of the material. In general, chitosan possesses the following favorable properties for use as a biomaterial:

- elicits minimal foreign body reaction,
- can be dissolved in water, depending on pH,

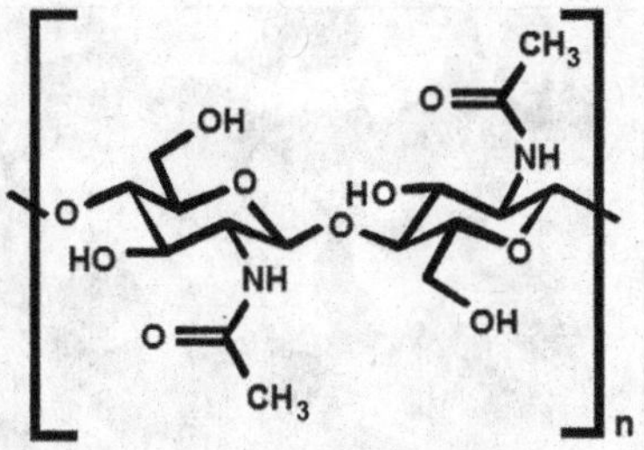

Figure 8.3

Chemical structure of a polymeric chain of chitin.

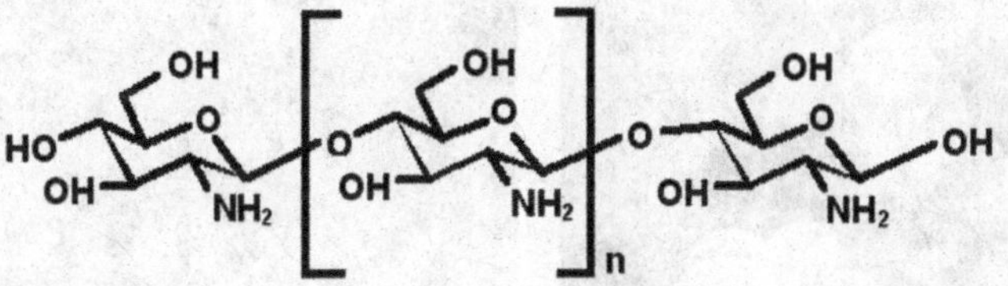

Figure 8.4

Chemical structure of chitosan that is 100% deacetylated.

- possesses available side groups for attachment of molecules such as growth factors, and
- mechanical and biodegradation properties are controllable through polymer length or porosity.

Processing of chitosan involves grinding of the shells of shrimp and crabs and the deproteination of the ground material in a strong base. Demineralization of the deproteinated material is performed in a strong acid, followed by deacetylation in a strong base at high temperature. Having a pKa value of approximately 6.5, the amino groups in chitosan become protonated at neutral to acidic pH, thus resulting in a positively-charged chitosan. Additionally, the pKa value of approximately 6.5 also means that chitosan is soluble in dilute acid or neutral solutions, with a charge density that is dependent on the pH and percent DDA. The positively-charged chitosan also provides the basis for many hydrogen bonding and ionic interactions. These interactions make chitosan a favorable bioadhesive and hemostatic material, allowing it to readily bind to the negatively-charged mucosal membranes. Additionally, the presence of many hydroxyl and amino groups in chitosan allows it to be chemically modified, whereas its solubility in aqueous solutions allows it to be processed into fibers, films, gels, and three-dimensional porous structures.

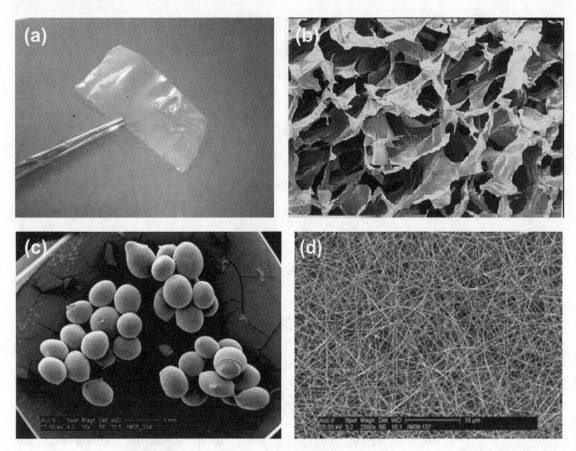

Figure 8.5

Laboratory-fabricated chitosan. (a) Photograph of chitosan film; (b) scanning electron micrograph of chitosan sponge; (c) scanning electron micrograph of chitosan microspheres (magnification: 16 ×); and (d) scanning electron micrograph of chitosan nanofibers (magnification: 2500 ×) (courtesy of Dr. Joel D. Bumgardner, University of Memphis).

Shown in Figure 8.5 are examples of laboratory-fabricated chitosan films, sponges, microspheres, and nanofibers.

Box 8.6

- Chitosan can be easily fabricated to form complex structures.
- In addition to biodegradability, chitosan also exhibits antimicrobial properties and promotes cell adhesion and migration.
- Although tougher and more flexible than ceramics, chitosan tends to swell when immersed in physiologic solution and is mechanically weak.

Broken down into simple sugars by lysozymes and by hydrolysis, the degradation of the chitosan in the body is generally dependent on its crystallinity and the degree of deacetylation. An increase in the degree of deacetylation increases the crystallinity of chitosan and decreases the rate of degradation. Degradation of chitosan is also affected by sterilization methods. Chain scissions and/or cross-links may be affected by gas sterilization, autoclaving and gamma irradiation, which in turn may affect the degradation behavior of the chitosan.

In addition to degradation rate, the degree of deacetylation also affects cell behavior, with higher cell viability reported on chitosan with a higher degree of deacetylation.[3] Chitosan that is highly deacetylated also induces more fibrosis, whereas a lower deacetylated chitosan has been associated with increased osteo-genesis.[4] Commercially chitosan medical products that are currently available or in different phases of clinical trials include BST-CarGel® (BioSyntech Canada, Inc., Québec, Canada), a gel to repair focal cartilage lesions resulting from trauma, physical activity, or osteoarthritis, and HemCon® Patch, HemCon® Bandage and ChitoFlex® Hemostatic Dressings (HemCon Medical Technologies, Inc.), a local-ized wound dressing to control blood loss and stop bleeding.

In addition to dressings, other applications for chitosan include tissue engineer-ing. Challenges with chitosan as a polymer scaffold in tissue-engineering include low strength and inconsistent behavior with seeded cells. Fortunately, chitosan may be easily combined with other materials in order to increase its strength and cell-attachment potential. Chitosan mixed with synthetic polymers such as poly (vinyl alcohol) and poly(ethylene glycol), ceramics, or natural polymers such as collagen have already been produced.[5] These combinations have demonstrated promise for improving the performance of the combined construct over the behavior of either component alone.

- Modification of chitosan can be achieved through ionic complexation, whereby negatively charged biomolecules of interest can be introduced to the positively charged chitosan.
- Physical adsorption or entrapment of biomolecules can also be used as an approach to modify chitosan.

8.5   Cellulose

Cellulose, with the formula $(C_6H_{10}O_5)_n$, is one of the many β-glucan compounds. Being a linear polysaccharide, cellulose is the most abundant organic polymer in

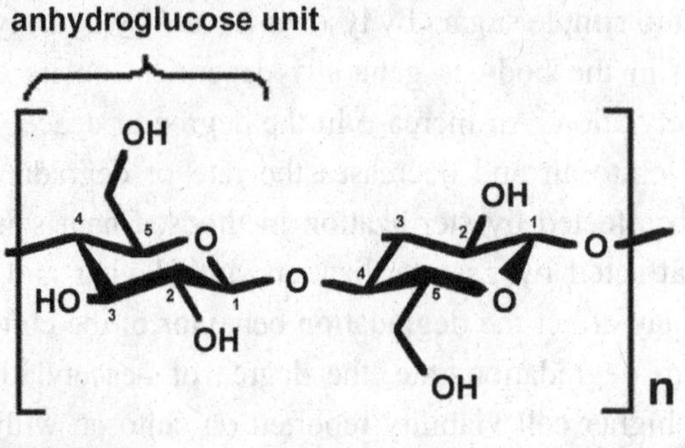

anhydroglucose unit

Figure 8.6

D-glucose units linked by β-1,4 glycosidic bonds linkage between the C-1 of one glucose unit and the C-4 of another glucose unit.

the world. It is found in all plant matter and is the main constituent of the cell wall in green plants. Cellulose is also secreted by some species of bacteria and prokaryotes such as acetobacter, rhizobium, agrobacterium, as well as fresh water and marine algae. Like other β-glucan compounds, cellulose is made up of several hundreds to thousands of D-glucose units. Collectively called pyranose, the six-membered ring of glucose consists of five carbon atoms and one oxygen atom. These rings are linked together between the C-1 of one glucose unit and the C-4 of another glucose unit by single oxygen atoms called the acetal linkages or the β-1,4 glycosidic bonds linkages (Figure 8.6). This reaction of the alcohol of one glucose unit with the hemiacetal of another glucose unit results in the formation of an acetal. The glucose units in cellulose polymers are thus referred to as anhydroglucose units.

- Cellulose is a low cost, readily available polymer.
- Cellulose is made of beta (β)-glucose units. These units exhibit a *trans* arrangement (the hydroxyl group at C-1 being on the opposite side of the ring as the C-6 carbon).
- Starch is made of alpha (α)-glucose units. These units exhibit a *cis* arrangement (the hydroxyl group at C-1 being on the same side of the ring as the C-6 carbon).

As shown in Figure 8.6, every other anhydroglucose unit within the cellulose chain is rotated by approximately 180° about its axis. Such rotation results in the glucan chain possessing disaccharides called cellobiose, with formula $(HOCH_2CHO(CHOH)_3)_2O$. With reference to Figure 8.7, each cellobiose unit contains eight free alcohol $(- C - O - H)$ groups and three ether $(- C - O - C -)$

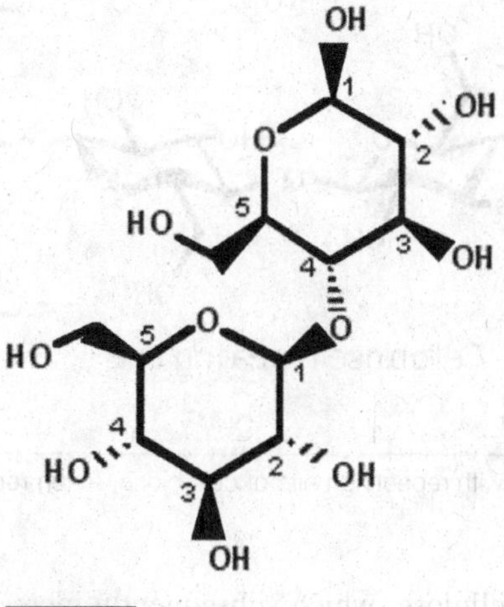

Figure 8.7

Molecular structure of disaccharides called cellobiose. Numbers in the figure indicate the position in the carbon ring.

linkages. The equatorial position of these free hydroxyl units in these free alcohol groups protrudes laterally along the extended molecule, allowing for strong intramolecular and intrastrand hydrogen bonding with oxygen molecules on neighboring chains, holding the chains firmly together side-by-side to form a highly ordered, crystalline structure within the microfibrils.

Shown in Figure 8.8, a cellulose molecule is made of repeating units of cellobiose and two non-equivalent chain ends. One of the two terminal hemiacetal groups acts as a reducing end aldehyde group and hence is called the reducing end. On the opposite end of the cellulose molecule is a closed ring structure termed as the non-reducing end. Additionally, the cellulose molecule is rigid and rod-like as a result of the orientation of the glycosidic bonds linking the glucose units that causes the six-membered rings of glucose to be arranged in a flip-flop manner.

Unlike glucose, which is water soluble, cellulose is insoluble in water. The water-insolubility of cellulose is due to the presence of intramolecular and intrastrand bonding within the crystalline structure that provides the polymer with high fiber strength. In regions of the cellulose fibers that are less ordered or less crystalline, the cellulose chains are farther apart and are more readily available for bonding with other molecules. As such, when immersed in water, the presence of these less crystalline regions permits cellulose to swell as a result of absorbing large quantities of water and thus making cellulose structures hygroscopic. Other cellulose properties such as viscosity are governed by its molecular weight. Variation in molecular weight is controlled by the degree of polymerization of the cellulose backbone, which in turn is dependent on the cellulose source and its isolation process. As a result, an increase in the degree of polymerization increases

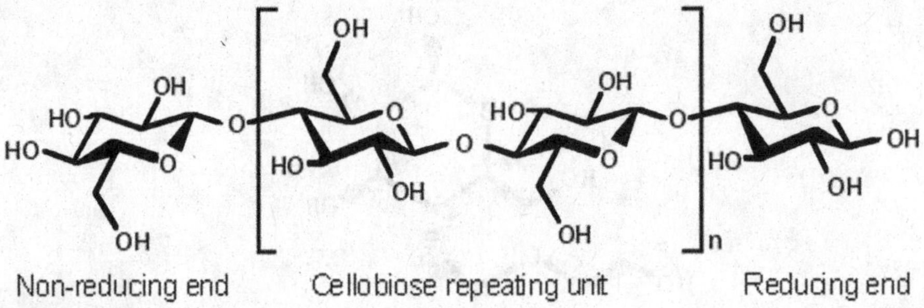

Non-reducing end   Cellobiose repeating unit   Reducing end

Figure 8.8

Molecular structure of cellulose with repeating units of cellobiose, a non-reducing end, and a reducing end.

the molecular weight of cellulose, which subsequently increases the viscosity of the cellulose polymer. Although properties of cellulose are dependent on its chain length, other general properties of cellulose are

- tasteless,
- odorless, and
- hydrophilic.

> - Although insoluble in water, cellulose is a hydrophilic polymer.
> - As a result of their high cohesive structure, cellulose fibers possess exceptional strength.

There are four general crystalline polymorphs of cellulose. Cellulose I is found naturally in most plants and contains more beta sheet (I_β) protein structures than alpha helix (I_α) protein structures. Unlike cellulose I, cellulose II contains more I_α protein structures than I_β protein structures and is found naturally only in some algae, mold, and bacteria such as *Sarcina ventriculi*. Additionally, cellulose I is metastable and consists of parallel β-1,4-linked glucan chains that are uni-axially arranged. Distinguishable from cellulose I using X-ray diffraction, nuclear magnetic resonance, Raman spectroscopy, and infrared analysis, cellulose II consists of randomly arranged β-1,4-linked glucan chains that are mostly anti-parallel and are linked with a larger number of hydrogen bonds. These high hydrogen bondings contribute to the thermodynamic stability of cellulose II. Aside from naturally produced cellulose II, it can also be synthesized in agitated bacterial culture as well as chemically converted from cellulose I. As a result, chemically converted or regenerated cellulose is cellulose II. Being less crystalline compared to cellulose I, regenerated cellulose is different from native cellulose owing to the occurrence of extensive degradation during the dissolution process. In the process of converting to regenerated cellulose, cellulose is allowed to age by treating with alkali, thereby subsequently reducing the molecular weight of cellulose as a result of possible oxidation degradation. Cellulose III is formed by having cellulose I and II undergoing a liquid ammonia treatment, whereas cellulose IV is formed by heat treating cellulose III.

The abundance and ease by which cellulose can be converted to derivatives makes it an attractive raw material. Purification and isolation of cellulose involves a variety of steps including a pulping process, partial hydrolysis, dissolution, re-precipitation, and extraction with organic solvents. These purification and isolation processes almost always degrade the cellulose as well as allow the cellulose to undergo oxidation by reacting with both acids and bases. There are a number of cellulose derivatives that can be synthesized including the following:

- cellulose esters,
- cellulose ethers,
- graft copolymers, and
- cross-linked cellulose.

Current commercially available cellulose products include the different grades of Natrosol® (hydroxyethylcellulose) and Klucel® HPC (hydroxypropylcellulose), both marketed by Ashland Inc. (Delaware, USA). As non-ionic water-soluble cellulose, both Natrosol® and Klucel® HPC exhibit aqueous or organic solvent solubility, thermoplasticity, and surface activity. Available in various viscosities, cellulose are used in pharmaceutical formulations for various purposes. The less viscous forms of cellulose are employed as tablet binders in the matrix formulations for immediate drug release, whereas medium and high viscosity forms of cellulose are employed in pharmaceutical matrix formulations for sustained drug release. Other commercially available cellulose products used in medical applications include Biofill® manufactured by Carki Erba S.A. (Duque Caxias, PR Brazil) for use as artificial skin in burn therapy and ulcers, Cuprophan™ manufactured by Allmed Medical GMBH (Hamburg, Germany) for use as hemodialysis membranes, and Surgicel® from Ethicon, Inc., a Johnson and Johnson company (New Jersey, USA), for control of hemorrhage during surgical procedures owing to its hygroscopic properties.

8.6 Alginate

Alginate is a family of non-branched anionic polysaccharide copolymers that are derived from the *Peudomonas* and *Axotobacter* bacteria as well as being found in marine algae such as brown seaweeds. Also known as alginic acid or algin, it is

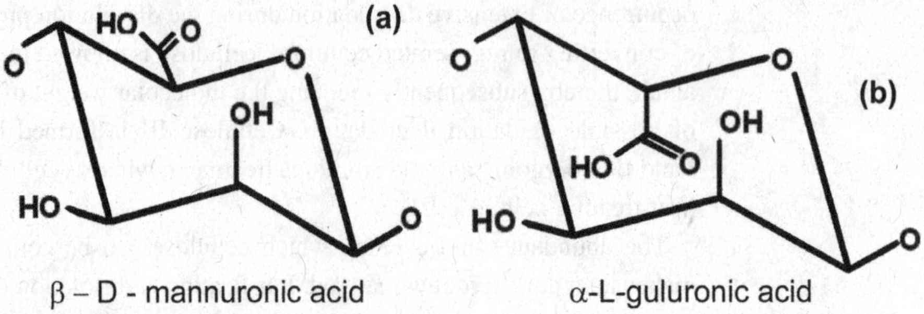

β – D – mannuronic acid   **α-L-guluronic acid**

Figure 8.9

Molecular structure of (a) β-D-mannuronic acid and (b) α-L-guluronic acid exhibiting a difference in stereochemistry at the C-5 position.

composed of β-D-mannuronic acid (M) and α-L-guluronic acid (G) monomers, linked by 1–4 glycoside bonds in a block-wise fashion. As shown in Figure 8.9, β-D-mannuronic acid and α-L-guluronic acid are epimers with the only difference in stereochemistry at the C-5 position.

Box 8.7

- Properties of alginates range from slimy and viscous solutions to pseudoplastic when they are cross-linked with divalent cations.
- Alginate properties are affected by the variability in the comonomer blocks and the degree of acetylation.
- The M residues are capable of undergoing acetylation, whereas the G residues are not acetylated.

The polymerization process to form polymannuronate or M-block homopolymer in the bacteria is dependent on the presence of guanosine diphosphate (GDP)-mannuronic acid as a precursor. This homopolymer can be further modified by acetylation (introduction of an acetyl functional group) at positions 2 and/or 3 hydroxyl groups of the D-mannuronic acid residues and by epimerization, leading to a variable content of acetyl groups and G residues, respectively. As a result, alginates isolated from natural sources have a wide variation in fractional contents and sequence distributions of the M and G residues. Alginates containing the G-blocks have also been reported to be experimentally formed *in vitro* by bacteria. As shown in Figure 8.10, the three possible sequences in the alginates are the homopolymeric structures of M-blocks (M–M–M) and G-blocks (G–G–G),

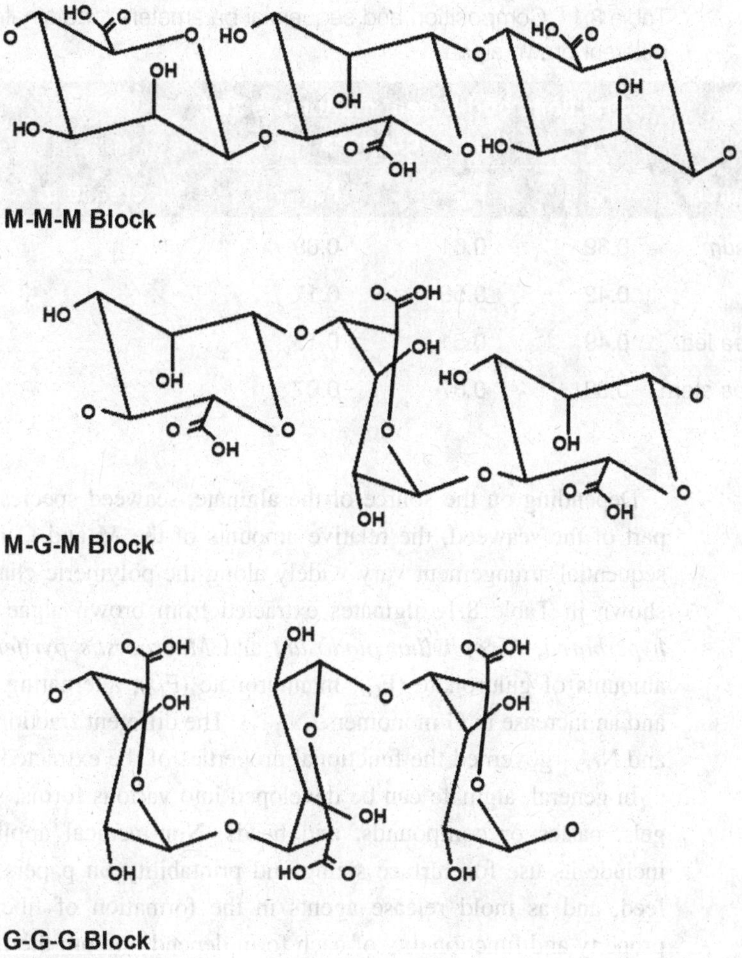

M-M-M Block

M-G-M Block

G-G-G Block

Figure 8.10

Sequential structures of alginate.

and the heteropolymeric structures of alternating MG-blocks (G–M–G–M). Being hydrophilic, alginate possesses the following favorable properties:

- fairly non-toxic,
- non-inflammatory,
- easily processed in water,
- ability to absorb water quickly,
- biodegradable,
- controllable porosity,
- can be linked to biologically active molecules, and
- a good mucoadhesive agent as a result of the presence of carboxyl end groups.

Table 8.1   Composition and sequential parameters for alginates extracted from different brown algae

Alginate source	Fractions of different monomers and their sequences			Increase in G monomers ($N_{G>1}$)
	F_G	F_M	F_{MGM}	
Asco. nodosum	0.39	0.61	0.09	5
M. pyrifera	0.42	0.58	0.17	6
L. hyperborea leaf	0.49	0.51	0.13	8
L. hyperborea stem	0.63	0.37	0.07	15

Depending on the source of the alginate, seaweed species, plant age, and the part of the seaweed, the relative amounts of the M and G monomers and their sequential arrangement vary widely along the polymeric chain. For example, as shown in Table 8.1, alginates extracted from brown algae such as *Laminaria hyperborea*, *Ascophyllum nodosum* and *Macrocystis pyrifera* contain different amounts of guluronate (F_G), mannuronate (F_M), alternating sequences (F_{MGM}), and an increase in G monomers ($N_{G>1}$). The different fractions of F_G, F_M, F_{MGM}, and $N_{G>1}$ governed the functional properties of the extracted alginates.

In general, alginate can be developed into various forms, such as cross-linked gels, pastes or compounds, and beads. Non-medical applications of alginate include its use for surface sizing and printability on papers, as binders for fish feed, and as mold release agents in the formation of fiberglass plastics. The property and functionality of each form depend on both the innate characteristics of the alginate (composition, sequential structure, molecular weight, and molecular conformation), as well as the way it is processed.

The uniqueness of alginate gels is that they are heat stable, and that they can be developed and set at physiologically relevant temperatures. Like calcium sulfate dihydrate, which is a divalent cation, alginate gels can be instantaneously formed through interactions with many other divalent and multivalent cations. Hydrogels are formed when the divalent or multivalent cations participate in the inter-chain binding between G-blocks of adjacent alginate chains. These inter-chain bindings create inter-chain bridges, giving rise to a three-dimensional (3D) network of alginate fibers that are held together by ionic interactions. As such, the cross-linking capacity of an alginate and the overall strength of the network are dependent on the G content, the number of consecutive Gs in the G-blocks, and the concentration and types of cations present. Unlike divalent or multivalent cations, interactions with monovalent cations do not induce gelation. Similarly, not all divalent cations induce gel formation, since interaction with magnesium ions (Mg^{2+}) do not result in gel formation. However, interactions with other divalent cations such as barium ions (Ba^{2+}) and strontium ions (Sr^{2+}) result in stronger alginate gels when compared to alginate gels produced by interacting with calcium ions (Ca^{2+}). Similarly, alginates with a high G content or higher $N_{G>1}$ values develop stiffer and more porous gels which maintain their integrity for longer periods of time. These alginates with high G content do not undergo excessive swelling and subsequent shrinking and thus have the ability to better retain their form or shape during cationic cross-linking. In contrast, alginates with high M contents form softer and less porous gels that tend to degrade with time. These alginates with high M content have less ability to retain their shape or form as a result of the high degree of swelling and shrinking during cationic cross-linking. The ability to retain shape for tissue regeneration is critical in applications such as tissue engineering, and this ability is dependent on the overall network strength, which in turn, affects nutrient diffusion and cell-to-cell contact.

- The G-residue content and distribution strongly affect the gel forming capacity and properties of alginates.
- The gel forming capacity and properties of alginates are also affected by the types of cations present.

Used in dental prosthetic applications, potassium, sodium, and ammonium salts of alginates have properties that make them suitable for use as pastes or compounds for impression making. Potassium alginate ($KC_6H_7O_6$) acts as a stabilizer, thickener, and emulsifier. Similarly, sodium alginate ($NaC_6H_7O_6$) is used as a stabilizer and an emulsifier. Impression compounds, such as Jetrate® Alginate from DENTS-PLY International Inc. (Pennsylvania, USA), Identic from DUX Dental (California, USA), and Orthoprint from Zhermack, Inc. (New Jersey, USA) are formed when solutions of potassium, sodium, or ammonium alginates react with calcium sulfate dihydrate in the presence of water. The calcium sulfate dihydrate dissolves and reacts with potassium alginate to produce a water insoluble calcium alginate. Such a reaction results in the production of an elastic gel and is irreversible since it is not possible to convert the calcium alginate to a sol once it is set. Equation 8.1 shows a typical reaction in the formation of the calcium alginate impression compounds:

$$\text{potassium alginate} + \text{calcium sulfate dihydrate} \xrightarrow{\text{(water)}} \text{calcium alginate gel} \\ + \text{potassium sulfate.} \quad (8.1)$$

In addition to the formation of calcium alginate impression compound, highly pure calcium alginate is also used in other medical applications. Calcium alginate fibers

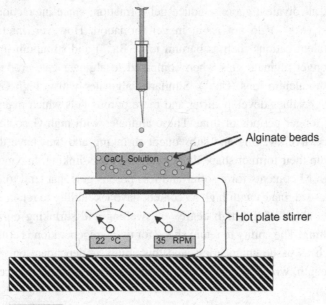

Schematic showing alginate beads produced by using the dripping method in calcium chloride solution.

are used in bandages, particularly for large wounds and burns. Commercially available alginate products sold for wound dressings applications include Restore® from Hollister Wound Care LLC (Illinois, USA), Kendall Curasorb™ from Coviden (Massachusetts, USA), and AlgiSite® M from Smith & Nephew Healthcare (Hull, UK). Being insoluble in water but soluble in simple salt solution, alginate-based bandages are more easily removed compared to cellulose-based bandages and minimize wound disruption.

In other applications, the production of calcium alginate beads uses the dripping method, as shown schematically in Figure 8.11. In principle, by dropping alginate solution from a large syringe into a bath of calcium-containing solution, beads are instantly produced by forming spheres of calcium containing solution with a calcium alginate coating or a solidified matrix.

However, solid beads or spheres can be formed if the alginate solution is forced into a droplet shape prior to dropping into the calcium containing solution. Bead sizes produced can range from micron range (minimum of 100 μm diameter) to macro range (2–4 mm diameter), depending on the instruments used. Since these beads have the capability to undergo reversible gelation in aqueous solution, they are used to immobilize or encapsulate cells or to deliver specific biomolecules such as proteins or drugs for applications such as tissue engineering. Shown in Figure 8.12 are examples of islets and kidney cells encapsulated in alginate beads.

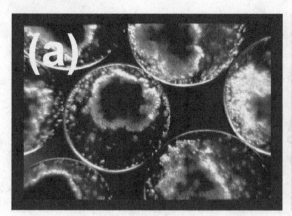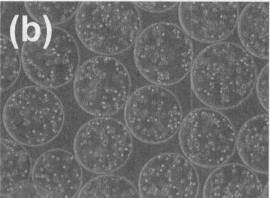

Figure 8.12

Alginate beads used for encapsulating (a) islets and (b) human embryonic kidney 293 cells.
(Reproduced with permission from NovaMatrix, Sandvika, Norway.)

If used in the development of a delivery system, it is critical to know that alginate-based materials are pH sensitive. At low pH, alginate theoretically shrinks and the encapsulated biomolecules are not released. However, rapid degradation of the alginate occurs theoretically at higher pH, thereby potentially resulting in burst release and denaturalization of the biomolecules by proteolytic enzymes. The ability to control the degradation kinetics of calcium alginate hydrogels, as a result of the loss of divalent cations releasing high and low molecular weight alginate units, is a problem. Other potential problems with this technology may include the loss of biomolecules during the processing as a result of leaching through the pores in the beads or hydrogels. Modifications of the beads through coating with other biopolymers and alginates have been made to try to overcome these limitations.

8.7 Hyaluronan

Hyaluronan, also known as hyaluronic acid, is a linear, hydrophilic, polyanionic polysaccharide. It is found naturally as non-sulfated, high molecular weight glycosaminoglycan in extracellular matrix, connective tissues, epithelial tissues, neural tissues, synovial fluid, vitreous humor, and umbilical cord. Synthesized in the plasma membrane, the polymer chain is made up of alternating disaccharide units of α-1,4-D-glucuronic acid and β-1,3-N-acetyl-D-glucosamine linked via alternative β-1,4 and β-1,3 glycosidic bonds (Figure 8.13). In the human body, hyaluronan has many physiological roles, including tissue and matrix water

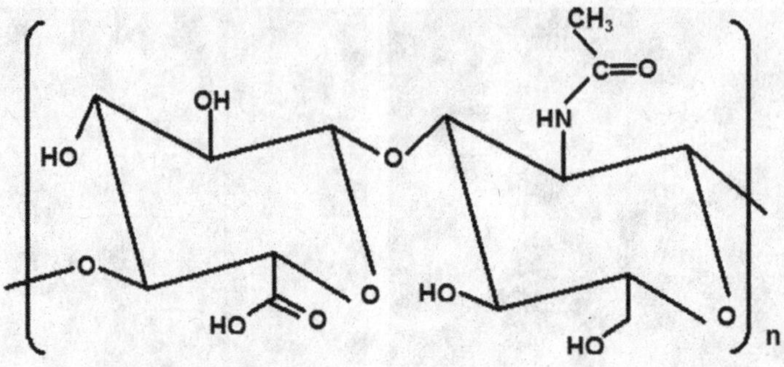

Figure 8.13

Molecular structure of hyaluronan comprising alternating disaccharide units of α-1,4-D-glucuronic acid and β-1,3-N-acetyl-D-glucosamine linked via alternative β-1,4 and β-1,3 glycosidic bonds.

regulation as in wound repair process of the skin and resistance to compression in cartilage. Present in the synovial fluid, the viscoelastic property of hyaluronan also allows it to function as a lubricant and shock absorber. In addition, the polyionic structure of hyaluronan allows it to scavenge free radicals, thus imparting an antioxidant effect which mediates inflammation.

Box 8.8

- Hyaluronan plays a major role in tissue regeneration, embryonic development, wound healing, and extracellular matrix hemostasis.
- Naturally derived, non-immunogenic, and possessing multiple sites for modifications, hyaluronan has been used widely in various applications such as orthopedic, cardiovascular, and dermatology.

There are several different sources for hyaluronan. Naturally extracted hyaluronan from umbilical cord, rooster comb, synovial fluid, or vitreous humor, is polydisperse in size. The hyaluronan in physiological solution will take up a stiffened helical or random coil configuration. The ability to swell in the presence of water allows hyaluronan to play several roles, including space filler, lubricant, and osmotic buffer. In addition to extraction from tissue, large amounts of hyaluronan can also be industrially produced through microbial fermentation from strains of bacteria such as *Streptococci*. In an attempt to maintain biocompatibility and biological activity, properties of such hyaluronan can be engineered through the use of derivatization and cross-linking approaches, thereby chemically modifying the hydroxyl, carboxyl, and acetamido groups of the hyaluronan.

Box 8.9

- Hyaluronan solution is highly viscoelastic.
- As a polymer, hyaluronan is highly hydrophilic.
- In the presence of water, hyaluronan molecules swell in volume.

Methods used for derivatization of the hyaluronan include esterification, carbodiimide-mediated modification, and sulfation. Hyaluronan derived from esterification are more hydrophobic, rigid and less susceptible to enzymatic degradation. Through an alkylation step with an alkyl halide, the available carboxyl group in hyaluronan can undergo different degrees of modifications. As the degree of esterification increases, the derivatized hyaluronan becomes more rigid, more hydrophobic, and less degradable by enzyme. Hyaluronan derivatized using carbodiimide-mediated modification permits covalent binding of the carboxyl group with an amine of another. Depending on the pH condition, carbodiimide reactions are sensitive and often result in the formation of acylurea, an unreactive intermediate from the carboxyl group. Additionally, yields of derivatized hyaluronan are also affected by the reaction conditions of carbodiimide-mediated modification. Hyaluronan derived from sulfation can occur by reacting the hydroxyl group with sulfur trioxide pyridine, resulting in the incorporation of one to four sulfur groups per dissacharide.

The cross-linking method to engineer hyaluronan's properties involves either a one-step procedure of exposing the hyaluronan to a cross-linker, or a two-step procedure of synthesizing a highly reactive hyaluronan derivative followed subsequently by cross-linking. The five common cross-linking approaches are as follows.

- **Diepoxy cross-linking**. Using diepoxy cross-linking compounds such as diepoxyoctane, ester linkages between carboxyl groups are formed at low pH, whereas ether linkages are formed between hydroxyl groups at high pH.
- **Carbodiimide-mediated cross-linking**. Carbodiimides are used to mediate reactions of the carboxyl groups on hyaluronan with amines, thereby inducing inter- and intramolecular cross-links.
- **Aldehyde cross-linking**. Examples of aldehyde cross-linkers include formaldehyde and glutaraldehyde, both of which have been popularly used to preserve tissues. Cross-linking of hyaluronan with formaldehyde results in a more viscous, elastic, and water soluble hyaluronan, whereas cross-linking with glutaraldehyde results in a more degradation resistant hyaluronan.
- **Divinyl sulfone cross-linking**. Cross-linking with divinyl sulfone results in an insoluble hyaluronan hydrogel. Depending on the reaction condition, the resulting hyaluronan hydrogel can range from soft gels to solids.

- **Photocross-linking**. Unlike the above cross-linking methods whereby cross-linking occurs when hyaluronan is mixed with the cross-linking compounds, photocross-linking of hyaluronan occurs when exposed to the appropriate wavelength of light. Occurring under physiological conditions, such a cross-linking approach allows for copolymerization of bioactive molecules, cell encapsulation, as well as patterning of the hyaluronan surfaces.

The derivatized and cross-linked hyaluronan are widely studied for tissue engineering and drug and gene delivery in dermal, orthopedic, cardiovascular, and ophthalmology applications. For orthopedic applications, hyaluronan products are used as lubricant and mechanical support for osteoarthritis patients. Commercially available products for orthopedic applications include Hyalgan® (Fidia Farmaceutici S.p.A., Padua, Italy), Artz® (Seikagaku Corporation, Tokyo, Japan), Orthovisc® (DePuy Mitek, Inc, Massachusetts, USA), and Synvisc-One® (Genzyme Biosurgery, New Jersey, USA). Commercially available hyaluronan products such as Bionect® (Fidia Farmaceutici S.p.A., Padua, Italy), and Jossalind® (Hexal AG, Holzkirchen, Germany) are also used as a viscoelastic gel or viscous cream for surgery and dressing to manage partial to full thickness dermal ulcers, wounds, and burns, whereas Healon® (Pharmacia & Upjohn Company, Michigan, USA), Opegan® (Seikagaku Corporation, Tokyo, Japan), and Opelead® (Shiseido Company Limited, Tokyo, Japan) are used as ophthalmic surgical aid in cataract extraction, intraocular lens (IOL) implantation, corneal transplant surgery, glaucoma filtration surgery, and retinal attachment surgery.

8.8 Chondroitin sulfate

Unlike hyaluronan, chondroitin sulfate, an unbranched polysaccharide, is a sulfated glycosaminoglycan. It is a major component of the extracellular matrix and is important for maintaining the structural integrity of tissue such as cartilage. A chondroitin chain consists of more than 100 individual sugars made of alternating disaccharides, N-acetyl-D-galactosamine and D-glucuronic acids. Mediated by specific sulfotransferases, sulfation of the N-acetyl-D-galactosamine occurs in most instances at the hydroxyl groups in the 4 and 6 positions of the monosaccharide. For example, chondroitin 4-sulfate or chondroitin sulfate type A is made up of alternating monosaccharides, N-acetyl-D-galactosamine and D-glucuronic acids, with sulfation occurring in position 4 of the N-acetyl-D-galactosamine, whereas sulfation of chondroitin 6-sulfate or chondroitin sulfate type C occurs in position 6 of the N-acetyl-D-galactosamine (Figure 8.14). Disaccharides with a different number and position of sulfate groups can be located in different percentages within the polysaccharide

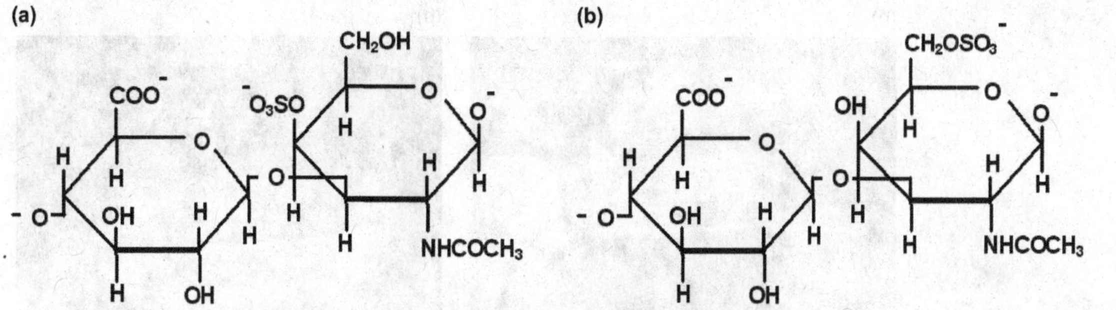

(a)   **(b)**

Figure 8.14

Typical disaccharide units found in chondroitin sulfate: (a) = chondroitin 4-sulfate or chondroitin sulfate type A; (b) = chondroitin 6-sulfate or chondroitin sulfate type C.

chains, and these parameters confer specific biological activities to the chondroitin chains. These chondroitin chains are usually found attached or linked to the hydroxyl groups of serine residues of proteins as part of a proteoglycan.

Chondroitin sulfate is produced from enzymatic digestion of bovine and marine animal tissues, specifically bovine nasal septum and trachea, and shark cartilage. Properties of chondroitin sulfate include:

- ability to bind and modulate growth factors and cytokines,
- inhibition of proteases,
- involvement in cell adhesion, migration, proliferation, and differentiation,
- non-immunogenic, and
- degrade to non-toxic oligosaccharides.

Since chondroitin sulfate is water soluble, tailoring of its properties is possible by cross-linking or combining it with other polymers such as chitosan, gelatin, hyaluronan, collagen, or poly-(lactic-co-glycolic acid). Commercially marketed as a nutritional supplement for osteoarthritic patients (Figure 8.15), chondroitin sulfate is also an attractive polymer for tissue engineering applications. As a delivery system, the negatively charged chondroitin sulfate can be used to interact with positively charged molecules such as polymers or growth factors. For example, the Integra® Dermal Regeneration Template (Integra™, New Jersey, USA) is a commercially available tissue-engineering product that utilizes chondroitin sulfate with porous matrix fibers of cross-linked collagen for the treatment of burns and reconstructive surgery. Other commercial products such as Viscoat® Ophthalmic Viscoelastic Device (Alcon Laboratories Inc., Texas, USA) is a chondroitin sulfate–sodium hyaluronate dispersive viscoelastic material and is used as a surgical aid to protect the corneal endothelium during anterior segment procedures such as cataract extraction and intraocular lens implantation.

(a)

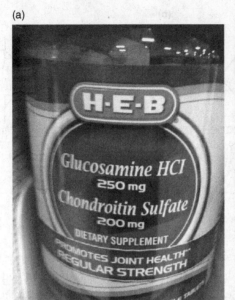

(b)

(c)

Figure 8.15

Bottles of different chondroitin compounds marketed as nutritional supplements.

8.9 Coral

Natural corals have been investigated as a bone substitute experimentally for more than 20 years. They have been clinically used for more than 10 years as a filling for defects or an onlay graft in orthopedics and traumatology, in craniofacial, dental and neurosurgery, and in spinal fusion. These corals belong to the subclass Hexacorallia of the Anthozoa class. Within this subclass are the coral reef builders which include the order Scleractinia (also called stony corals) and the sea anemones. For medical applications, the coral graft substitutes are naturally derived from the exoskeleton of stony corals belonging to the genus *Madrepora*, *Porites*, *Acropora*, *Lobophyllia*, *Goniopora*, *Polyphyllia* and *Pocillopora*, with many of these corals coming from the Caribbean Sea, New Caledonia, the Red Sea, the East Coast of Africa, the Thailand Sea, the coast of Hainan island, and the coastline of Australia.

Box 8.10

- Natural coral is biocompatible and osteoconductive.
- The mechanical properties of coral are comparable to those of cancellous bone.

The head of the coral is formed by individual polyps, with each polyp being only a few millimeters in diameter. This colony of polyps functions as a single organism, sharing nutrients via a well-developed gastrovascular network. The outer layer of coral is made up of calicoblasts that secrete the mineral phase made principally of aragonite, a naturally occurring crystalline form of calcium carbonate ($CaCO_3$). Impurities present in the corals include magnesium (0.05%–0.2%), potassium (0.02%–0.03%), sodium (0.4%–0.5%), amino acids (0.07%), and trace amounts of strontium, fluorine, and phosphorus. The presence of these impurities is important in aiding in the calcification process. Once calcified, the secreted $CaCO_3$ forms the coral scaffold with an architecture characteristic to each species.

- Pores in coral species typically range between 100 μm and 800 μm.
- Coral resorption and bone regeneration rates are dependent on the size and interconnectivity of the coral pores.

As shown in Figure 8.16, the porous structure found in corals offers substantial surface exchange area, with the size and interconnectivity of the pores playing a critical role in the corals' resorption rate. Used as a bone substitute graft material, these parameters are also crucial in determining vascularization and successful bone regeneration. Corals from the genus *Goniopora* and *Porites* resemble trabecular or spongy bone and have an average pore size of 500 μm and 150 μm, respectively and an average open volumetric porosity of 80% and 50%, respectively. Similarly, the pores of corals derived from genus *Acropora* are open and interconnected, and these corals possess an average pore size and an

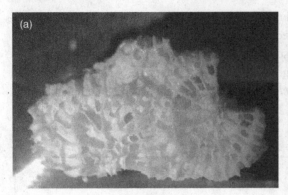 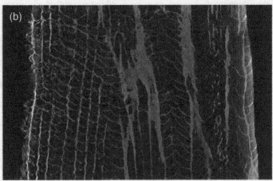

Figure 8.16

(a) Stereoscope image showing the porous interconnected structure of the coral (*Goniopora*).
(b) Micro-CT cross-sectional images of the coral showing the 3D interconnected organization of axially aligned columns and radial cells.

average volumetric porosity of 200 μm and 20%–30%, respectively. Although corals derived from the genus *Polyphyllia* have an average volumetric porosity of 60% porosity, the pores are not completely interconnected. This incomplete interconnectively lowers the resorption rate and they thus have slower bone formation when used for medical applications. This difference in resorption rate allows corals derived from the genus *Polyphyllia* to provide greater volumetric stability, and they are used as anchors for bone.

Untreated, these corals are unstable when exposed to physiological fluids. In general, a hydrothermal exchange process is used to improve coral stability. In this process, the carbonates found within the calcium carbonate coral are substituted with phosphates under high temperature (250–260 °C) and pressure (15 000 psi) for 24–48 hours. The end-product of the hydrothermal exchange process yields coralline heterogeneous apatites that have core carbonated hydroxyapatites on inner $CaCO_3$ struts. Although such a process retains the macroscopic interconnective porosity, biocompatibility, and bioactivity of the coral, the resulting heterogeneity of the coralline apatites produces inconsistent degradation rates and reactivity levels in biological environments. In addition, the coralline apatite is also brittle. As such, coralline apatite lacks the adequate functional mechanical integrity needed for many of the load bearing applications. Other improvements in the process and in coral properties include a series of cleansing steps to remove impurities from the corals prior to the hydrothermal conversion and the varying of phosphate concentrations or phosphate solutions during the hydrothermal conversion. Current commercially available coral bone graft substitute products include Pro Osteon® from Interpore Cross International, Inc. (California, USA) and Biocoral® from Biocoral France (La Garenne Colombes, France).

8.10   Summary

This chapter provides a summary of the different natural materials that have applications in medicine. In general, natural polymers can be divided into protein-based or polysaccharide-based materials. These natural polymers perform diverse functions in their natural environments such as intracellular communications, providing structure, storage, and acting as catalysts. In addition to the natural polymers, other natural materials include corals that have been harvested and processed for orthopedic, spinal, and craniofacial applications. Although there are drawbacks in the use of natural materials as biomaterials such as batch-to-batch variability, low mechanical properties, and limited processability, natural materials are still very attractive candidates for biomedical applications because of their biocompatibility, degradability, low cost, availability, similarity with the extracellular matrix, and intrinsic cellular interaction.

References

1. Buehler, M. J. (2006). Nature design tough collagen: explaining the nanostructure of collagen fibrils. *Proc. Nat. Acad. Sci.*, **103**, 12 285–12 290.
2. Megeed, Z., Cappello, J. and Ghandehari, H. (2002). Genetically engineered silk-elastinlike protein polymers for controlled drug delivery. *Adv. Drug Deliv. Rev.*, **54**, 1075–1092.
3. Freier, T., Koh, H. S., Kazazian, K. and Shoichet, M. S. (2005). Controlling cell adhesion and degradation of chitosan films by N-acetylation. *Biomater.*, **26**, 5872–5878.
4. Hidaka, Y., Ito, M., Mori, K., Yagasaki, H. and Kafrawy, A. H. (1999). Histopathological and immunohistochemical studies of membranes of deacetylated chitin derivatives implanted over rat calvaria. *J. Biomed. Mater. Res.*, **46**, 418–423.
5. Chesnutt, B. M., Yuan, Y., Brahmandam, N. *et al.* (2007). Characterization of biomimetic calcium phosphate on phosphorylated chitosan films. *J. Biomed. Mater. Res.*, **82A**, 343–353.

Suggested reading

• Khor, E. (2001). *Chitin: Fulfilling a Biomaterials Promise.* Elsevier Science, ISBN 9780080440187.

Problems

1. List the different criteria that natural materials need to satisfy in order for them to be used in biomedical applications.
2. What is the difference between protein-based and polysaccharide-based materials?
3. In general, what is the molecular difference between collagen and elastin?
4. Describe the make up of tropocollagen.
5. What is the general function of elastin in the body?
6. Define inverse transition temperature.
7. What is the composition of silk?
8. What makes up fibroin?
9. Which region of the fibroin is responsible for the high strength and toughness of the silk fiber?
10. How is chitosan formed?
11. Define degree of deacetylation in chitosan.
12. What is the effect of deacetylation on the properties of chitosan?
13. What is cellulose made of?
14. How is cellulose different from starch?
15. Although glucose is water soluble, why is cellulose insoluble in water?
16. What are the four general crystalline polymorphs of cellulose?
17. What is alginate made of?
18. How are the properties of alginate varied?
19. What is the effect of the different cations on alginate gel formation?
20. What is the composition of hyaluronan?
21. What is the function of hyaluronan in the body?
22. List the five common cross-linking approaches of hyaluronan.
23. Name the two typical disaccharide units found in chondroitin sulfate.
24. What factors regulate coral resorption and bone regeneration rates?
25. Why do corals need to undergo a hydrothermal exchange process?

9 Surface modification

Goals

After reading this chapter, students will understand the following.

- Main categories of the different surface modification techniques commonly used to engineer biomaterial surfaces.
- Principles underlying these surface modification techniques.
- Advantages and limitations of each technique.

Despite the advent of advanced manufacturing tools and newly developed materials with unique properties, why is the attention of biomedical scientists and engineers focused on biomaterial surfaces? Unlike the material's bulk, which generally governs the mechanical integrity of medical devices, the material's surface governs tissue–biomaterial interactions and these usually occur within a narrow depth of less than 1 nm on the material's surface. Surface modification of biomaterials allows the tailoring of surface properties without affecting bulk material properties. These altered surface properties influence tissue–biomaterial interactions, which ultimately determine the success or failure of a device placed in the human body. In addition to improving tissue–biomaterial interactions, modification of biomaterial surfaces may also be performed for the purpose of improving surface mechanical properties such as wear resistance. This can be achieved through surface or subsurface alloying, or heat treatment. Enhancement of surface oxide thickness or the presence of a dense protective coating on a metallic surface can provide surface barriers which minimize the metal's chemical reactions with its surroundings, thereby improving its corrosion resistance. Thus, through surface modification, the native surfaces of biomaterials can be physically or chemically transformed with the primary goal of engineering desired surface chemistry, topography, reactivity, biocompatibility, hydrophilicity, or charge.

- Surface modification is employed to modify surface properties, with minimal effect on bulk properties.
- The parameters used in the surface modification processes significantly impact how surface properties are altered.

Since surface modification for medical implants is primarily used to improve or optimize the properties of a biomaterial, the various modification techniques should generally be simple, robust, and cost effective so as to promote their inclusion in the manufacturing process. These techniques may include passivation, anodization, etching, deposition of coatings, and sterilization. Although there are many different approaches to engineering biomaterial surfaces, most surface modification techniques can generally be classified into two categories, namely

- subtractive modifications, and
- additive modifications.

Subtractive surface modifications refer to a change in the surface as a result of removal of the surface layer. Such modifications are also known as physical modifications. Means to achieve subtractive surface modification include abrasive blasting and plasma treatments. In contrast, additive surface modifications refer to a change in the surface as a result of depositing films or biological factors on the biomaterial surface. Also known as over-coating or chemical modifications, common methods for achieving additive surface modifications include physical and chemical vapor deposition, grafting, attachment of self-assembly molecules, and layer-by-layer deposition.

With any modification process, it is important to note that the parameters used for altering biomaterial surfaces, such as gas used, voltage, current density, environmental pressure, and temperature, play important roles in governing the chemical, structural, and mechanical properties of the surface and near-surface. Surface properties altered could include chemistry, grain size, crystallographic orientation, defects, stress concentration, film adhesion, surface topography, and biology–biomaterials interactions. Discussed in the following sections of this chapter are principles governing the different surface modification techniques used in the research and manufacture of medical devices.

9.1   Abrasive blasting

Abrasive blasting techniques are commonly used for preparing material surfaces for further treatment, cleaning or finishing of a surface, and shot peening, which is

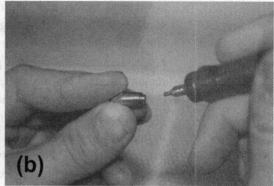

Figure 9.1

Photographs of an abrasive blasting system, showing (a) a self-contained chamber where blasting is performed, and (b) a close-up view of an abrasive blasting nozzle (left) and a sample (right) being modified.

a process used to increase strength and durability by replacing tensile surface stresses with compressive stresses. In the case of medical devices, abrasive blasting is performed to remove surface contaminants, roughen implant surfaces to increase the surface area of the implants, enhance cellular–biomaterials interactions, and/or improve the adhesion of coatings on implant surfaces. Shown in Figure 9.1 is the set up for performing abrasive blasting. There are several types of abrasive blasting techniques, including hydro-blasting and bead blasting. The most common abrasive blasting technique used for the modification of medical devices is micro-abrasive blasting (also known as pencil blasting). The fundamental components of a blast system include:

- blast media,
- accessories in the blasting equipment such as nozzles, and
- air compressor.

- Surface irregularities created as a result of abrasive blasting provide a mechanical interlocking mechanism for coatings to adhere strongly to the substrate surfaces.
- A change in surface area after abrasive blasting provides a greater surface area for cell attachment.

The blasting process involves the use of an air compressor to direct a fine stream of abrasive media under high pressure and dry conditions to the biomaterials' surface. The parameters determining the flow conditions of the abrasive medium are critical factors that control the removal of the substrate layer. With most abrasive media particles ranging from 10 μm to 150 μm in size, the media commonly used for modifying medical devices include sodium bicarbonate, glass beads, alumina, and calcium phosphates. The nozzles used are typically small, ranging from 0.25 mm to 1.5 mm in diameter. Additionally, the amount of pressure necessary to propel the abrasive media depends on the pipe fittings within the blasting equipment. A volume of air through a small diameter fitting can result in high pressure air flow. However, the flow pressure will not be the same if the same volume of air flows through a large fitting. The air compressor is primarily responsible for providing the required pressure for optimal blasting. In general, the different surface alterations of biomaterials achieved by abrasive blasting are dependent on several parameters such as:

- type of abrasive media, since each medium type possesses different intrinsic properties,
- size and shape of abrasive media,
- substrate type, since blasting of a softer substrate such as titanium results in a rougher surface compared to blasting of the harder cobalt chrome surfaces,
- gas or air pressure,
- type of nozzle used, and
- nozzle to substrate distance.

Abrasive blasting of surfaces of medical devices is common, especially in the arena of stents, and dental and orthopedic implants. Roughening of implant surfaces with abrasive blasting provides surface irregularity for improved coating–substrate interlocking and also greater surface area for enhanced tissue–biomaterial interactions. Additionally, since calcium phosphates such as hydroxyapatite, tricalcium phosphate, or mixtures of different calcium phosphates, are often biocompatible and may possess osteoconductive properties, particles from this family of materials are used to clean surfaces of orthopedic or dental implants; any particles left behind embedded in the implant surfaces may offer beneficial properties. Dental implants with embedded calcium phosphate particles are often referred to as resorbable blasted media (RBM) implants. An example of an RBM dental implant is shown in Figure 9.2.

Although abrasive blasting can be utilized as a stand-alone approach to modify and engineer biomaterial surfaces, this modification technique is often used in combination with other types of surface modifications. For instance, with orthopedic implants such as total hip prostheses, surfaces are often roughened to optimize the adhesion of plasma-sprayed hydroxyapatite coatings. Crevices and pits are created on the implant surface as a result of abrasive blasting, thereby providing a means to mechanically interlock the plasma-sprayed coatings to the substrate. Additionally, the combined use of abrasive blasting with other

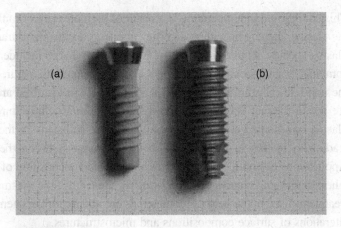

Figure 9.2

Dental implant surfaces roughened with an abrasive blasting process. (a) An SLA implant, that is, exposure of implant surface to acid etching after abrasive blasting. (b) An RBM implant, that is, implant surface that is abrasive blasted with calcium phosphate particles. (Courtesy of Dr. Weihui Chen, Union Hospital, Fujian University of Medical Sciences, P.R. China.)

modification techniques such as acid etching, have allowed for the production of Sand-blasted, Large-grit, Acid-etched (SLA) dental implants. Problems associated with abrasive blasting include the embedment of abrasive media in the substrates. This embedment of abrasive media can introduce stress concentration points on the surface of the substrates resulting in weakening of the mechanical interlocking between coatings and the substrates. Removal of these embedded media through post treatments is typically performed to enhance the coating–metal adhesion.

Box 9.1

- In addition to changes in surface area, surface roughness, and surface topography, abrasive blasting also affects the wettability of the surface.
- The adhesion properties of coatings to abrasive-blasted surfaces are affected by post treatments, including steam treatment and ultrasonic cleaning.

9.2 Plasma glow discharge treatments

There are two different types of gas discharge plasma, namely the local thermal equilibrium (LTE), and the non-LTE. In LTE plasma, all plasma species in localized areas are in thermal equilibrium, that is, they exhibit similar temperatures.

This equilibrium in temperature is a result of high pressure or multiple collisions in the plasma, thereby providing an efficient energy exchange or transfer between the plasma species. Applications for LTE plasma discharge include cutting, welding, spraying, or analytical analysis where high temperature is required. In contrast, for the non-LTE plasma, the different plasma species in localized areas of the plasma are not in thermal equilibrium and exhibit dissimilar temperatures. The non-LTE plasma is achieved at low gas pressure or few collisions in the plasma, thereby leading to inefficient energy transfer or exchange between the plasma species. Applications of non-LTE plasma discharge include deposition of layers or etching where high temperatures are not required. Generally, the plasma glow discharge treatments used in the field of biomaterials are mainly for the chemical and physical alterations of surface compositions and microstructures.

Box 9.2

- Glow discharge plasma treatment is frequently used for cleaning, preparation, and modification of biomaterial and implant surfaces.
- Effect of glow discharge plasma treatment on surfaces is strongly dependent on the process parameters.

A simple glow discharge chamber or cell can be made by inserting a cathode and an anode in the cell or by having the walls of the cell act as these electrodes. Typically, the material to be modified acts as the cathode. An inert or a reactive gas at pressures ranging from a few mTorr to atmospheric pressure is then used to fill up the cell or chamber. Ionization of atoms occurs when a potential difference is applied between the two electrodes, resulting in gas breakdown and yielding free electrons and positively charged ions. The positively charged ions move towards and bombard the cathode, thereby resulting in the release of electrons, called secondary electron emission. In contrast, electrons arrive in the plasma and collide with gas particles, thereby resulting in ionizing and exciting the gas particles. In the absence of an outside excitation source, the gas molecules in a glow discharge chamber are electrically neutral.

- In a simple glow discharge, the material to be modified is the cathode.
- A reactive or inert gas is used to create the plasma.

In the application of an outside excitation source between the two electrodes, such as an electric current or via intermolecular interactions or collisions with one another at high temperatures, plasma is created when the atoms or gas molecules gain enough energy. Ionization occurs, resulting in the physical removal of electrons from an atom or molecule in the gaseous state at low pressure and these free electrons are accelerated toward the anode. As a result of the electron's low mass (9.11×10^{-31} kg), these free electrons rapidly achieve high velocity or kinetic energy. Since kinetic energy can be related to temperature, the electrons are also said to have achieved extremely high temperatures rapidly in a cell filled with reactive or inert gas molecules. As a result of the high kinetic energy, these electrons are "hot," and the high electron temperature results in the need for inelastic electron collisions in order to sustain the plasma.

Gas molecules such as argon have high mass. Having an atomic weight of 39.948 amu, argon moves slowly towards the cathode. These slow-moving gas molecules are also termed "cold" gas molecules. Very little of the electron's energy is depleted during elastic collisions between the electron and gas molecules. As a result of the great mass difference between the gas molecules and electrons, gas molecules are not significantly influenced by elastic collisions. However, during inelastic collisions of the electrons and the gas molecules, two primary events can occur in the cell or chamber.

- Ionization: this results in the production of positive ions and electrons, and
- excitation or elevation of energy level of the gas molecules: the excitation state is usually short-lived and the excited molecules returned to the state with the lower energy level as a result of decay or relaxation. The excitation–decay process of the gas molecules is thus responsible for characteristic light of the glow discharge.

9.2.1   Direct current glow discharge

When the outside excitation source is a direct current (d.c.), the plasma is created when a potential difference of a few hundred volts to a few kilovolts is applied between the two electrodes. As shown schematically in Figure 9.3, the application of the potential difference results in the emission and the acceleration of electrons from the cathode towards the anode.

On the way to the anode, the accelerated electrons collide with the gas atoms or molecules. Through these collisions, a plasma is formed when electrons are forced out of the neutral gas molecules, thereby resulting in the production of ion pairs consisting of the resultant positive ion and detached negative electron. Some electrons are attached to the neutral gas molecules forming negative ions. As such, the plasma formed will consist of equal concentrations of positive and negative ions as well as a large number of neutral gas molecules. Additionally,

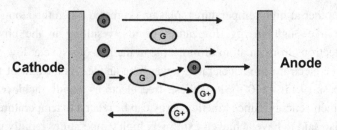

Figure 9.3

Schematic showing the basic ionization process in a d.c. glow discharge: e = electrons; G = gas molecules, G+ = positive ions. Arrows indicate direction of flow of ions.

the plasma created is self-sustaining. This is because the positively charged ions created by the ionization collisions accelerate toward the cathode or the negatively biased sample surface. Also known as cathodic sputtering, these positively charged ions strike the sample surface with sufficient energy, disrupting atomic bonds, ejecting atoms and releasing secondary electrons. These secondary electrons are accelerated away from the cathode, resulting in more ionization collisions. As a result of the various collision processes, the plasma can contain several kinds of radicals and excited species, in addition to position and negative ions, electrons, and neutral gas molecules or atoms. The interactions of these different species thus make the glow discharge plasma a complicated gas mixture.

In a d.c. glow discharge, the application of a continuous potential difference between the cathode and the anode results in a constant current. Problems with the d.c. glow discharge can include the following.

- The inability to use a non-conducting electrode in the chamber or cell. Since no current will flow through a non-conductive material, the use of a non-conductive electrode or having non-conductive samples as one of the electrode leads to a short breakdown, followed by the creation of a surface charge when d.c. voltage is applied.
- Buildup of the electrons due to the constant current, thereby resulting in the burn-out of the glow discharge.

9.2.2 Alternating current glow discharge

To overcome the problems of the d.c. glow discharge, magnetic fields and alternating current (a.c.) are some of the commonly used excitation sources for the ionization process. The application of magnetic fields, as in the magnetron glow discharge, allows electrons to circulate in helices around the magnetic field lines, thereby increasing ionization. The application of a.c. causes a time-varying potential difference, thereby resulting in short consecutive discharges. During these short discharges, the two electrodes alternatively play the role of the cathode and the anode, with electrons oscillating in the plasma between the electrodes. Variations of the a.c. glow discharge can include

- the dielectric barrier glow discharge, wherein the electrodes are covered by a dielectric barrier,
- the pulsed glow discharge, where the glow discharge is short, followed by a longer period of afterglow, and
- the radiofrequency (RF) glow discharge.

Since materials used for medical devices can be non-conductive or conductive, the RF glow discharge is commonly used as one of methods for modifying biomaterials surfaces, with typical frequencies of a.c. voltages applied in the RF range of 1 kHz to 10^3 MHz. Clean, surface-activated surfaces, with thinner and more stable oxide films are produced on metallic biomaterials surfaces with RF glow discharge. There are two general types of RF glow discharge, namely the capacitively coupled RF glow discharge and the inductively coupled RF glow discharge.

9.2.3 Capacitively coupled radiofrequency glow discharge

In the capacitively coupled RF glow discharge, the cathode is typically connected to a RF power supply. The electrodes and their sheaths form a capacitor (Figure 9.4) in a vacuum chamber. An inert or a reactive gas at pressures ranging from 10 mTorr to 100 mTorr is then used to fill up the cell or chamber. With the frequency of a.c. voltage applied at about 13.56 Hz to 15.56 MHz between the two electrodes, the charge is allowed to accumulate during one half of the cycle to be

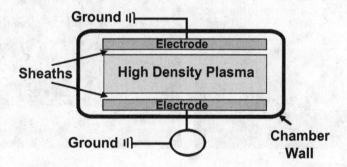

Schematic showing configuration of a capacitively coupled radiofrequency glow discharge, with one of the electrodes (cathode) connected to the radiofrequency power supply.

neutralized by the opposite charge accumulated during the next half cycle.[1] In this configuration, the energy from the electric field is transferred to both electrons and ions in the chamber. Depending on the gas pressure, the high plasma density produced can vary from 10^9 ion/cm^3 to 10^{11} ion/cm^3. Additionally, heat is generated across the sheath as well as by collision of the ions and electrons in the plasma. These ions and electrons behave differently due to differences in masses and are thus not at thermal equilibrium. Although capacitively coupled RF glow discharge efficiently converts power from the RF supply into the plasma, a major drawback with the system is its inability to independently control the ion density and ion energy.

9.2.4 Inductively coupled radiofrequency glow discharge

The inductively coupled RF glow discharge is an electrodeless discharge. Figure 9.5 shows a set up of the RF glow discharge, which is basically a compact electrodeless device.

The chamber of the inductively coupled RF glow discharge is surrounded by the coil thus allowing the RF power to be inductively coupled to the plasma. Application of the time-varying current or RF current induces a time-varying magnetic field or magnetic flux to penetrate the plasma region. This flux thus induces a RF electric field which then accelerates the free electrons and sustains the plasma. A difference between the inductively coupled RF glow discharge and the capacitively coupled RF glow discharge is that a high-energy ion bombardment

Figure 9.5

Photographs showing (a) the set up of a radiofrequency plasma glow discharge system with dual gas feeds connected to a gas flow mixer and a vacuum pump, and (b) a close-up view of the plasma chamber showing the viewing window and RF coils inside the chamber.

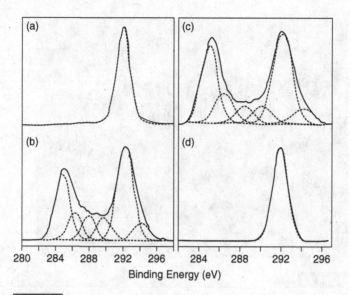

Figure 9.6

High-resolution X-ray photoelectron (XPS) showing C_{1s} spectra of polytetrafluoroethylene that is (a) untreated, (b) exposed to an oxygen plasma treatment for 30 seconds, (c) exposed to an oxygen plasma treatment for 120 s, and (d) exposed to an oxygen plasma treatment for 900 s. Changes in morphology of the XPS indicated changes in surface chemistries.

can be achieved with capacitive coupling when compared to inductive coupling. Additionally, while low and medium plasma densities are produced with capacitive coupling, a high plasma density can be achieved with inductive coupling. An example of the effect of RF glow discharge on surface properties is shown in Figure 9.6 where the surface chemistry of polytetrafluoroethylene is altered after exposure to RF glow discharge treatment in an oxygen-filled chamber.

9.3   Thermal spraying

Of the many techniques used for modifying orthopedic and dental implant surfaces, thermal spraying is one of the common commercially accepted methods for depositing hydroxyapatite and titanium. Shown in Figure 9.7 is an example of a non-coated and a thermal-sprayed hydroxyapatite dental implant.

Although the thickness of the coatings on medical implants ranges from 30 μm to 120 μm, coatings as thick as a few millimeters have been produced for other applications. The thermal spraying process is a line-of-sight process and is generally associated with flame spraying, plasma spraying and high velocity oxy fuel (HVOF) processes. The line-of-sight effect from the thermal spraying is

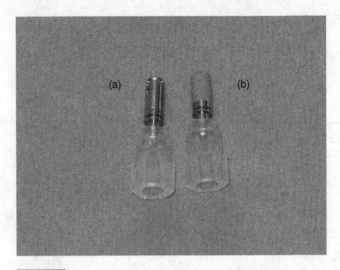

Figure 9.7

Dental implants with vial cap as holders to aid clinicians place implants into surgical site without contaminating the surfaces. These implants have surfaces that are (a) non-treated and (b) plasma-sprayed with hydroxyapatite.

shown in Figure 9.8, indicating non-uniformity or lack of coatings in areas that are not in line with the thermal spray.

Box 9.3

- Thermal spraying process uses a gas stream to carry powders, which are then passed through a plasma.
- Advantages of coatings produced by thermal spraying are the rapid deposition rate and a low cost of production.

All thermal spraying processes have the same fundamental principle. As shown in the schematic in Figure 9.9, thermal spraying requires the presence of the following:

- an electric heat source as in plasma spraying, or a chemical heat source as in flame spraying to melt the powder particles;
- a media supply, that is gaseous or liquid to generate the flame or plasma jet. The gas used can be nitrogen, argon, helium, hydrogen, or combinations of these gases;
- a feeder or mechanism to introduce the powder particles to the heat source.

Plasma spraying can be performed in air (air plasma spraying), in an enclosed chamber filled with inert gas (controlled atmosphere plasma spraying), or in vacuum

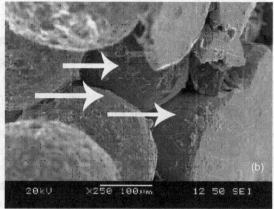

Figure 9.8

Scanning electron micrographs of cross-sectioned plasma-sprayed hydroxyapatite coatings on titanium beads. (a) Micrograph shows the surface of the beads coated with hydroxyapatite coatings at × 100. (b) Micrograph of (a) at × 250 showing beads below the surface region exhibiting non-uniform coatings as well as non-coated areas (arrows) of the beads that are not in line with the spray.

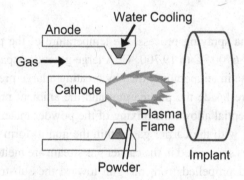

Figure 9.9

Schematic showing the setup of a plasma spraying process showing an electric heat source, gas supplied to generate the plasma flame, and a powder feeder. Powder is fed into the plasma flame through a feed and is propelled to the implant surface as semi-molten particles.

(vacuum plasma spraying). Properties of the coatings can vary, depending on the spray environment. For example, coatings with a lower percent porosity and impurities can be obtained using vacuum plasma spraying compared to air plasma spraying.

Aside from the spraying environment, the plasma jets also play an important role. There are generally two types of plasma jets that are used for the plasma spraying process.

- *Direct current (d.c.) plasma*. The plasma gun used for generating d.c. plasma consists of a water-cooled tungsten cathode and a water-cooled copper anode. The plasma gas flows around the cathode and through the anode which is shaped like a constricting nozzle. Local ionization occurs when a high voltage is applied, resulting in an electric arc between the two electrodes. Resistance heating occurs as a result of the electric arc, thereby enabling the gas to reach extremely temperatures. These high temperatures cause the dissociation and ionization of the gas leading to the formation of a d.c. plasma.

- *Induction plasma or radiofrequency plasma.* An induction plasma torch used for generating induction plasma or radiofrequency plasma consists of an induction coil, a confinement tube which serves to contain the plasma, and a gas distributor or a torch head. Depending on the alternating, radiofrequency power, the induction coil is made of several spiral turns, wrapping the confinement tube. Cooling of the coil by water is a requirement as a result of high coil current. The confinement tube is commonly made of quartz, either water-cooled or air-cooled, and is open on one end for continuous gas flow. There are three gas lines in the torch head, with one line for the carrier gas, a line for the plasma forming gas, and one line for the gases that will help stabilize the plasma discharge as well as to act as a cooling medium to protect the confinement tube. Upon the application of an alternating, radiofrequency current to the torch coil, an alternating magnetic field or magnetic flux inside the coil is induced. This flux thus induces a time-varying or a radiofrequency electric field which then accelerates the free electrons and sustains the plasma.

During the plasma spraying process, the temperature of the plasma can range from approximately 6700 °C to 19 700 °C. A large gaseous expansion occurs as a result of this increase in temperature, thereby creating a large pressure differential between the pressure inside the plasma gun and the ambient pressure outside it. This pressure differential allows the mixing of the powder materials and for them to be sprayed along with the carrier gas within the gun to form a plasma stream. The powder particles suspended in the carrier gas stream are melted as they are fed into the plasma and propelled from the gun toward the substrates at velocities ranging from 200 m/s to 400 m/s. Upon reaching the substrate surface, the molten droplets rapidly solidify and form an adherent coating. Properties of the coatings depend on several factors including

- the distance between plasma nozzle and substrates,
- powder size,
- speed and temperature of the plasma,
- rate of cooling,
- gas used and flow rate,
- energy input, and
- spray time.

Depending on the gun distance to the substrate material (typically ranging from 25 mm to 150 mm), the substrate temperature is usually in the range of

Plasma-sprayed Particles

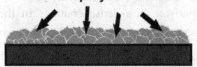

Figure 9.10

Schematic showing the piling and flattened effects of plasma-sprayed droplets or particles are they are being deposited on implant or substrate surfaces.

Figure 9.11

Representative scanning electron micrograph showing microcracks (arrows) on the surface of plasma-sprayed hydroxyapatite coatings on titanium substrate.

38 °C to 260 °C. The sprayed droplets become flattened lamellae upon depositing on the substrate surface. The thickness and lateral dimensions of the flattened droplets depend on the size of the powders. The rapid solidification of the droplet, the piling of the flattened lamellae on each other, and the impinging flattened lamellae on each other during the coating buildup can result in the formation of intralamellar or interlamellar cracks. Figure 9.10 shows a schematic of the piling and flattening effects of the droplets as the coatings are deposited on the substrate.

The substrate and the coating material may have different coefficients of thermal expansion. For example, a metallic material has a higher coefficient of thermal expansion when compared to a ceramic material. As a result of such differences between the metallic substrate and the ceramics coatings, cracks in the coatings can occur. As shown in the scanning electron micrograph of a plasma-sprayed hydroxyapatite coating on a titanium substrate (Figure 9.11), microcracks are observed on the surface of the sprayed coatings. The presence of these microcracks can weaken the overall mechanical integrity of the coating.

Percent porosities found in the coatings can vary, depending on the spraying environment. The amount of porosity in the coating critically determines its quality. For instance, a highly porous coating is more susceptible to dissolution and delamination in comparison to a dense coating. In the thermal spraying process, the formation of porosity is generally attributed to

- low velocity unmelted particles resulting in low energy deposits,
- shadowing effects as a result of either spray angle, and
- shrinkage and stress relief effects of the coatings as the result of cooling when deposited on the substrate.

The plasma heat content as well as the ability to increase the powder's temperature depends strongly on the gas used. The use of argon gas during plasma spraying generates a higher plasma temperature compared to nitrogen gas. Additionally, plasma temperature is also influenced by the amount of time the powders reside in the plasma, with extended exposure of powder particles in the plasma resulting in higher plasma temperature. Although some phase transition is unavoidable during the exposure of powder particles to the plasma, it is ideal that a thin outer layer of each powder particle gets into a molten plastic state in order to ensure dense and adhesive coatings. As the coatings are deposited on the substrate surfaces, the cooling rate of the coatings highly influences the formation of metastable phases and crystallinity, and thus affects the dissolution behavior of the coatings. High cooling rates for the powder particles can result in the production of amorphous coatings. In biomedical devices where hydroxyapatite coatings are deposited on implant surfaces, an amorphous coating generally results in a higher dissolution rate when compared to coatings that are more crystalline.

Box 9.4

- Plasma generated for thermal spraying usually incorporates one of or a mixture of argon, helium, nitrogen and hydrogen gases.
- The gas used to create the plasma governs the effective heating properties of the plasma.
- The energy for normal thermal spraying is also governed by the type of gas used.
- Helium gas can be heated to 13 000 K without forming a plasma, whereas nitrogen gas undergoes ionization and dissociation to form a plasma at 10 000 K.

In addition to the rate of cooling, plasma gas composition is an important parameter influencing the crystallinity and composition of the coatings. Argon and nitrogen are the primary plasma gases, whereas hydrogen and helium are used mainly as secondary gases. The addition of the secondary gas to the primary plasma gas is done to alter the plasma characteristics and energy level. Properties such as the coating crystallinity can be controlled using argon mixed with some hydrogen gas. The use of higher hydrogen gas concentration mixed with argon results in more plasma enthalpy and a lower plasma velocity, thus leading to more melting and decomposition of the powder particles. In the absence of hydrogen gas, the high velocity and viscosity of the argon gas cause the powder particles to bounce back from the plasma, instead of entering it. Thus, the type of gas used in plasma spraying allows for the production of either amorphous coatings, partially crystalline coatings, or crystalline coatings. In addition to the plasma spray time, the plasma gas composition also influences the thickness of the coatings. Unlike argon and helium, which are monoatomic gases, nitrogen and hydrogen are diatomic gases and thus have higher energy content at a given temperature due to the energy associated with the dissociation of the molecules. As such, the use of nitrogen gas results in the production of a thicker coating compared to argon gas. Although thicker coatings often result in greater protective coverage of the substrate, thereby improving the corrosion resistance of metallic implants, a thicker coating can also significantly reduce the adhesion integrity of the coating to the substrate.

- Plasma temperature is affected by the composition of the plasma gas used for the thermal spraying process.
- The composition of gas used for the plasma affects the composition and structure of the thermal-sprayed coatings

Cleaning and abrasive blasting are important for preparing the substrate surface in order to bond the coating. A cleaned surface provides a more chemically and physically active surface needed for chemical bonding. The roughened substrate surface resulting from the abrasive blasting provides an increase in surface area for bonding as well as promoting mechanical interlocking of the coating to the substrate, thereby enhancing coating bond strength. A higher temperature during deposition on the substrate may promote diffusion-related bonding between the coating and the substrate. However, in the case of metallic substrates, a higher coating temperature may also increase the substrate temperature, thereby increasing substrate oxidation that may be detrimental to the coating–substrate bond strength. Shrinkage and stress relief of the coatings as a result of cooling highly influence the bond strength between the coatings and the substrates. Rapid cooling and solidification during the coating process result in the buildup of tensile stress within the coatings while the substrate experiences compressive stress. If the tensile stress within the thick coating exceeds the cohesive strength or the bond strength of the coating–substrate interface, the coating fails. Additionally, the coating–substrate bond strengths are also significantly influenced by the presence of porosity and cracks in the coatings. Thus, the bond strength between the coatings and substrates is dependent on several factors including

- substrate preparation,
- cooling rate,
- coating thickness, and
- coating imperfections such as voids, pores, cracks.

- Rapid cooling of the thermal-sprayed coatings results in the formation of an amorphous coating.
- Rapid solidification of the coatings during thermal spraying results in the formation of cracks on the coatings.

The major difference between plasma spraying and flame spraying is that the flame spraying process uses an oxy-acetylene burner as the heat source. The temperature of the flame during the flame spraying process is approximately 2700 °C. Unlike the deposition rate of up to 5 kg/h for the plasma-spraying process, the flame-spraying process has a higher deposition rate of up to 10 kg/h thereby resulting in thicker coatings. Unlike plasma spraying, particle velocity in flame spraying is low, ranging from 40 m/s to 200 m/s. This low particle velocity typically results in low coating–substrate bond strengths, a higher porosity and more impurities within the coatings as compared to plasma-sprayed coatings (200 m/s to 400 m/s).

Box 9.5

- High-temperature thermal spray process has been used to coat hydroxyapatite on titanium implant surfaces since the 1980s.
- Problems with thermal-sprayed hydroxyapatite implants include variation in coating–substrate adhesion strength, and alterations of structure and composition of the hydroxyapatite.

The high velocity oxy fuel (HVOF) spraying process is a modified flame-spraying process which uses an advanced nozzle design technology. Using a mixture of hydrogen, oxygen, and methane gas, powder particles can achieve velocities exceeding 600 m/s. Although the deposition rate of the HVOF process is similar to that of plasma spraying, the coatings produced can be denser with

Table 9.1   Differences in the plasma temperature, deposition rate, and particle velocity in the plasma using the flame spraying, the plasma spraying, and the HVOF spraying process

	Flame spraying	Plasma spraying	HVOF spraying
Plasma temperature	2700 °C	6700 °C to 19 700 °C	6700 °C to 19 700 °C
Deposition rate	10 kg/h	5 kg/h	5 kg/h
Particle velocity in the plasma	40 m/s to 200 m/s	200 m/s to 400 m/s	Exceeds 600 m/s

high coating–substrate bond strength when compared to coatings achieved using the flame and plasma spraying process. Table 9.1 summarizes the differences in plasma temperature, deposition rate, and particle velocity in the plasma using the flame spraying, the plasma spraying and the HVOF spraying process.

9.4   Physical vapor deposition (PVD)

Physical vapor deposition (PVD) is a general term that refers to a family of vaporization coating techniques for depositing thin films on substrates. Like the thermal spraying process, PVD is a line-of-sight process in a vacuum chamber involving the pure physical transfer of material at the atomic level. In general, the four steps involved in PVD are

- evaporation,
- transportation,
- reaction, and
- deposition.

> - Like the thermal spraying process, the physical vapor deposition process is a line-of-sight process.
> - The physical vapor deposition process involves the transfer of material on an atomic level in a vacuum chamber.

During evaporation, a target making up the material to be deposited is converted into vapor by physical means such as beams of accelerated, high-energy electrons or ions. The bombardment of these high-energy electrons or ions onto target materials causes surface atoms from the target to be dislodged with sufficient energy. The energized, dislodged atoms from the target are then transported to the substrate surfaces in a straight line. Depending on the gas used in the PVD, the dislodged atoms may react with reactive gases such as oxygen, nitrogen, and methane during the transportation process. No reactions occur if a noble gas is used. Deposition is said to occur when the atoms are deposited on the substrate surface and build a coating layer.

A number of different PVD systems are currently used either in engineering research or commercially for modifying medical device surfaces. These different PVD systems used include

- evaporative deposition,
- pulsed laser deposition, and
- sputter deposition.

9.4.1 Evaporative deposition

For evaporative deposition to occur, the target material must be evaporated and condensed onto the substrate surface. Under high vacuum, the target material is melted using an electric resistance heater. A schematic of evaporative deposition is shown in Figure 9.12.

The vapor pressure of the melted material is raised, thereby allowing the vapor to be deposited directly onto the substrate surface without reacting with or colliding against other gas atoms in the vacuum chamber. To avoid contamination of the deposited film, target materials with a vapor pressure higher than the electric resistance heating element can be deposited using this process. As an alternative to

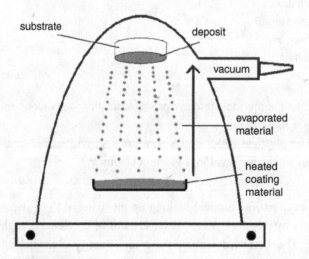

Figure 9.12

Schematic of experimental set up for evaporative deposition under high vacuum. The arrow indicates the direction of vapor flow towards the substrate.

the electric resistance heater, an electron gun can also be used for melting the target material under high vacuum. A high-energy beam of up to 15 keV, provided by the electron gun, can be used on a small spot of the target material. An advantage of using the electron gun over the electric resistance heater is that heating using the electron gun is not uniform and thus allows for lower vapor pressure target materials to be deposited. In general, the quality of the coating depends on the vacuum pressure and the purity of the target material.

9.4.2 Pulsed laser deposition

Another type of PVD is the pulsed laser deposition process. Pulsed laser deposition requires a highly focused, power pulsed laser beam to strike a target material in a vacuum chamber filled with a gas such as argon or oxygen. A schematic of the pulsed laser deposition process is shown in Figure 9.13.

Penetration of the laser pulse into the surface of the target results in electronic excitation of the target due to the removal of electrons from its atoms. Oscillating within the electromagnetic field of the laser, these free electrons collide with target atoms as well as the gas atoms in the chamber. The collision of electrons with the gas

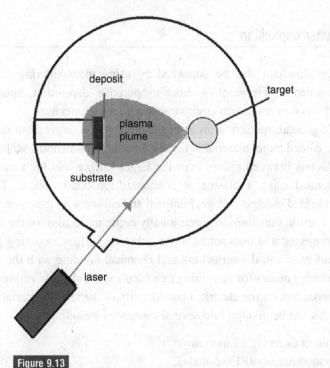

Figure 9.13

Schematic of pulsed laser deposition process under high vacuum. The arrow represents a highly focused laser beam from a laser source striking the target during the process.

induces the formation of plasma in the chamber through ionization and excitation of the gas molecules, whereas the collision of electrons with the target heats up the target surface, thereby resulting in vaporization and expansion of the target material content in the plasma. As in other deposition processes, the addition of target materials in the plasma results in the acceleration of these materials and the eventual bombardment of the substrate surfaces with these target materials. As a result, deposition of the target material and cleaning of the substrate surfaces occur simultaneously. A collision region within the plasma plume is consequently formed where species from the substrate that are emitted as a resulting of cleaning, collide with the arriving target materials, thereby generating a condensation of particles. A thermal equilibrium is achieved when the condensation rate is sufficiently high to result in a net deposition of target materials on substrate surfaces as thin films. Factors affecting the thickness of the deposited film include

- composition of the target material,
- energy of the laser beam,
- distance between the target and the substrate, and
- gas used in the vacuum chamber.

9.4.3 Sputter deposition

Thin films can also be deposited by using the sputtering deposition process, illustrated in Figure 9.14. Like evaporative deposition, sputter deposition is performed in a vacuum environment at room temperature.

In general, sputtering involves the erosion of a target material from its surface. The eroded target material is largely driven by momentum exchange as a result of collisions between atoms from the target surface and high-energy gaseous ions generated using a plasma or a separate ion beam source. The gaseous ions, accelerated by high voltage, bombard and collide with atoms on the target surface. As a result, the atoms, or occasionally entire molecules, of the target material are then ejected and transported to the substrate surface, resulting in a thin film that forms mechanical interlocking and chemical bonding with the substrate. Several sputtering modes for depositing thin films exist and the selection of a sputter mode is dependent on the electrical conductivity of the target material. These sputtering modes can be divided into several categories including

- direct current (d.c.) sputtering,
- radiofrequency (RF) sputtering,
- RF magnetron sputtering, and
- ion beam sputtering.

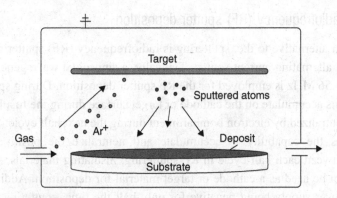

Figure 9.14

Schematic of a sputter deposition process under high vacuum. Ions are observed to strike the target surface, thereby allowing the erosion of the eroded target materials to be deposited on the substrate surface.

Direct current (d.c.) sputter deposition

The d.c. sputtering mode is used for depositing conductive materials. In this sputtering configuration, the cathode is made of the target material and the anode consists of the substrate to be coated. In a vacuum chamber filled with a noble gas such as argon, the application of a high negative potential difference of approximately 1000 V between the cathode and anode results in the release of electrons from the cathode. These electrons collide elastically or inelastically with the gas atoms in the vacuum chamber and ionize them, creating positive ions and more electrons. The principle of gas ionization has been described in the previous sections of this chapter. Positive ions are then accelerated towards the target or cathode surface, striking it with high energy and ejecting the target material. Traveling in a straight line, the target material is then deposited onto the anode or the substrate surface. Since a high negative potential difference between the cathode and the anode needs to be generated in the d.c. sputtering mode, the choice of a target material becomes critical. Ceramics, such as hydroxyapatite and tricalcium phosphates, are electrically insulating materials and thus are not materials of choice for use as cathodes in the d.c. sputtering mode.

- Direct current sputter deposition requires a continuous potential difference between the cathode and the anode.
- The use of non-conductive materials in the direct current sputter deposition process leads to a short breakdown, followed by the creation of a surface charge on the cathode.

Radiofrequency (RF) sputter deposition

An alternative to d.c. sputtering is radiofrequency (RF) sputtering, which utilizes an alternating current source. Typically, a sinusoidal wave generator operating at 13.56 MHz is employed for the RF sputter deposition. During sputtering, positive ions accumulate on the cathode or target surface during the first half cycle and are neutralized by electron bombardment during the next half cycle. Since the cathode has the capability to accumulate and neutralize the positively charged ions between each half cycle in RF sputtering, insulating materials, such as ceramics, can be used as a cathode or target material for deposition. Additionally, with the power supply being negative for only half the time, sputtering rates using a RF sputtering mode are lower than the sputtering rates in a d.c. sputtering mode.

- Radiofrequency sputter deposition is used for the deposition of non-conductive target materials.
- Sputtering rate using the radiofrequency sputter deposition process is generally lower than the use of direct current sputter deposition process.

Radiofrequency (RF) magnetron sputter deposition

A modification of the RF sputtering mode is the inclusion of a magnetic source. Also known as the planar RF magnetron sputtering, the efficiency of the RF sputtering is enhanced by the presence of a hydraulically moving magnetic source in the sputtering system. Shown in Figure 9.15 is an example of a RF magnetron sputtering system that is found in research laboratories. The system usually consists of a RF power supply, a deposition chamber and electronic control panels to monitor chamber pressure and to regulate RF energy.

Box 9.6

- For optimal sputtering, the magnetic field in the radiofrequency magnetron sputtering system is generally located no more than 2.5 cm above the target.
- The combined use of magnetron with sputtering allows for more efficient sputtering rate, minimal growth of an insulating film on the target surface, and independent control of bombardment of the growing film.
- Although the sputtering efficiency is increased with a magnetron sputtering system, the sputtering rate from a multi-component target material is lower when compared to a single material target.

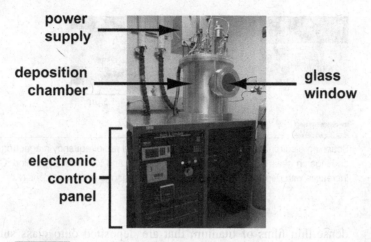

power
supply

deposition
chamber

glass
window

electronic
control
panel

Figure 9.15

Photograph of a radiofrequency magnetron sputter coater showing the chamber and the electronic control panel. The primary components in the electronic control panel consist of the switches to regulate the radiofrequency power, turbo pump, and roughing pump. A glass window is available for viewing the sputtering process in the chamber.

Within the RF magnetron sputtering system, the moving magnets are located in close proximity to the target material. A closed-loop magnetic line of flux is created, extending from one pole of a magnet to the opposite pole. This magnetic field is above and parallel to the target surface. The spiral path around the magnetic field lines provides a re-directed, much longer path length for the free electrons to the target surface before they are lost to the anode. This re-direction of free electrons greatly increases their chance of colliding with argon atoms and ionizing them, thereby enhancing the effectiveness of plasma discharge. Additionally, the use of the magnetic fields in re-directing the electrons allows the production of an ion density that is similar to a d.c. sputtering mode, except that this ion density is created at a much lower pressure. The increased discharge effectiveness results in higher deposition rates, which thus makes the planar RF magnetron sputtering a preferred mode for depositing conductive and non-conductive materials over the d.c. sputtering and the RF sputtering modes. However, despite the advantage of having a higher discharge, the production of the desired thin film with optimum coating properties requires careful control of the coating parameters. In addition to the target composition, it is known that the coating properties are also affected by the ionized gas pressure, substrate temperature, voltage across the different sheaths, and the amount of time used for deposition. Shown in Figure 9.16 are

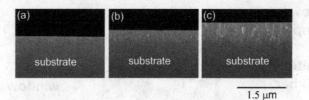

Figure 9.16

Scanning electron micrographs of cross-sectioned radiofrequency magnetron sputtered titanium coatings on glass substrates after (a) 1 h, (b) 2 h, and (c) 3 h of deposition. Coating thickness increases with deposition time with relatively no change in coating density or quality.

dense thin films of titanium that are deposited onto glass substrates using an RF magnetron sputtering process.

Ion beam sputter deposition

Ion beam sputtering, like the other sputtering processes, is a low temperature process. However, unlike d.c. sputtering, RF sputtering, and planar RF sputtering where the plasma is localized to the target, ion beam sputtering or ion beam deposition uses a separate ion beam source to generate a neutralized beam. There are a variety of ion beam sources depending on the application, and they can range from d.c. to RF and either be gridded or without a grid. All gridded ion beam sources require an electron or ion source to create and sustain a plasma discharge. There are several electron or ion sources, including:

- d.c. filament source – this source is simple to operate and service;
- d.c. hollow cathode source – the advantage of using a d.c. hollow cathode electron source is that it allows for a long process or deposition time;
- RF source – inductively coupled, a RF electron source is used for operating in reactive gases, such as oxygen.

In ion beam sputtering, the gas flows through the ion source between the anode and the cathode. In general, ionization of the gas occurs upon application of a positive voltage of up to 10 keV to the anode, thereby generating a plasma discharge. Energies used for ion beam sputter deposition are typically in the range of 500 eV to 1500 eV, resulting in little damage to the target. When a RF source is used, plasma discharge is generated when the positive voltage applied is combined with the magnetic field. Similar to RF sputter deposition (described previously), the magnetic field from the RF source is used to confine and direct electrons to collide and ionize the gas atoms. The positive ions generated are then repelled by the anode electric field that originates from a grid, thereby creating an ion beam of accelerated ions towards a target. As these accelerated positive ions leave the ion source, they are neutralized by electrons generated from a second external filament. Since the flux that strikes the target is composed of neutral atoms, either insulating or conducting targets can be sputtered. Unlike the other sputtering modes, the target used in ion beam sputtering is not used as a cathode and thus the electrical conductivity of the target is no longer of importance. Examples of modified ion beam sputtering are

- ion beam-assisted deposition (IBAD), and
- dual ion beam sputter deposition.

Ion beam-assisted deposition (IBAD)

In IBAD, deposition of the target materials is carried out by using a combination of ion beam sputtering with another PVD process. As shown schematically in Figure 9.17, the principle behind IBAD deposition is the interaction of the accelerated ions generated by the ion source with the coating atoms from the PVD process as they are being deposited on the substrate surface. This results in the production of a low-stress, uniform, and adherent coating. Coatings deposited with the IBAD process are much thinner and denser than those deposited using the other sputtering processes.

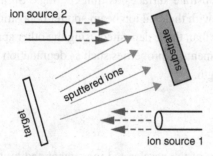

Figure 9.17

Schematic of an ion beam-assisted deposition process showing an ion source 1 accelerating ions to the target surface. Target atoms are eroded from the target surface and subsequently are deposited on the substrate surface. Ion source 2 accelerates ions to the coated substrate surface, thereby interacting with the deposited coatings.

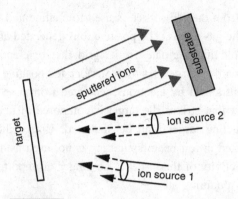

Figure 9.18

Schematic of a dual ion beam sputter deposition process. Ion sources 1 and 2 accelerate ions to the target surface, thereby eroding the target material as atoms that subsequently are deposited on the substrate surface. The energy of ion source 2 is less than ion source 1.

Dual ion beam sputter deposition

Unlike IBAD, which uses a separate ion source to direct the ion beam towards the coating atoms, dual ion beam sputter deposition utilizes two ion sources to bombard and erode the target surface. A schematic of the dual ion beam sputter deposition process is shown in Figure 9.18. The second beam in the dual ion beam sputtering is of lower energy, typically ranging from 4 eV to 20 eV. Unlike the heavy sputtering ions, such as those from argon gas that are generated from the first ion source, other ions from gases such as nitrogen, oxygen, and hydrogen may be generated from the second ion source. The resulting combination of low-energy ions with atoms on the target surface or the eroded atoms from the target surface are then deposited on substrate surfaces as thin coatings. Similar to IBAD coatings, coatings deposited with the dual ion beam sputter deposition process are also much thinner and denser than those deposited using the other sputtering processes, and thus render improvements in properties such as degradation rates and coating–substrate adhesion strength.

Ion implantation

Depending on the energy of the accelerated ions generated by the ion source, ion implantation is a process whereby high-energy accelerated ions can penetrate or embed into the surface of the substrate material. Energies in the range 1 keV to 10 keV may result in the penetration of ions a few nanometers deep into the target. Energies of up to 5 MeV are commonly generated with the help of an accelerator, and such high energies provide a broad distribution of ions to penetrate into the substrate surface. The penetration of ions results in structural damage to the substrate, but the net composition change at any point in the target is small when high energies are used. In principle, the energy of a penetrating ion is greatly diminished with penetration depth, and the ion eventually comes to rest within a certain depth of the substrate material. In addition to ion penetration, other cascades of activities, such as surface sputtering, can also occur during the collision between the substrate atoms and the accelerated ions. Depending on the ion energy and the composition of the substrate, surface sputtering during the ion implantation process can limit the amount of ions implanted in the substrate.

- Ion implantation results in penetration of ions into the substrate, causing structural damages to the surface.
- Surface compositional alterations occur after penetration of ions into the substrate surface.

In general, ion implantation offers the capability of forming alloys, with the new phases formed at the near-surface and surface regions of a substrate. This formation of new phases alters the mechanical properties of the near-surface and the surface. In metallic and ceramic substrates, the atomic and nuclear collisions induced by ion implantation lead to the formation of highly disordered and sometimes amorphous structures. For example, ion implantation of titanium and titanium alloys with nitrogen species improves the wear resistance. The induced change in the crystallinity of the near-surface and surface regions of metals also alters their friction properties. In polymeric substrates, ion implantation induces chain cross-linking and scission in the near-surface and surface regions. For example, ion implantation of polycarbonate with nitrogen species results in the formation of a complex three-dimensional, cross-linked surface, thereby increasing solvent resistance and improving the wear and friction characteristics.

9.5 Chemical vapor deposition (CVD)

Although use of chemical vapor deposition (CVD) has many applications in a wide range of industries such as in the fabrication of semiconductor devices and the production of ceramic matrix composites, CVD can also be used for surface modification in the biomedical arena. For biomaterials, CVD can be utilized to produce diamond and diamond-like carbon (DLC) coatings. In general, CVD is an extremely versatile process that can be used for depositing almost any metallic or ceramic compound. Unlike PVD which is a pure physical deposition process requiring a high temperature, the generation of a plasma plume, and the bombardment of ions on the target surface, CVD is a chemical process using thermal energy to heat up the gases in the coating chamber and drive the deposition reaction. The temperature during the CVD process is at least 1000 °C. Like PVD, which consists of four general steps in the deposition process, the CVD process consists of the following five general steps.

- Introduction of gas reactants into the vacuum or reaction chamber at room temperature. The gas reactants contain mixtures of precursor gases and a carrier gas.
- Moving of gas reactants toward the heated substrates.
- Absorbing of the gas reactants on the heated substrate surfaces.
- High temperature of the heated substrates allowing for chemical reactions of gas reactants to form a coating.
- Desorption and evacuation of gaseous by-products from the reaction chamber.

- In the chemical vapor deposition process, gas reactants absorb and react with the substrate surface at high temperatures, thereby resulting in the formation of a film on the substrate surface.
- Gaseous by-products are pumped out of the reaction chamber.

Precursor gases in the CVD process consist of materials that are to be deposited on substrate surfaces. Since chemical reactions of the gas reactants occur at high temperatures in the reaction or vacuum chamber, the precursors must be volatile, but at the same time stable enough to be able to be delivered to the heated substrate surfaces. Two types of reactions can occur within the reaction or vacuum chamber, and these are

- homogeneous reactions, and
- heterogeneous reactions.

Homogeneous reactions occur between the gas reactants as they move toward the heated substrates and result in the formation of gas-phase aggregates. These aggregates adhere poorly to the substrate surface and form a porous or low density coating. On the other hand, heterogeneous reactions occur when the gas reactants first adsorb on the heated substrate surfaces and react with these surfaces to form dense coatings. Obviously, of these two reactions, the heterogeneous reactions are preferred over the homogeneous reactions because of the resulting quality of the coatings. In general, CVD coatings are produced at a fairly slow deposition rate of a few hundred micrometers per hour. These high purity coatings are fine grained, impervious, and are usually only a few micrometers thick. Even though the carrier

Table 9.2   Advantages and disadvantages of the different chemical vapor deposition processes

CVD process	Advantages	Disadvantages
Atmospheric pressure chemical vapor deposition (APCVD)	Simple process, fast deposition, low temperature	Low coating coverage, low purity
Low pressure chemical vapor deposition (LPCVD)	High purity, good coating coverage	Slow deposition, high temperature
Ultrahigh vacuum CVD (UHVCVD)	High purity, good coating coverage, low temperature	Slow deposition, high cost, difficult process

gases such as argon and nitrogen are inert, one of the disadvantages of using the CVD process is that many of the reactant gases used such as ammonia, silane, arsine, oxygen, hydrogen, nitrous oxide, and phosphine can be toxic, corrosive, flammable, explosive or pyrophoric.

There are a variety of different types of CVD processes, depending on the chemical reactions and the process conditions. In an atomic layer CVD process, coatings with distinct layers can be formed on the substrates as gas reactants containing different but complimentary precursor gases are introduced into the vacuum or reaction chamber. Examples of other CVD processes are

- atmospheric pressure chemical vapor deposition (APCVD),
- low pressure chemical vapor deposition (LPCVD),
- ultrahigh vacuum CVD (UHVCVD), and
- plasma enhanced chemical vapor deposition (PECVD).

The atmospheric pressure chemical vapor deposition (APCVD) process performs at atmospheric pressure, whereas the low pressure chemical vapor deposition (LPCVD) and ultrahigh vacuum CVD (UHVCVD) processes perform at sub-atmospheric pressure. The LPCVD process operates at medium pressure ranging from 30 Pa to 250 Pa, whereas the UHVCVD process operates at ultralow pressure, typically below 10^{-6} Pa. This reduced pressure in the reaction chambers during LPCVD and UHVCVD processes enhances gas diffusivity, thereby reducing unwanted gas-phase reactions and improving coating uniformity across the substrate. Listed in the Table 9.2 are some of the advantages and disadvantages of the APCVD, LPCVD, and UHVCVD processes.

Unlike the APCVD, LPCVD, and UHVCVD processes, which are dependent on the operating pressure, the plasma enhanced chemical vapor deposition (PECVD) process relies on the presence of plasma generated to enhance the chemical reaction rates of the precursors. This plasma is created during the application of an RF field to a low pressure gas, thereby creating free electrons that are energized by the electric field. Collisions of the gas reactants (containing precursors) and the energetic electrons induce gas-phase dissociation and ionization to occur, thereby allowing energetic species to be adsorbed on the surface to be coated. These surfaces are subjected to ion and electron bombardment, rearrangements, reactions with other species, new bond formation, and film formation and growth. In general, the PECVD process allows for the deposition of coatings at lower temperatures, which is often critical in the manufacture of semiconductors or temperature-sensitive substrates such as low temperature insulators over metals.

Box 9.7

- Diamond-like carbon coatings produced by PECVD are also known as hydrogenated amorphous carbon coatings.
- In general, the diamond-like carbon coatings possess high hardness, high electrical resistivity, and low coefficient of friction.

The discharge gas used in PECVD is a reactive gas, whereby chemical reactions in the plasma result in the formation of radicals and ions, and these radicals and ions can diffuse to a substrate and be deposited. The use of plasma containing mixtures of methane (CH_4) and hydrogen (H_2) in a PECVD process results in the deposition of diamond-like carbon layers.[2] Other PECVD applications include the use of silane–hydrogen (SiH_4–H_2) plasma to deposit amorphous hydrogenated silicon layers for the fabrication of thin film transistors, xerographic materials, and solar cells. Although a good coating coverage can be achieved, problems with the PECVD process are linked to chemical and particle contamination during deposition, resulting in the production of low purity coatings.

9.6   Grafting

Grafting is one of the many popular surface modification techniques utilized in the biomedical engineering industry to alter surface properties for selective protein adsorption as well as for enhancing biocompatibility. Modification of surfaces by grafting biological signals or ligands such as arginine–glycine–aspartic acid (RGD) peptides allows for the optimization of the biomaterial surface for cell adhesion and proliferation. In general, the grafting process requires the reaction of one or more polymeric species or functional groups to the main chain of the polymeric macromolecules. Typically, these functional groups or species have properties that differ from the main polymeric chain and their attachment is achieved by a variety of methods such as covalent coupling, surface graft polymerization, surface segregation, and interpenetration of the surface. Factors affecting the properties of a grafted polymer layer include

- the molecular weight of the polymer,
- the degree of chain branching,
- the density of functional and charged groups on the substrate and polymer, and
- the solubility of the polymer.

Box 9.8

- The development of the new generation of grafted materials is mainly driven by the textile industry.
- In general, grafting provides the potential to tailor custom made properties on existing polymers.

Schematically shown in Figure 9.19, the strategy of introducing functional groups or polymeric species usually involves chemical activation or radiation activation using external energy sources to create active sites on the substrates and to initiate the reaction. In both chemical and radiation grafting, free radicals are generated either through the treatment of substrate surfaces with a free radical generating agent such as an organic peroxide or irradiation. The monomers or graft polymer chains are then covalently bonded to the substrate.

During chemical grafting, polymeric substrates containing active hydrogen are generally treated in an aqueous solution containing functional species and graft initiators or free radical polymerization catalysts such as silver nitrate. The graft initiators or catalysts act on the substrate to remove the active hydrogen. Such removal allows for the creation of active sites or free radicals along the molecular chain, thereby allowing the monomers to covalently bond to the substrate at the site of the removed hydrogen as side chains. Initiation and propagation of the covalently bonded monomers occur through linking carbon–carbon bonds, thereby modifying the substrate and imparting new and desirable properties to it. The chemical grafting process is dependent on the substrate possessing active hydrogen atoms, such as the tertiary hydrogen in polypropylene, the amide hydrogen in proteins, and the hydroxyl hydrogen in polysaccharides. A problem with the chemical grafting process is the need to remove oxygen from the aqueous solution, and this can make the process difficult and costly.

Unlike chemical grafting, which is dependent on graft initiators or catalysts, radiation grafting involves the creation of active sites or free radicals on the

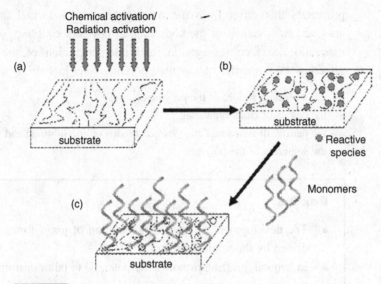

Chemical activation/
Radiation activation

(a)

substrate

(b)

substrate

● Reactive
species

Monomers

(c)

substrate

Figure 9.19

Schematic representation of the grafting process showing (a) activation of the substrate surface using chemical, radiation, or photochemical activation, (b) active sites or reactive species created on the substrate surface, and (c) the grafting of the monomers to the substrate.

substrate either by pre-irradiation of the substrate or by radiating the substrate while immersed in an aqueous solution containing grafting monomers. Active sites on the substrate are created by the absorption of radiant energy from gamma radiation, UV radiation, corona discharge, radiofrequency glow discharge, electron beam radiation, or other high-energy radiation generated from external sources. In either the pre-irradiation method or the irradiation of the substrate after immersion in aqueous solution, the rate and efficiency of the initiation is dependent on the type of radiation, the radiation dose or the total energy absorbed, the dose rate or the rate at which energy is absorbed, and the radiation sensitivity of materials involved. Like chemical grafting, problems with irradiation grafting are that

- grafting efficiency using radiant energy is substrate dependent, and
- the need to eliminate oxygen from any monomer solution used for the grafting process can greatly add to the expense and difficulty of the process.

9.7 Self-assembled monolayer (SAM)

The self-assembled monolayer (SAM) technique deposits a thin film consisting of a monolayer of organic molecules on surfaces to improve surface properties for controlling corrosion rates, reducing friction, or enhancing wetting or adhesion.

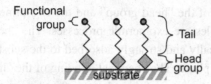

Figure 9.20

Representation of a functionalized self-assembled monolayer structure deposited on a substrate.

The technique is sometimes employed in the fabrication of biochips and biosensors for applications such as pH sensing. It has also been used to develop implantable drug delivery systems. Additionally, the SAM surface provides a model system for the simulation and evaluation of complex biological events, such as protein adsorption or cell adhesion, occurring *in vivo* or under quasi-physiological conditions. Immobilization of biomolecules such as growth factors on SAM surfaces has the potential to influence tissue–biomaterials interactions, whereas the use of nano-patterned SAM can provide high throughput cell-based platforms for mass screening of drugs and proteins. One advantage of using a SAM is that it can form on planar surfaces as well as on curved surfaces and even on a nanoparticle. Once deposited on the surface of nanoparticles, the reactive sites on the surface can be stabilized as well as functionalized for different applications.

In general, the thin organic film formed is made of chemical molecules containing both hydrophilic and hydrophobic properties. Such molecules are also known as amphiphilic molecules. As shown in Figure 9.20, the hydrophilic end of the SAM molecule is a "head group" that is adsorbed on the substrate surface. Also shown in Figure 9.20 is the tail of the SAM molecule that is made up of the backbone chain of either long alkyl or derivatized alkyl backbone chain. The alkyl chain can have a carbon number ranging from 1 to 20 whereas the derivatized alkyl chain can contain groups such as aromatic rings and unsaturated alkenes. At the other end of the backbone chain is the "tail end" which is capable of being functionalized by either adding an amine, hydroxyl, or carboxylic acid group. Figure 9.20 shows the schematic of a functionalized group at the end of the tail. Depending on the desired property, the addition of a particular functional group, such as cell binding peptides, to the "tail end" will modify the surface properties of the substrate surface.

In general, the process of SAM formation is dependent on the head group and the backbone chain of the SAM molecules. The following two critical conditions have to occur for SAM formation to take place:

- the chemisorption of the "head group" to the substrate surface, and
- the two-dimensional organization of the SAM molecules through the lateral inter-chain interactions of the hydrocarbon backbone.

Both the chemisorptions of the "head group" and the two-dimensional organization of the SAM molecules are exothermic processes. The "head group" of the SAM molecule is chemically and strongly adsorbed to the substrate surface from either the vapor or liquid phase. The chemical bonding of the "head group" to the substrate lowers the surface energy of the substrate, thereby creating a SAM that is more stable in a variety of solvents, temperature range, and potential when compared to physical adsorption.

The hydrophilic heads of the SAM molecules initially adsorbed on the substrate form either a short-range disordered mass of molecules that are randomly dispersed on the substrate surface or are randomly lying flat on the substrate surface. Areas of close-packed molecules nucleate and grow until the surface of the substrate is covered in a single monolayer over time. As more of the SAM molecules are packed into the monolayer, the monolayer undergoes changes, rearranging itself to its optimal configuration. Additionally, with more SAM molecules adsorbed on the substrate surface, the free energy of the SAM molecules is reduced as a result of increased van der Waals interactions between the hydrocarbon backbone chains of the SAM molecules. Over time, the SAM molecules become tightly packed, forming well-ordered crystalline or semi-crystalline structures on the substrate surface. The order of the monolayer generally increases with an increase in chain length.

- The van der Waals forces between the carbon chains allowed the SAM molecules to form densely packed, semi-crystalline or crystalline monolayers.
- Monolayer formation is governed by the time of annealing, cleanliness of the substrate surface, purity of the solution, and properties of the head group and tail end.

The process of the newly formed monolayers undergoing changes and rearranging themselves to their optimal configuration is also known as annealing of the monolayers. This annealing process can last from several hours to several days. An equilibrium state of the monolayer's assembly, wherein the majority of the molecules are arranged in their final, optimal configuration, is usually achieved within 1–2 days. In general, the process of SAM formation is dependent on several factors including the

- time available for coating,
- medium or solvent containing the SAM molecules,
- length and composition of the hydrocarbon backbone,

- "head group" and the "tail end," and
- substrate material.

Although SAM formation occurs rapidly, in a matter of minutes, the amount of time required for a given level of order within a monolayer will depend on the initial solution concentration and temperature. In comparison to concentrated solutions of SAM molecules, a longer period is required for the monolayer to reach a well-ordered state when assembled from dilute solutions. The purity of the assembled monolayers formed on a substrate is dependent on the purity of the medium or solvent containing the SAM molecules. Solutions containing a mixture or trace amounts of different SAM molecules may create competitive adsorption between the different groups for the substrate surfaces.

Kinetics of the SAM formation and the order and stability of the monolayer are affected by the chain length of the SAM molecules. For example, in an impure solution containing SAM molecules with different alkyl backbone chain lengths, SAM molecules with longer alkyl chain lengths will generally assemble more rapidly than SAM molecules with shorter alkyl chain lengths. Since there are attractive inter-chain forces between the backbone chains of adjacent SAM molecules adsorbed on the substrate surface, the ordering of the molecules within the monolayer is affected by the chemical nature of these backbone chains. Additionally, properties of the "tail end" define the surface properties of the self-assembled monolayer since it is this end that is present at the outer surface. The ordering of the assembled monolayer and the stability of the monolayer depend on the properties and size of the "tail end," with bulky tail end or repulsive tail end groups disrupting the assembled monolayer. In addition, selection of the "head group" depends on the substrate material that it has to react with as well as the end application of the SAM. Cleanliness of the substrate surface is also critical for the adsorption of the "head group." Listed below are some of the different "head group" and substrate combinations for SAM formation:

- organosulfur on metal surfaces, and
- organosilicons on hydroxylated surfaces.

Among the different organosulfur compounds used for SAM formation, alkanethiols are the most commonly used SAM molecules. The alkanethiol molecules are made of an n-alkyl or $(C–C)^n$ backbone chain, a "tail end," and a sulfur (S)–hydrogen (H) "head group." Other organosulfur compounds include dialkyl disulfide, dialkyl sulfide, aromatic thiol, alkyl xanthate, and dialkyl thiocarbamate. These organosulfur compounds have high affinity for surfaces of transition metals.[3] Although alkanethiols can react with other metals such as silver, platinum, mercury, iron, and copper as well as non-metallic oxide surfaces such as silanes, the surface of gold (Au) is most frequently studied. Under ambient conditions, the immersion of Au in ethanol solution, containing alkanethiol molecules results in a reaction between the thiol and gold. The S on the "head group" has a strong affinity to the Au surfaces. The oxidative reaction of S–H bonds in the "head group," followed by the reductive elimination of the hydrogen results in the formation of thiolate through the robust formation of a covalent bonding at the S–Au interface.

The organosilicon compound is a family of molecules having a general formula R_nSiX_{3-n}, where R is an aliphatic or an aromatic hydrocarbon backbone chain and X is the chlorine (Cl) atom. In addition to the backbone chain and a silicon (Si)–Cl or alkoxy "head group," it also contains a functional group at the "tail end," like the alkanethiol molecules. Examples of organosilicon compounds that form monolayers are the dodecyltrichlorosilane ($CH_3–(CH_2)_{11}–SiCl_3$), n-alkyltrichlorosilane ($CH_3–(CH_2)_n–SiCl_3$), and the octadecyltrichlorosilane ($CH_3–(CH_2)_{17}–SiCl_3$). The substrate surfaces have to be hydroxylated to react with organosilicon compounds. These substrate surfaces include hydroxyl (OH) on groups such as SiO_2 found in glass, SnO_2, and TiO_2.

Unlike the formation of alkanethiol monolayers, the process for forming organosilicon monolayers occurs through two pathways. In a solution containing organosilicon compounds such as n-alkyltrichlorosilane, the presence of water molecules on the substrate surfaces allows for hydrolysis reaction to occur, transforming the Si–Cl$_3$ on the "head group" to trimethoxysilane ($–Si(OH)_3$) groups. The presence of water molecules also allows for the lateral rearrangement of the adsorbed monolayer through lateral van der Waals interactions between adjacent molecules, thus transforming the adsorbed molecules into a dense, homogeneous monolayer over time. Once the dense layer is formed, a condensation reaction of the silanol (SiOH) groups between neighboring molecules results in the formation of siloxane (Si–O–Si) linkages.

In the second pathway, condensation reactions occur between the SAM molecules and the OH groups on the substrate resulting in the formation of R–Si–O–Si covalent bonds between the molecules and the substrate surface. The final state of the SAMs is a dense plane of extended hydrocarbon chains chemically linked into a two-dimensional network and grafted to the solid substrate at a certain number of hydroxyl sites. Typical reaction time leading to well-ordered organosilicon SAM formation ranges from 1 h to 90 min. However, given that organosilicon SAM formation occurs through two different processes, it is more disordered when compared to alkanethiol SAMs, since alkanethiol molecules have more freedom to establish long-range order. With a monolayer completely covering the substrate surface and forming Si–O bonds, there is not enough space for hydrocarbon chains to organize a closely packed, ordered structure.

9.7.1   Patterning of self-assembled monolayers

Regardless of the types of molecules used for SAM formation, patterning of the SAM is critical for many applications such as biosensors used to detect changes in biological responses. Using localized deposition strategy, the major techniques for patterning SAM formation include

- micro-contact printing,
- dip-pen nanolithography, and
- localized removal strategy using ultraviolet irradiation, electron beam, or high resolution microscope.

In the micro-contact printing method, the SAM molecules are inked onto a pre-shaped elastomeric stamp with a solvent and transferred to the substrate surface by stamping. The SAM solution is applied to the entire stamp, and the SAM is formed only in the areas where the substrate surface is in contact with the stamp. The nano-patterned SAM formed by using this strategy is dependent on

- the SAM molecule,
- concentration of SAM molecules in solution,
- contact time between the substrate surface and the mold, and
- the pressure applied during printing.

In general, there are several different types of micro-contact printing, including the use of a planar stamp on a planar surface and the use of a rolling stamp on a planar surface. A schematic depiction of using the planar stamp and the rolling stamp on a planar surface is shown in Figure 9.21.

During micro-contact printing, the pre-shaped elastomeric stamp is made of polydimethylsiloxane (PDMS), one of the many organosilicon compounds. It has

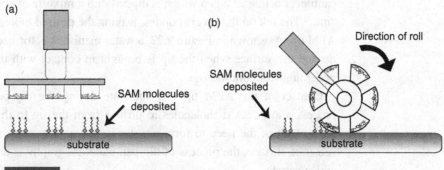

Figure 9.21

Schematics of micro-contact printing using (a) a planar stamp on a planar substrate surface, and (b) a rolling stamp on a planar substrate surface.

a chemical formula of $CH_3[Si(CH_3)_2O]_nSi(CH_3)_3$, with n being the repeating monomer $[SiO(CH_3)_2]$ units. Easily moldable, the PDMS can be molded to be a planar stamp or a rolling stamp for micro-contact printing. One of the many reasons for selecting PDMS as a material for the stamp is due to its elastomeric properties ($E = 1.8$ MPa), thereby allowing it to fit and conform to the contour or features of micro-surfaces, including non-planar surfaces. Additionally, the low surface energy of the PDMS ($\gamma = 21.6$ dyn/cm^2) and its chemical inertness keep the substrates from adhering to the PDMS stamp. The advantages of the micro-contact printing are that it has the potential to pattern a large area of substrate in a short time and the ability to correctly align the PDMS stamp with submicron features in order to create high resolution patterns.

Box 9.9

- Fabrication of the polydimethylsiloxane stamp is generally achieved using standard microelectronics photolithography methods to create the desired grooves, pits, or patterns.
- Successful pattern transfer using micro-contact printing is dependent on the rinsing or cleaning of the stamp, the concentration of the SAM solution, inking time, and the period that the stamp is in contact with the substrate.

Another strategy for creating patterning SAM is dip-pen nanolithography. This strategy utilizes an atomic force microscope (AFM) or a scanning probe micro-cope (SPM) to transfer SAM molecules on the AFM tip to a substrate surface via a solvent meniscus. Surface patterning on scales of less than 100 nm can be produced by using this strategy. The process begins with the tip of the AFM cantilever acting as a pen which is dipped into a mixture of solution known as the "ink." The ink on the tip evaporates, leaving the desired molecules attached to the AFM tip. As shown in Figure 9.22, a water meniscus is formed between the AFM tip and the surface when the tip is brought in contact with the substrate surface under ambient conditions.

Contact of the AFM tip and substrate during the dip-pen nanolithography process allows SAM molecules to diffuse from the tip to the substrate surface. As a result of the need to form a water meniscus between the AFM tip and the substrate surface, the process of dip-pen nanolithography is dependent on the ink and substrate.

Depending on the AFM tip used, the SAM molecules can be accurately formed on a specific location on the substrate surface using dip-pen nanolithography.

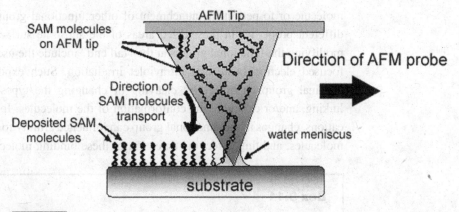

Figure 9.22

Schematic of a dip-pen nanolithography process showing the formation of a water meniscus between the AFM tip and the surface when the tip is brought in contact with the substrate surface under ambient conditions.

The ability to precisely form a SAM on a specific location allows the ability to create protein nano-arrays in the field of proteonomics or grow nanocrystals.

Box 9.10

- Limitation of the dip-pen nanolithography process is the need of special equipment.
- Although precision and control of nano-patterning is possible with the dip-pen nanolithography process, it requires significant time to cover a small area.

Besides using high resolution localized deposition strategy, nano-patterning of SAM can also be achieved using the strategy of localized removal or through the modification of the functional group at the "tail end." With the substrate surface covered with SAM, the localized removal strategy involves the use of either ultraviolet (UV) irradiation, low-energy electron beam, or a high resolution microscope such as scanning tunneling microscope, AFM, or SPM to remove individual SAM molecules from undesired sites of the substrate surface. UV irradiations involve the use of patterned apertures to project UV light on SAM-covered substrate surface for up to 20 min, thereby photo-oxidizing the SAM molecules.[4] Similarly, with high resolution microscopes, localized removal of SAM molecules can be achieved through mechanical dragging of the tip across the SAM-covered substrate surface. SAM patterning can also be achieved by deactivating the functional group at the "tail end" so as to create an inert SAM molecule or to permit the attachment of other functional groups so as to render different properties in preselected areas of the substrate surface. Techniques for modifying the functional groups at the "tail end" include the use of low-energy or focused electron beam and ultraviolet irradiation. Such exposures change the terminal group chemistry by cleaving or changing the types of bonds, cross-linking, and/or changing the conformation of the molecules. In biosensing applications, changes in the functional group can influence affinity for cells, proteins, or molecules, and thus affect the detection of these binding molecules.

Box 9.11

- With ultraviolet irradiation, the rate of oxidation is dependent on the chain length.
- Shorter chain length oxidizes faster, since longer chain length presents a greater barrier to oxygen diffusion, which slows down oxidation rate.

9.8   Layer-by-layer (LbL) assembly

Similar to SAM formation, layer-by-layer (LbL) assembly, also known as electrostatic self-assembly, is a simple, economical, versatile, and powerful bottom-up surface modification technique involving the sequential adsorption of water soluble anionic and cationic polyelectrolytes. With the capability to easily self-assemble nano-scaled multiple layers of polyelectrolytes, LbL assembly has been used in the fabrication of biosensors as well as vehicles for delivering drugs. An added advantage of the LbL assembly is that the coating process is not a line-of-sight process and thus allows the process to be applied to a broad range of substrates with different geometry and sizes. Depending on the nature of the porosity of the substrates, the polyelectrolytes used for LbL assembly can penetrate into three-dimensional porous surfaces thereby coating the inside of the pores. Additionally, LbL films are capable of incorporating sensitive biological drugs such as cell adhesion proteins and peptides, drugs, DNA, and other nucleic acid-based therapeutics on biomaterial surfaces.

- Layer-by-layer film assembly is a bottom-up surface modification technique.
- Film is formed by alternatively exposing the substrates to soluble anionic and cationic polyelectrolytes.

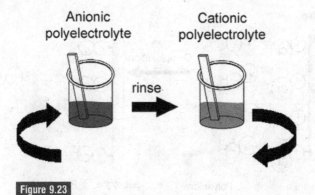

Figure 9.23

Simplified schematic of the layer-by-layer (LbL) dipping process. The substrate is dipped in polyanionic and cationic polyelectrolytes.

> • Proteins and growth factors can be incorporated in the layer-by-layer film assembly during the alternate dipping.

Layer-by-layer films are formed by alternately dipping of the substrates into soluble anionic and cationic polyelectrolytes. A schematic of the dipping process is shown in Figure 9.23. The process of alternate dipping allows for electrostatic interaction, covalent bonding, hydrogen bonding, and specific interactions between each charged layer. The monolayers formed are rinsed with pure solvent in between each dipping of the substrates in anionic and cationic polyelectrolyte solutions to remove any non-adhering polymer solution. Factors affecting LbL film assembly include the

- chemical structure and molecular weight of polyelectrolytes,
- ionic strength of the polyelectrolytes, and
- pH of the polyelectrolytes.

The polyelectrolytes are polymers with ionic groups contained in the repeating polymeric units. In aqueous solutions, these groups dissociate, thereby conferring charged properties to the polymers. These charged properties play an important role in determining structure, stability, and the interactions with the subsequent layer during LbL deposition. Like polymers, the polyelectrolytes are viscous and have properties similar to the polymer and electrolyte. Similarly, like acids and bases, polyelectrolytes can be divided into weak and strong polyelectrolytes. A strong polyelectrolyte completely dissociates in solution, whereas a weak polyelectrolyte will partially dissociate at intermediate pH.

First Step

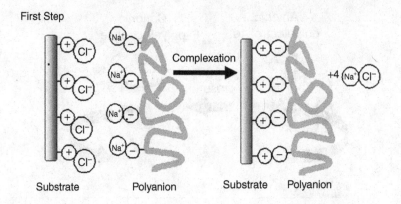

Second Step

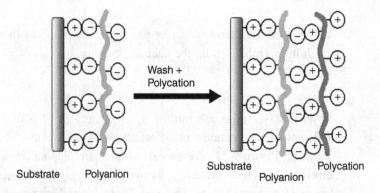

Figure 9.24

Simplified molecular picture showing the complexation of the weak polyelectrolyte to the positively charged substrate during the first adsorption steps of the LbL assembly process. During the first step, the counter-ions from the substrate (Cl$^-$) and the polyelectrolyte (Na$^+$) are being expelled during the process, forming NaCl. The second step of the process involves the relaxation of the adsorbed polyelectrolyte. This is followed by subsequently washing and immersing of the substrate in solution containing polycations for building of the next multilayer coating.

In general, during the LbL assembly, immersion of the substrate into the oppositely charged weak polyelectrolyte solution results in complexation of the polyions by the charged substrate surface, thereby forming ion pairs. The energetically favorable ion pairing of the polymer segments results in the loss of water and the liberation of undissociated low molar mass counter-ions to maintain electrostatic neutrality. As such, during ion pairing, the oppositely charged polymer segment and counter-ions compete to balance the charge density. This competition to balance the charge during the LbL assembly is a two-step process, with the first step involving the initial formation of the polyelectrolyte–substrate complex by having the molecules come together to expel their counter-ions as they complex. Figure 9.24 shows a schematic of the first and second adsorption steps, depicting film deposition starting with a positively charged substrate. As such, as the charged substrate is absorbing an oppositely charged polyelectrolyte, the counter-ions from the substrate surface are being displaced by the adsorbing polymer segments from the polyelectrolyte. The second step shows the relaxation of the adsorbed polyanion polymer followed by subsequent building of the next layer of polycation coating.

Each monolayer adsorbed on the substrate surface is amorphous, has distinct order and thickness, and is profusely interpenetrating with neighboring layers. This interpenetration allows the newly formed polyelectrolyte layer to interact with up to seven to eight previously formed monolayers. With the different types of interactions driving the assembly process, the end result of the LbL assembly is a deposition of nano-scale ultrathin multilayered films on substrate surfaces. The thickness of each monolayer varies from 1 nm up to several tens of nanometers, depending on the time provided for the adsorbed polymers to relax. Additionally, the deposited films are stable even in harsh chemical environments due to the covalent bonding of the irreversibly bound oppositely charged layers. In circumstances where non-covalent intermolecular inter-actions occur between the oppositely charged layers, additional treatments such as cross-linking and thermal treatment can be introduced to enhance layer bonding.

9.8.1 Different layer-by-layer (LbL) assembly techniques

Several variations of the LbL assembly technique are available, besides the dipping method discussed above. Other common methods of the LbL assembly include

- spin-coating LbL assembly, and
- spray-LbL assembly.

As depicted in the schematic in Figure 9.25, the spin-coating LbL assembly involves having aliquots of the polyelectrolyte solution on the substrate surface. The substrate is then rotated at a constant speed and the centrifugal forces spread the polyelectrolyte solution thinly on the substrate. Evaporation of the solvents results in polymers adsorbed on the substrate surfaces.

Unlike the dipping method whereby the monolayers formed are influenced by the interactions between the adsorbed polymers from the polyelectrolyte solution and the oppositely charged substrates and the amount of time required for the

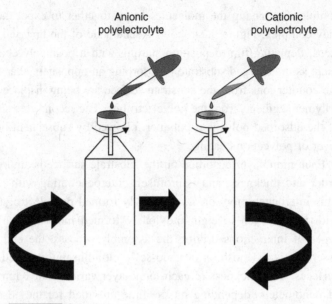

Anionic
polyelectrolyte

Cationic
polyelectrolyte

Figure 9.25

Schematic of experimental sequence for multilayer film deposition using the spin-coating LbL process.

adsorbed polymers to rearrange themselves on the substrate surface, the spinning speed used during spin-coating LbL assembly influences the adsorption and rearrangement of adsorbed polymers on the substrate surface and the simultaneous removal of weakly bound polymer from the substrate. Since the presence of water molecules in the LbL assembly screens the electrostatic attraction, the rapid removal of water molecules as a result of the spinning process also enhances the electrostatic attraction. In addition, the rapid removal of the water molecules also increases the molar concentration of the polyelectrolyte solution, thereby increasing the thickness of the monolayers formed. The disadvantages of the spin-coating LbL assembly method include the

- restriction in the substrate size, depending on the spinning component, and
- inability to effectively deposit monolayers on substrates that are non-planar.

As an alternative to spin-coating LbL assembly process, the spray-LbL assembly method is convenient, fast, and applicable for depositing monolayers on large substrate surfaces. Shown in Figure 9.26, the deposition process involves spraying of atomized mists of polyelectrolyte solutions on substrates to rapidly form homogeneous monolayers.

The speed of the spray process is increased when excess polyelectrolyte solutions arriving at the substrate surface are constantly removed by drainage, thereby avoiding the rinsing step used in the dipping method. In comparison to the

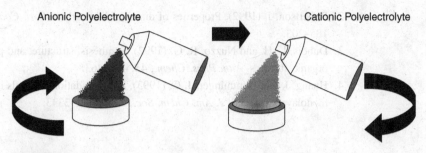

Anionic Polyelectrolyte   Cationic Polyelectrolyte

Figure 9.26

Schematic of experimental sequence for multilayer film deposition using the spray-LbL assembly process.

dipping method, the monolayers deposited are thinner in the spray-LbL assembly method. There are several factors influencing the monolayers deposited using the spray-LbL assembly method including

- spraying time,
- polyelectrolyte concentration, and
- distance and angle of the spray.

9.9   Summary

The different surface modification techniques described here represent many of the strategies that are used for modifying surfaces of medical devices. However, these techniques do not cover the entire spectrum of strategies available to the medical device industry. In many instances, a combination of different processes are utilized by manufacturers to optimize the overall properties of a biomaterial. Examples of such strategies include the use of abrasive blasting with thermal spraying or the use of abrasive blasting with acid etching. Advances in science and technology result in improvements to existing technologies and help optimize the surface modification processes. Overall, engineers and scientists should take into account the substrate properties, the final desired outcome, ease of the method, limitations of the process, and cost when selecting a modification technique.

References

1. Bogaerts, A., Neyts, E., Gijbels, R. and van der Mullen, J. (2002). Gas discharge plasmas and their applications. *Spectrochimica Acta Part B*, **57**, 609–658.

2. Robertson, J. (1992). Properties of diamond-like carbon. *Surf. Coat. Technol.*, **50**, 185–203.
3. Dubois, L. H. and Nuzzo, R. G. (1992). Synthesis, structure, and properties of model organic surfaces. *Ann. Phys. Chem.*, **43**, 437–463.
4. Huang, J. and Hemminger, J. C. (1993). Photooxidation of thiols in self-assembled monolayers on gold. *J. Am. Chem. Soc.*, **115**, 3342–3343.

Suggested reading

• Kim, K.-H., Narayanan, R. and Rautray, T. R. (2010). *Surface Modification of Titanium for Biomaterial Applications*. Nova Science Publishers, ISBN 1608765393.
• Williams, R. (2010). *Surface Modification of Biomaterials: Methods Analysis ad Applications*. Woodhead Publishing, ISBN 1845696409.

Problems

1. Define the term "surface modification."
2. What are the two general classifications of surface modification?
3. Name two alterations of biomaterials surfaces that are impacted by abrasive blasting.
4. What are the problems associated with the use of d.c. glow discharge treatment?
5. What is the advantage of using an a.c. glow discharge over a d.c. glow discharge treatment?
6. With the use of a.c. as an excitation source for ionization, what happens to the two electrodes?
7. Thermal spraying requires high temperature (minimum of 2700 °C). Because of the high temperature involved, what detrimental effect can happen to the ceramic coatings after being deposited on metallic substrates?
8. How does high temperature in the thermal spraying process affect the adhesion of the coatings to the metal substrates?
9. Physical vapor deposition process is a family of vaporization processes for depositing thin films. How is this process different from all the other processes?
10. How is chemical vapor deposition different from physical vapor deposition?
11. Grafting requires the reaction of one or more polymeric species to the main chain of the polymeric macromolecules. Name the two types of activation that are commonly used for the grafting process.

12. For self-assembled monolayers (SAM) to form, SAM molecules need to be amphiphilic. Define what is meant by amphiphilic.

13. During SAM formation, how is the SAM molecule attached to the substrate surface?

14. What are the factors affecting layer-by-layer film assembly?

15. What problems are associated with the use of layer-by-layer film assembly?

16. Name two surface modification techniques that are considered line-of-sight and non-line-of-sight processes.

10 Sterilization of biomedical implants

Goals

After reading this chapter, students should understand the following.
- Fundamentals and importance of sterilization.
- Different types of sterilization methods commonly used.
- Principles behind determining the type of sterilization method suitable for an application.

The world around us is full of microorganisms such as bacteria, fungi, and viruses. These microorganisms are present in the atmosphere, on the surface of all objects, and even on our own skin. In the process of manufacturing an implantable medical device, there is always a possibility of contaminating the device surface with microorganisms. As required by regulatory agencies such as the Food and Drug Administration (FDA), it is mandatory to sterilize all devices prior to implantation. The goal of sterilization is to render the implant devoid of all potential infection-causing organisms, making it one of the most important steps in the manufacturing process of biomedical devices. In this chapter, the basic terminology associated with sterilization such as *bioburden* and *sterility assurance level* is explained. In addition, commonly employed sterilization methods such as steam sterilization, ethylene oxide sterilization, and gamma radiation sterilization are described along with various other new and old sterilization methods. The suitability of these methods for sterilizing different biomaterials is also discussed.

10.1 Common terminology

Bioburden is the term used to represent the number of microorganisms that are adhered to the implant surface prior to sterilization. The duration of any sterilization process depends on the number of microorganisms that are present on the implant surface. The process is quicker if few microorganisms are present. So, it is important to control the adherence of microorganisms on the implant surface during the manufacturing process. Bioburden is typically determined by randomly choosing product samples (10 to 30) from different production batches.[1] These samples are then immersed in a sterile solution, followed by mechanically agitating them (typically by shaking or sonicating) to remove the microorganisms from the surface. The extracted microorganisms are counted using a standard plate method and conventional microbiological techniques.

The probability of the occurrence of one unsterile product during a sterilization process is known as sterility assurance level (SAL). The generally acceptable value of SAL for biomedical devices is 10^{-6}, which means that no more than one device in a million can be unsterile after a sterilization process. The determination of SAL involves the initial determination of bioburden. It is followed by exposing the implant to specified sterilization conditions for a fraction of the proposed total sterilization time and recording the corresponding SAL. This process is continued until the SAL reaches 10^{-6}. The time needed to reach this level is the minimum time required to achieve acceptable sterilization. An example plot of $\log_{10}$ microorganisms versus time is shown in Figure 10.1, with the SAL reaching 10^{-6} at 6 hours. It is recommended to keep the total sterilization time slightly more than the minimum time required to reach the SAL. This is done in order to avoid any natural variations in the number of microorganisms on different product samples or differences in bacterial resistance to sterilization, which could result in failure to achieve the required 10^{-6} SAL level. It is also important to thoroughly characterize the implants after sterilization to confirm that no changes in the surface and bulk properties have occurred during the sterilization process. Some of the techniques used for characterizing implants and medical devices are described in Chapter 4.

- Bioburden refers to the number of microorganisms adhered on the implant surface prior to sterilization.
- Sterilization assurance level (SAL) is the probability of an occurrence of one unsterile product during a sterilization process.

10.2 Steam sterilization

Steam sterilization, a method traditionally used for sterilization of implants, utilizes pressurized heated steam to eliminate microorganisms. An autoclave,

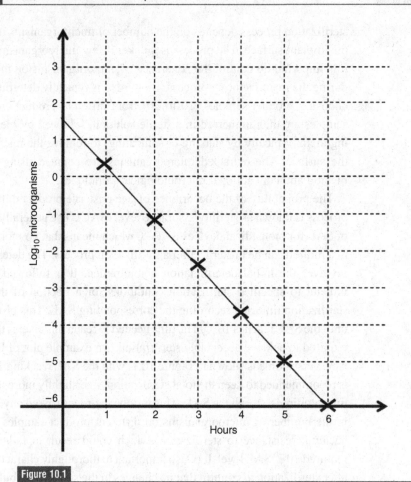

Figure 10.1

An example plot of $\log_{10}$ microorganisms versus time. SAL reaches 10^{-6} at 6 h.

a pressurized steam heated chamber as shown in Figure 10.2, is commonly used for this purpose. During sterilization, the medical device is exposed to saturated steam under high pressure of approximately 15 psi and at a temperature of 121 °C for 15 min. The temperature used can be increased up to 134 °C, provided that the medical device can withstand such a high temperature. This increase in sterilization temperature may be accompanied by decrease in sterilization exposure time. The Medical Research Council recommends temperatures of 121 °C, 126 °C, and 134 °C for sterilization exposure times of 15 min, 10 min, and 3 min, respectively.[2] Steam sterilization kills the microorganisms mainly by irreversibly denaturing and coagulating proteins and enzymes essential for their viability and reproduction. However, great care must be taken to ensure that the steam passes freely through the packaging material and contacts all surfaces of the device. The required exposure time of 15 min starts only after the entire medical device

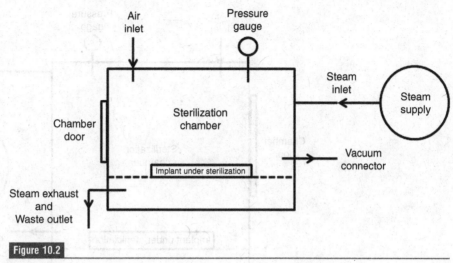

Figure 10.2
Schematic of an autoclave.

reaches a temperature of 121 °C. The advantages of steam sterilization are that it is simple to use, can be performed rapidly, is efficient, and leaves no toxic residues. Although steam sterilization is the preferred method for sterilizing implants, surgical instruments, and devices which are metallic or made of other heat-resistant materials, it cannot be used for many polymeric biomaterials or any materials that cannot withstand the heat employed during this process.

- Steam sterilization is preferred for metal-based implants, devices, and surgical instruments.
- Steam sterilization is not compatible with the majority of polymers and heat-sensitive materials.

10.3 Ethylene oxide sterilization

Ethylene oxide (EtO) gas is the most commonly used chemical agent for gas sterilization. EtO is a colorless flammable liquid with a boiling point of 10.7 °C. Above this temperature, it is a gas that is highly reactive and explosive. The gas is toxic and is a potential human carcinogen. It is rarely used in its pure form because of its explosive nature and is often mixed with other inert gases such as nitrogen or carbon dioxide. Although chlorofluorocarbon compounds have been mixed with

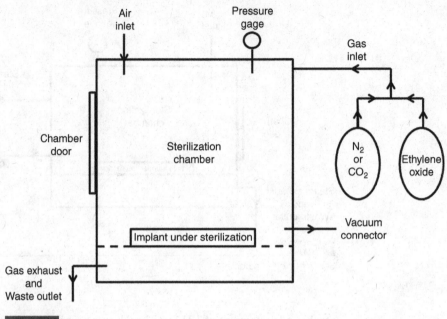

Figure 10.3

Schematic of an ethylene oxide sterilization chamber.

EtO gas, the use of such compounds has been greatly restricted because they can deplete ozone in the atmosphere.

A schematic of an ethylene oxide sterilization chamber is shown in Figure 10.3. The medical device to be sterilized is enclosed in packaging material which is permeable to EtO gas and is placed in the sterilization chamber. The air inside the chamber is evacuated and a relative humidity of 60%–80% is maintained by passing steam through the system. EtO gas is injected into the chamber and its concentration is normally maintained in the range of 700–800 mg/l.[2] The temperature of the chamber is maintained between 30–50 °C, and a time of 2–16 h is typically used to achieve the required SAL. The four most important parameters that govern the efficiency of EtO gas sterilization are the relative humidity, EtO concentration, temperature, and time. After sterilization, the chamber is evacuated to remove the EtO, and the chamber is flushed with sterile air or nitrogen several times to bring the EtO levels below the acceptable limit. In addition, the material is often aerated outside the sterilization chamber to remove any traces of EtO present.

- Ethylene oxide gas sterilization is a low temperature process that is compatible with most biomaterials.

- The possible deposits of residues and by-products of ethylene oxide on implant surfaces is a concern.

Ethylene oxide gas sterilization kills microorganisms by altering the chemical structure of nucleic acids (DNA and RNA) or proteins in the microorganisms. Alkylation is the primary mechanism involved in EtO gas sterilization.[3] EtO is a strong alkylating agent, and it replaces labile hydrogen atoms of amine groups (the reactive functional groups commonly present in nucleic acids and proteins) with hydroxyethyl groups. This chemical alteration affects the viability and proliferation of microorganisms.

Since EtO sterilization is carried out at a low temperature, it can be used for sterilizing a wide variety of implants including vascular grafts, heart valves, intraocular lenses, and bioabsorbable fracture fixation devices. This technique is also highly effective for sterilizing porous materials since the EtO gas can easily penetrate through the pores. The presence of toxic residues of EtO and its by-product, ethylene chlorohydrin, on the implant surface are the main limitations of this method. Additionally, laboratory personnel are required to use appropriate personal protective equipment (PPE) to minimize EtO exposure during the sterilization process.

10.4 Gamma radiation sterilization

Radiation sterilization of medical devices is performed by exposing the material to gamma radiation. Cobalt-60 (^{60}Co), a radioactive isotope of cobalt with a half-life of 5.27 y, is used as the source for producing gamma radiation. A schematic of gamma radiation sterilization is shown in Figure 10.4. The gamma irradiator consists of a room with thick concrete walls for holding the ^{60}Co radiation source. These radiation-shielding walls are necessary to contain the gamma rays and to prevent them from escaping into the environment. Since the isotope decays continuously, the source is shielded under water in a large underground pool when not in use in order to avoid radiation exposure for personnel. The product to be sterilized is transferred to the radiation room via a conveyor belt and removed after irradiation. A radiation dose of 25 kGy is typically used to sterilize medical materials. Radiation dosimeters, which measure ionizing radiation, are placed along with the medical devices to ascertain if the material has received the minimum dosage required for effective sterilization and has not received greater than the maximum dosage that the material could withstand.

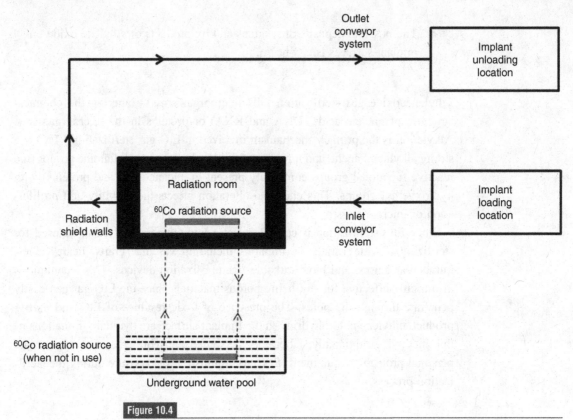

Figure 10.4

Schematic of gamma radiation sterilization.

The advantages of gamma radiation sterilization are its efficacy and compatibility with a wide range of materials. A variety of polymers, including polyethylene, polyesters, polystyrene, polysulfone, and polycaprolactone, can be sterilized using gamma irradiation. A limitation of this mode of sterilization is the high initial capital investment needed for setting up a gamma irradiator. Also, polymers such as polytetrafluoroethylene (PTFE), polypropylene, ultrahigh molecular weight polyethylene (UHMWPE), poly(lactic acid), poly(glycolic acid), and cyanoacrylates can be sterilized using gamma irradiation but with some precautions and prior consideration. PTFE and polypropylene are highly sensitive to radiation, while UHMWPE undergoes oxidation under gamma irradiation in air which is detrimental to its wear properties. However, the use of inert gases such as nitrogen and argon instead of air greatly reduces the formation of oxidative species on UHMWPE and enables the use of gamma sterilization. Exposing poly(lactic acid) and poly(glycolic acid) to gamma radiation sterilization causes the polymers to undergo radiation degradation and suffer a decrease in their molecular weight.

This factor has to be taken into consideration in determining the initial molecular weight during the design of the device. In the case of cyanoacrylates, a loss in elongation and tensile properties occurs when they are exposed to gamma radiation sterilization. Once again, appropriate accommodations in the initial pre-sterilization properties of the polymers can be made in the design phase by engineers to compensate for the effects of radiation.

Ionizing radiations such as gamma radiation kill microorganisms by damaging their DNA either directly or indirectly. The direct effects of radiation on DNA molecules include single- and double-strand breaks, base pair damage, and inter-strand cross-links. Similar damage to DNA molecules can be caused indirectly by the free radicals generated during radiation sterilization.

- Gamma radiation effectively sterilizes a wide range of materials.
- Setting up a gamma radiation facility is expensive.

10.5   Other sterilization methods

Apart from the commonly used sterilization methods described above, there are several other methods which can be used for sterilization. These include dry heat sterilization, formaldehyde and glutaraldehyde treatments, phenolic and hypochloride solution treatments, UV radiation sterilization, and electron beam sterilization. These methods are not commonly used owing to the use of high temperature (dry heat sterilization), their toxic nature (formaldehyde and glutaraldehyde treatments), their hazardous nature (phenolic and hypochloride solution treatments), or their ineffective penetration capability (UV radiation and electron beam sterilization).

10.5.1   Dry heat sterilization

As the name indicates, this sterilization is performed using dry heat with negligible amount of moisture in the air. The temperature used for this sterilization mode varies from 120 °C to 180 °C, with less exposure time required when high temperatures are used. The recommended temperatures and exposure time used in dry heat sterilization are 120 °C, 160 °C, and 180 °C for 6 h, 2 h, and 30 min, respectively. This sterilization is only suitable for metallic implants and devices.

10.5.2   Formaldehyde and glutaraldehyde treatments

Chemical agents such as formaldehyde and glutaraldehyde have been used for sterilizing materials. These chemicals are well known for their microbicidal effect on a variety of bacteria, fungi, and viruses. Their microbicidal property is the result of their cross-linking with proteins and nucleic acids of the microorganisms to inactivate them. A longer immersion time is usually required for glutaraldehyde treatment. These sterilization procedures are occasionally carried out in clinics but are employed only if there is a necessity. For example, formaldehyde or glutaraldehyde treatments are performed if the material cannot be sterilized using steam or ethylene oxide sterilization. The use of these chemicals is limited due to their toxic nature. Additionally, they require a great deal of labor to confirm that the required SAL level has been achieved.

10.5.3   Phenolic and hypochloride solution treatments

Chemical agents such as phenolic compounds (derivatives of phenol) and hypochloride solutions have been used for sterilizing polymers. Although these treatments involve low temperature, the immersion time in these solutions is generally longer and some polymers may be damaged under these conditions. Additionally, these treatments are not economical and not very commonly used due to the hazardous nature of the chemicals and ambiguity in achieving the desired SAL.

10.5.4   Ultraviolet (UV) radiation

Ultraviolet radiation is typically used to sterilize biological safety cabinets and instruments used for *in vitro* cell culture studies. With a wavelength of 260 nm, UV radiation can effectively kill most microorganisms by causing the dimerization of thymine molecules in their DNA and preventing their replication and transcription. Although UV can sterilize the surface of a material effectively, it cannot effectively penetrate glass, dirt or water. As a result, the main limitation in using UV radiation as a sterilization mode is its poor penetration capability. Additionally, care must be taken to avoid prolonged UV exposure as it may have deleterious effects. Since UV radiation can degrade organic molecules, this effect is a serious concern when using UV sterilization for polymers and implants with organic coatings.

10.5.5   Electron beam sterilization

In this sterilization method, an electron beam linear accelerator machine with power ranging from 1.7 to 20 kW is used to generate and accelerate electrons.[2] The accelerator is located inside a concrete room to limit the exposure of stray electrons (the electrons which have deviated from their normal path) to the outside environment. Unlike gamma radiation sterilization, no underground water pool is necessary since no radioactive material is used in this method. An electron beam dose of 25 kGy is normally used to sterilize implant materials with an exposure time of 5 min.

Although the advantage of electron beam irradiation is the short exposure time, the main limitation of this method is the low penetration depth of an electron beam. This depth of penetration depends on the accelerating voltage of the electron beam and the density of the material to be sterilized. For example, a material with a density of 0.1–0.2 g/cm^3 permits a beam penetration depth of 40–80 cm with a 10 MeV electron beam.[2]

10.6   Recently developed methods

Two newly developed methods which are promising for sterilizing biomedical implants are the low temperature gas plasma and gaseous chlorine dioxide treatments.

10.6.1   Low temperature gas plasma treatment

The hydrogen peroxide (H_2O_2) gas plasma system is developed as an alternate to steam autoclaving and ethylene oxide gas treatment. The significant advantage of H_2O_2 gas plasma treatment over steam autoclaving is that it can be performed at relatively low temperatures of less than 50 °C.[4] When compared to EtO sterilization, the processing time of H_2O_2 gas plasma treatment is significantly lower. Medical devices undergoing H_2O_2 gas plasma treatment require an exposure time of less than 75 min, whereas EtO gas sterilization requires an exposure time of 2–16 h to achieve the required SAL.[4] Additionally, unlike EtO treatment where aeration is mandatory to remove residues, there is no post-treatment aeration required to remove residual H_2O_2 deposited on the material surface. Instruments and devices are ready for use as soon as the H_2O_2 gas plasma sterilization process is completed. The only limitation in using H_2O_2 gas plasma treatment is that H_2O_2 is an oxidizing agent, and care must be taken to confirm that material properties are not affected during the treatment.

10.6.2 Gaseous chlorine dioxide treatment

Gaseous chlorine dioxide has been successfully used to sterilize biomedical implants. Similar to the low temperature H_2O_2 gas plasma treatment, this is also a rapid procedure requiring an exposure time of only 1.5–3 h, and no post-sterilization aeration.[5] The green color of the chlorine dioxide gas can be used to measure its concentration inside the sterilization chamber using a spectrophotometer.[5] This proves to be a significant advantage for this treatment since the concentration of the gas inside the sterilization chamber can be easily monitored and controlled. Similar to H_2O_2, chlorine dioxide is also an oxidizing agent. As such, careful monitoring is necessary to confirm that no change in material properties occurs during this treatment.

> Criteria for an ideal sterilization method include the following.
> - Low temperature process.
> - No toxic residues or by-products on the implant surface.
> - Short processing time.
> - No change in the properties of the material under sterilization.
> - Good penetration capability to sterilize implant in full.
> - Economical equipment.

10.7 Summary

Sterilization is a very critical procedure in the manufacture of biomedical implants. In general, an ideal sterilization method should:

- be a low temperature process,
- not leave toxic or hazardous residues on the surface of the medical device after sterilization,
- have a rapid processing time,
- not induce oxidation of the material,
- not affect the bulk and surface properties of the material,
- possess adequate penetration capability of the sterilizing agent in order to fully sterilize the implants, and
- be economical to set up.

As new biomedical devices are developed and manufactured, utmost attention should be given to determining the type of sterilization needed to achieve the required SAL. Commonly used sterilization techniques in today's biomedical industry include steam sterilization, ethylene oxide sterilization, and gamma radiation sterilization. Other less often used methods such as dry heat sterilization, formaldehyde and glutaraldehyde treatments, phenolic and hypochloride solution treatments, UV radiation sterilization, and electron beam sterilization can also be considered. As discussed in this chapter, there are advantages and limitations to each of the above sterilization techniques. New methods are continuously being developed in an attempt to overcome the limitations of the currently available methods.

References

1. Kowalski, J. B. and Morrissey, R. F. (2004). Sterilization of Implants and Devices, in Ratner, B. D., Hoffman, A. S., Schoen, F. J. and Lemons, J. E., editors, *Biomaterials Science: An Introduction to Materials in Medicine*. London, Elsevier Academic Press, pp. 754–760.
2. Giardino, R. and Aldini, N. N. (2002). Infection and sterilization, in Barbucci, R., editor, *Integrated Biomaterials Science*. New York, Kluwer Academic/Plenum Publishers, pp. 815–832.
3. Gad, S. C. (2002). *Safety Evaluation of Medical Devices*, 2nd edition. New York, Marcel Dekker, Inc.
4. Favero, M. (2000). Hydrogen peroxide gas plasma low temperature sterilization. *Infect. Control Today*, **4**, 44–46.
5. Kowalski, J. B. (1998). Sterilization of medical devices, pharmaceutical components, and barrier isolation systems with gaseous chlorine dioxide, in Morrissey, R. F. and Kowalski, J. B., editors, *Sterilization of Medical Products*. Champlain, Polyscience Publications, pp. 311–323.

Suggested reading

• Booth, F. (1999). *Sterilization of Medical Devices*. Buffalo Grove, Interpharm Press, Inc.
• Gad, S. C. (2002). *Safety Evaluation of Medical Devices*, 2nd edition. New York, Marcel Dekker, Inc.
• Matthews, P., Gibson, C. and Samuel, A. H. (1994). Sterilization of implantable devices. *Clinical Mater.*, **15**(3), 191–215.
• Kowalski, J. B. and Morrissey, R. F. (2004). Sterilization of implants and devices, in Ratner, B. D., Hoffman, A. S., Schoen, F. J. and Lemons, J. E., editors, *Biomaterials Science: An Introduction to Materials in Medicine*. London, Elsevier Academic Press, pp. 754–760.

• Hill, D. (1998). *Design Engineering of Biomaterials For Medical Devices*, 1st edition. New York, Wiley.
• Willey, J., Sherwood, L. and Woolverton, C. (2008). *Prescott's Principles of Microbiology*. New York, McGraw-Hill Higher Education.
• Giardino, R. and Aldini, N. N. (2002). Infection and sterilization, in Barbucci, R., editor, *Integrated Biomaterials Science*. New York, Kluwer Academic/Plenum Publishers, pp. 815–832.

Problems

1. What is bioburden? How is it determined?
2. What is the sterility assurance level (SAL)? What is the generally accepted value of SAL? Explain.
3. How does steam sterilization kill microorganisms?
4. What are the advantages and limitations of steam sterilization?
5. What are the advantages and limitations of ethylene oxide sterilization?
6. What are some of the polymers that require special consideration when sterilized using gamma irradiation? Explain.
7. What is the main limitation of UV irradiation sterilization?
8. What are the main advantages and limitations of electron beam sterilization?
9. What are the advantages of low temperature gas plasma treatment over steam sterilization and ethylene oxide gas treatment?
10. What are the advantages of gaseous chlorine dioxide treatment?

11 Cell–biomaterial interactions

Goals

After reading this chapter, students will understand the following.
- Key components of the extracellular space.
- Principal proteins and pathways that cells utilize to interact with both cellular and non-cellular environments.
- Adhesion mechanisms that bind cells to substrates and types of junctions found near biomaterials.
- The role of this cell matrix environment in the success or failure of biomaterial integration.

How do cells, containing the same genetic information, diversify and give rise to so many types of tissue? This fundamental question of cell biology has a surprisingly simple answer: environment. Despite the incredible complexity of internal genetic control, cells rely equally on their surroundings to define their form and function. In the study of cell biology, the plasma membrane traditionally defines the boundary between the functional unit of the cell and its environment. The interactions that occur at this interface represent an exceedingly complex and highly organized series of reactions that permits the cell to send and receive biochemical signals across the membrane. Most eukaryotic cells define their structure and function based on these signals. Even for those cells that are not substrate bound, such as those of the circulating immune system, it is essential to sense and respond to biochemical gradients and interactions in the body. The availability, intensity, and duration of these gradients are the signals which direct cells into their most common activities, that is, migration, division, and differentiation. These activities provide the complexity of all cell response in the body and can specifically define the reaction to any material implanted.

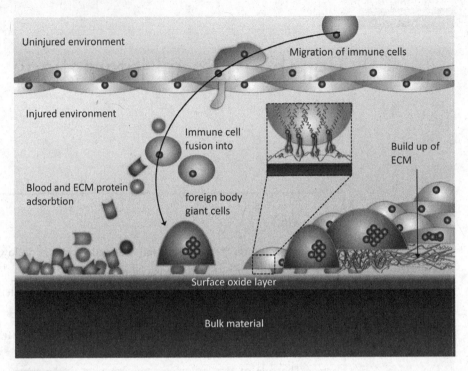

Uninjured environment

Migration of immune cells

Injured environment

Immune cell fusion into

Build up of ECM

Blood and ECM protein adsorbtion

foreign body giant cells

Surface oxide layer

Bulk material

Figure 11.1

A time course representation of extracellular proteins, various cells and mature tissues that illustrate the variety of interactions which occur throughout the lifetime of an implanted material. In this diagram early proteins are bound to a material surface from the blood environment eliciting immune type cells to bind and modify the surface, later connective tissue and mature bone cells alter the surface of the material with time. A magnified view of the maturing cells demonstrates the cell surface receptors mediating the connection from the extracellular space to intracellular components such as the actin cytoskeleton.

The native mammalian extracellular matrix and precisely how the cells mediate their interactions with this environment are defined in this chapter. In Figure 11.1, a representation is shown of the connective tissues surrounding an implanted biodegradable material and demonstrates the diverse connections that can be found between cells, their surrounding protein matrixes, and materials. Many of the topics covered in this chapter are illustrated in this figure, including the protein layers initially formed on a biomaterial surface and their subsequent incorporation into a thin matrix. These initial proteins represent the starting point for early to long-term response to a material which includes contacts with immunological, fibroblastic, and ultimately mature cell types such as the bone observed here. A closer inspection of these interactions will indicate how receptor proteins in the plasma membrane bind the cell to its extracellular environment (shown in the lower panel insert) and mediate this signal via their connections to intracellular proteins and structures.

11.1   The extracellular environment

The illustration in Figure 11.1 indicates a thin layer of protein separating cells from their environment and residing just outside of the plasma membrane. This thin protein layer is called the glycocalyx and represents an organized layer of secreted proteins which are able to interact with the carbohydrates protruding from the plasma membrane. The size of this layer can vary significantly, depending on cell type. For example, skin cells maintain a basement membrane on the order of microns (1×10^{-6} m) when transitioning from the epidermis to dermis. Another example of the glycocalyx is the digestive lining surfaces where ectodermal cells maintain a well-defined protein layer but of reduced scale in nanometers (1×10^{-9} m). This thin lining can be considered a protective coating to the cell membrane and a highly specific mechanism for cells to engage in extracellular communications.

Directly outside the glycocalyx begins an organized network of extracellular matrix (ECM) in which most mammalian cells reside. One of the best defined ECM environments is that (shown in Figure 11.2) of an osteocyte, a specialized

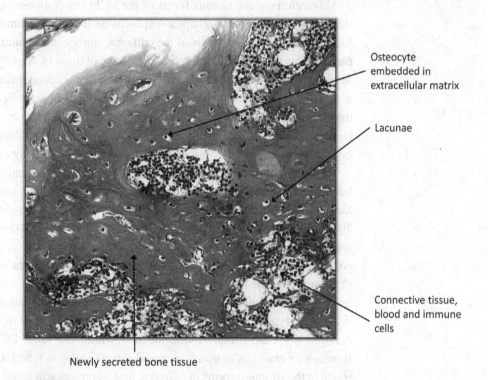

Osteocyte embedded in extracellular matrix

Lacunae

Connective tissue, blood and immune cells

Newly secreted bone tissue

Figure 11.2

Developing bone tissue showing an embedded osteocyte cell residing in a structure called a lacuna. The osteocyte can be seen organizing a local ECM between itself and the calcified bone tissue. Approximate size of the osteocyte is 20 μm in diameter, with extensions spanning 200 μm in length.

bone cell that maintains and responds to mechanical forces in the skeleton. Note how the cell has created a compartment for itself within the surrounding hard bone tissue. A highly active interplay is found in this local ECM where the osteocyte secretes proteins in response to injury. In cell culture studies where the ECM is removed, cells often will de-differentiate or lose their specialized function, which can be restored only through the return of their extracellular matrix.

Another well-defined ECM is the basement membrane. Often called the basal lamina, the basement membrane is found in many cell types including the epithelium, respiratory cells, inner endothelial layer of blood vessels and the digestive tract. These ECMs are dense structures, varying in thickness (10–300 nm), and they often serve to clearly compartmentalize differing tissue types. Basement membranes also serve as the attachment anchors for cells, create a platform for cell migration, and act as a barrier for most molecules and proteins. Because of their dense arrangement of protein, the basal lamina can be used to prevent diffusion, as in the case of keeping circulating proteins in the blood supply despite the porous characteristics of capillary vessel walls.

Although there are various forms of the ECM, its composition is quite similar throughout the body. This chapter will focus on the most common proteins in the ECM which include members of the collagen, fibronectin, laminin and proteoglycan families. As illustrated in Figure 11.3, many of these ECM proteins are found in fibrous strands rather than the form of protein usually found inside cells of single, globular shapes. By arranging the ECM into fibrous meshes, cells are able to create mechanically strong surroundings which also serve as a local environment for the incorporation of many signaling proteins. Additionally, any change to this native organization can have serious consequences that result in many disease states.

The most common ECM protein as well as the most abundant protein in the human body is collagen. Details of collagen can be found in Chapter 8. This family of proteins can be found throughout the ECM and represents approximately 30% of all protein. They are also one of the most resistant biological materials to applied tension or pulling force. This mechanical feature helps explain why collagen is found in so many of the tissues in our body and the incredible variety of forms in which it is found. To date, there are at least 28 different molecular types of collagen identified, and these distinct forms are described in Table 11.1. While each type of collagen can have specificity to individual tissues in the body, multiple types are usually found together within the same ECM environment. Bone, as an example, contains an abundance of collagen type I. Collagen type III is one of the strongest forms of collagen, and its presence in small amounts in bone tissue may contribute to the density of bone. A similar property can be found in nearly all biological tissue environments where each protein type uniquely contributes to the character of the tissue.

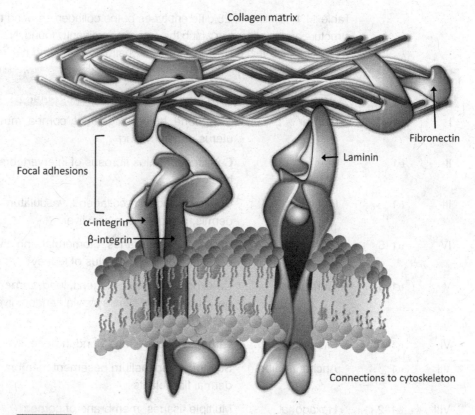

Collagen matrix

Fibronectin

Laminin

Focal adhesions

α-integrin

β-integrin

Connections to cytoskeleton

Figure 11.3

A representative illustration of a cell's surface with corresponding connections to a fibrous ECM. Of the proteins shown in the figure, note the inter connections between the collagen, laminin, fibronectin and how they provide anchoring and attachment places for cell surface receptors such as integrins and tyrosine kinases.

Despite their seeming heterogeneity, the key feature of the collagen family is that they possess the fundamental unit of the collagen strand. Depicted in Figure 11.4, this fundamental unit comprises three collagen chains (α-chains) wrapped around one another to make a triple helix design.

Classified in scale from small to large, the organization of the collagen design can be classified as follows:

- polypeptide α-chains of approximately 300 nm in length,
- three α-chains arranging into a *monomer*,
- multiple monomers stacked together into a *fibril*, and
- multiple fibrils stacked together into a *fiber*.

As shown in Figure 11.4, the so-called 'fibrillar' arrangement of monomers is a result of their slightly off-axis stacking, since the small gaps in lengthwise

Table 11.1 Listing of 28 different types of the collagen showing their identified structure and the tissue in which they are predominantly found. Note the co-listing of various types within a single tissue demonstrates the hybrid nature of most tissues

Type	Subunits	Structure / group	Tissue distribution (not comprehensive)
I	α1–2	Fibrillar	Skin, bone, tendon, ligament, cornea, muscle, liver, uterus, amnion, lung
II	α1	Fibrillar	Cartilage, annulus fibrosus of intervertebral disc, vitreous humor
III	α1	Fibrillar	Co-expressed with collagen I, vasculature and skin, fetal dermis, placenta, synovia, liver
IV	α1–6	3D Network	Basement membranes, glomerular and alveolar BM, lens capsule, aorta, glomerulus of kidney
V	α1–4	Fibrillar	Co-expressed with collagen I, lung, corneal stroma, bone, nervous system, skin, synovia, tendon, liver, muscle, amnion
VI	α1–6	Microfibril	Large distribution, not found in bone, liver, uterus, aorta
VII	α1	Anchoring fibril	Squamous epithelium basement membrane, placenta, dermal fibroblasts
VIII	α1–2	Hexagonal lattice	Multiple tissues, membrane of cornea
IX	α1–3	Facit	Identified with type II fibrils of cartilage and cornea, human fetus, hyaline cartilage, embryo, eye vitreous humor
X	α1	Hexagonal lattice	Hypertrophic cartilage
XI	α1–3	Fibrillar	Co-expressed with collagen II, cartilage, annulus fibrosus, nucleus pulposus (early development)
XII	α1	Facit	Found with type I fibrils of the perchondrium, ligament, tendon, embryonic tissue
XIII	α1	Transmembrane	Low concentrations in many fetal tissues, lens capsule membrane
XIV	α1	Facit	Identified with type I fibrils in multiple tissues
XV	α1	Multiplexin	Basement membranes
XVI	α1	Facit	Identified with type II fibrils in hyaline cartilage and microfibrils in skin

Table 11.1 (cont.)

Type	Subunits	Structure / group	Tissue distribution (not comprehensive)
XVII	α1	Transmembrane	Skin and intestinal epithelia
XVIII	α1	Multiplexin	Endothelial and epithelial membranes
XIX	α1	Facit-like	Low concentration in some membranes, in developing muscle
XX	α1	Facit	Identified with type I fibrils of cartilage, cornea, tendon
XXI	α1	Facit	Identified with type I fibrils in vessel walls
XXII	α1	Facit-like	Identified with microfibrils at tissue junctions
XXIII	α1	Transmembrane	Heart, lung and brain, metastatic cells
XXIV	α1	Fibrillar	Co-expressed with collagen I in bone and cornea
XXV	α1	Transmembrane	Neurons
XXVI	α1	Facit-like	Testis and ovary
XXVII	α1	Fibrillar	Co-expressed with collagen II in cartilage and epithelia
XXVIII	α1	Fibrillar	Peripheral nerves

arrangement seldom align together and can be physically seen as the banding pattern. These patterns are distinctly characteristic of collagen fibers. Additionally, such arrangement of monomers also accounts for their mechanical strength.

Chemical cross-links between neighboring collagen monomers tend to increase with the age of all tissues. Since there is reduced flexibility of muscles, tendons, and ligaments with age, the age-induced increase in chemical cross-links between neighboring collagen monomers may account for one explanation for the decreased elasticity of many tissues. Note that the covalent cross-linking between monomers is not exclusive to collagen, and this concept is equally applied to other ECM proteins and as well as biomaterial interfaces.

Owing to the abundant distribution of collagens in most ECMs, the properties of many tissues are highly dependent on its organization in terms of both structure and mechanical properties. The aforementioned cases of muscles, tendons, and ligaments share a similar organization of collagen with parallel arrangement to their long axis. In the case of muscle, this is seen where the collagen is arranged along the same axis as that of contraction. A highly organized arrangement can be seen in some of our most mechanically strong tissues of the body such as bone. At the surface of most long bones, like the tibia and femur, there is a stacked layer of calcified bone tissue composed of 65% collagen type I. Quite unique to bone

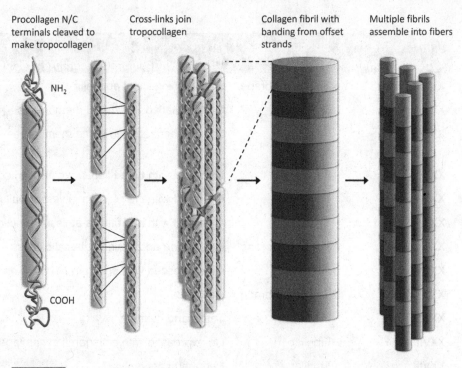

Procollagen N/C
terminals cleaved to
make tropocollagen

Cross-links join
tropocollagen

Collagen fibril with
banding from offset
strands

Multiple fibrils
assemble into fibers

NH$_2$

COOH

Figure 11.4

A structural view of collagen type I. From left to right, observe the tightly packed triple helix of collagen molecules representing three distinct monomer α-chains. Variation in the types of collagen can be identified through differences of individual monomers, for example a homotrimer design would contain the same three α-chains while variation in any one of the chains could create heterotrimer designs. Each momomer of collagen is found arranged in slightly off-axis stacking to create a stacked structure called a *fibril*. The small spacing between molecules accounts for the banding structure observed on the far right with a periodicity of 66 nm. Numerous arrangements of fibrils are often assembled into *fibers* that may be observed by the light microscope and can vary in scale from 100 nm to 300 nm.

tissue is a lamellar organization, where each layer has a highly organized direction of collagen, with each sheet varying fiber orientation from 0° to 90°, as shown in Figure 11.5. Using this arrangement, a plywood-like structure is formed of exceptional mechanical strength. This organization has been optimized to withstand the high compressive, tension, bending and torsion forces experienced by our bones and is repeated both on the outside layers and interior osteons of bone tissue.

In contrast to highly organized collagen stacking is the type of bone formed immediately after trauma. For example, in bone breaks or implantation of orthopedic materials, a 'woven' type of bone is often formed. In such bone type, the collagen is arranged in a more random orientation. Correspondingly, this less

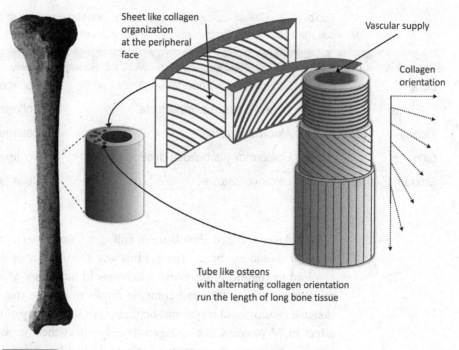

Sheet like collagen
organization
at the peripheral
face

Vascular supply

Collagen
orientation

Tube like osteons
with alternating collagen orientation
run the length of long bone tissue

Figure 11.5

The organization of collagen into different types of bone tissue. On the outer face of long bones note the layer structure of collagen type I arranged into sheets of bone with orientation varying from 0 ° to 90 ° to create a plywood-like stacking with exceptional mechanical strength. The interior of bone tissue is comprised of tube-like osteon arrangements and strut-like trabecular bone that also shares a lamellar organization.

organized arrangement of collagen lacks much of the strength of lamellar bone but can form at a faster rate to repair an injury. Of particular interest to the biomaterial engineer would be considering these arrangements of ECM on biomaterials, not only for initial integration but also for their long-term implant success.

Box 11.1 Conceptual discussion point

From our understanding of the alignment and cross-linking present in native ECMs such as collagen, what can we describe about the role of the ECM deposited on biomaterial surfaces? Taking into account the incredible variety in both form and function of these native ECMs, how would the ECM deposited on a biomaterial surface be affected in terms of chemical adhesion, formation, and strength? Should this be considered a time-dependent concern?

Table 11.2   Classifications of structural or genetic damage to members of the collagen family resulting in abnormal trauma or disease

Tissue	Disease or condition	Damage to ECM protein
Bone	Osteogenesis imperfecta	Collagen I
Cartilage	Chondrogenesis imperfecta	Collagen II
Kidney	Alport Syndome	Collagen IV ($\alpha5$)
Skin	Epidermolysis bullosa dystrophica	Collagen VII
Smooth muscle	Leiomyomatosis	Collagen IV ($\alpha5$)

With the established abundance of collagen found in so many different kinds of tissue, it should not be unexpected that any abnormality or defect in this protein can lead to significant disorders. Mutations in the genes which encode all ECM proteins are among the most common causes of disease that affect the musculo-skeletal, cardiovascular systems, brain, eye, or skin. Many of the mutations which affect ECM proteins like collagen directly affect their structural properties and give rise to severe abnormality. Table 11.2 identifies several major mutations or damage to collagen subtypes and the resulting disease, trauma, or syndrome.

The matrix of the ECM truly represents a variety of proteins to provide unique character reflected in each tissue type. Like collagen, fibronectin is a highly abundant polypeptide organized into a chain-like structure. In the case of fibronectin, the structural arrangement resembles a wishbone configuration, as shown in Figure 11.6. Two distinct chains form a series of potential cross-linking locations with other molecules. Prospective connections include high affinity binding for collagen, fibrin, heparin, and cell–receptor interactions.

The numerous binding locations for cells and other ECM components reinforce the perception of fibronectin as a fundamental linking protein. During embryogenesis, fibronectin plays a vital role directing cell migration patterns, particularly related to nerve tissue formation. Absence of most fibronectin types during embryogenesis results in a lethal condition. In another example, macrophages routinely migrate through tissues using fibronectin as an attachment point. Recent evidence has identified a potent chemotactic signal from damaged ECM involving fragmentation of fibronectin. These fragments represent a powerful stimulus to wound repair mechanisms and are a current target for biomaterial modification.

The primary component of the basement membrane or basal lamina is a unique glycoprotein called laminin. This foundation protein forms the base of cells and anchorage for organs. They are also a highly conserved member of the ECM and are the first members to form in development as distinguishable structural

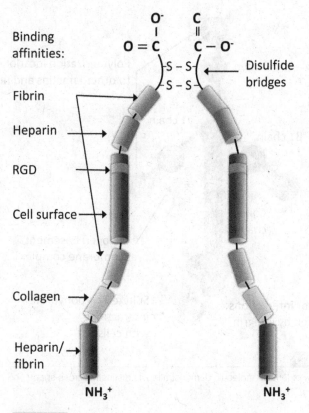

Figure 11.6

Structural organization of the extracellular matrix protein, fibronectin. Two distinct polypeptides of the protein contain ~17 repeating modules folded into several distinct domains with binding locales for collagen, fibrin, heparin, and cell–receptor interactions and other peptide interactions. These two chains are linked by a pair of disulfide bonds found at the C-terminal end in types I and II fibronectin but are absent in type III, which forms a barrel-like structure. Of particular focus later in this chapter is the ability of these cell–receptor interactions to bind with plasma membrane proteins and initiate signaling mechanisms.

components. They provide highly effective barriers between cells, in addition to their role as general mechanical support. The basement membrane underlies endothelial cells of blood vessels, epithelial cells of the skin, and encircles muscle tissue. Laminins are heterotrimeric chain polypeptides classified according to their composition of α, β, and γ chains, respectively. The complete protein appears as a cross-like molecule, with one long arm and up to three short arms. Currently, there are five known laminin-α, 4 known laminin-β, and 3 known laminin-γ chains in mammals that give rise to multiple pairings. An overview of laminin structure is shown in Figure 11.7, where each arm has selective affinity for unique binding partners.

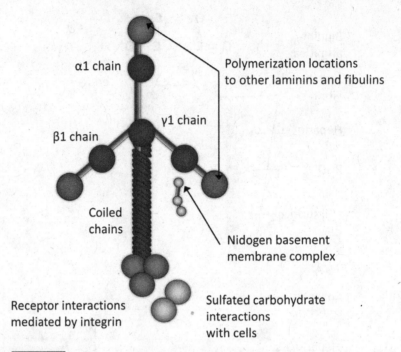

α1 chain

Polymerization locations
to other laminins and fibulins

γ1 chain

β1 chain

Coiled
chains

Nidogen basement
membrane complex

Receptor interactions
mediated by integrin

Sulfated carbohydrate
interactions
with cells

Figure 11.7

Overview of laminin molecule demonstrating its distinctive cross-shape and potential interacting members.

As seen in the structure of fibronectin, multiple recognition locations are present on the laminin molecule with affinity to cell surface receptors, other laminin molecules, collagen, proteoglycans, and other basement membrane components. In particular, laminin is often arranged into large polygonal grids, woven together with collagen IV in a lattice configuration. The notable difference between these two proteins in the basement membrane is that laminin is strongly tethered to the cell via its affinity with integrin in the long chain. Collagen IV has not shown this cell-based dependence on its organization. These networks are essential in a multitude of normal tissue functions and their importance is clearly observed during development. Limb syndactyly, fusion of two or more digits during development, has been linked with a discontinuous basement membrane and correlated with loss of laminin-10 during limb formation.

As previously mentioned, many proteins of the ECM are networked with a class of protein–polysaccharides called proteoglycans. Shown in Figure 11.8, this superfamily of more than 30 molecules is defined by their structure, which contains a long series of carbohydrate polymers with covalently attached glyco-saminoglycan (GAG) side chains. Like other members of the ECM, proteoglycans

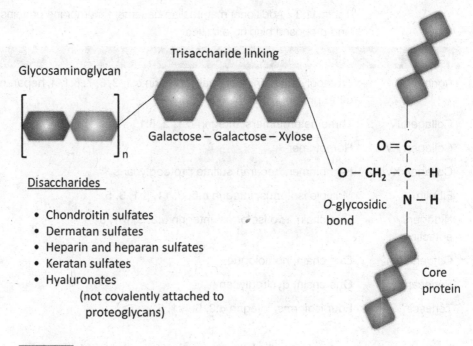

Trisaccharide linking

Glycosaminoglycan

Galactose – Galactose – Xylose

$O = C$

$O - CH_2 - C - H$

O-glycosidic bond

$N - H$

Core protein

Disaccharides

- Chondroitin sulfates
- Dermatan sulfates
- Heparin and heparan sulfates
- Keratan sulfates
- Hyaluronates
  (not covalently attached to proteoglycans)

Figure 11.8

Schematic and structural representation of the "core" protein and side-chain glycosaminoglycans (GAG) which comprise the structure of a proteoglycan.

have multiple binding domains for the interconnection of cells and polymers. However, the vast number of side chains attached to these proteins gives them unrivalled diversity and function. For example, multiple classes of GAG chains including chondroitin sulfate (CS), dermatan sulfate (DS), heparan sulfate (HS), hyaluronan, and keratan sulfate (KS) are available for tissue variety. Details of hyaluronan and CS are provided in Chapter 8. As a surface layer on epithelial cells, proteoglycans are found with millions of syndecan molecules that mediate a diverse range of matrix interactions. As one of the most well understood and simple GAGs, hyaluronan cannot truly be classified as a proteoglycan since it does not covalently associate with a core protein. However, hyaluronan is still often categorized with proteoglycans based on function as a major linker of core proteins. Hyaluronan is a unique non-sulfated GAG, forms in the plasma membrane and can reach massive size with molecular weights in the millions. In cartilage, hyaluronan GAGs and proteoglycans protectively encapsulate chondrocytes, attracting vast numbers of water molecules to dampen mechanical compression by slowing the movement of liquid in a confined space similar to the steady viscous resistance of force provided by a shock absorber.

Table 11.3 Additional mammalian basement membrane proteins, characteristics and proposed binding affinities

Protein	Characteristics, affinity
Agrin	Proteoglycan monomer chain, integrin α:1–3, 6, 7, β:1, 4, heparan, sulfatides and other proteoglycans
Collagen IV	Three heterotrimers, integrin α:1, 2, β:1
Collagen XV	Homotrimer
Collagen XVIII	Homotrimer, heparan sulfate proteoglycans
Fibulin	Multiple isoforms, integrin α:5, 9, 11, β:1, 3, 5
Nidogen/ entactin	One chain, two isoforms, integrin α:3, 5, β:1, 3
Ostoenectin	One chain, homologues
Perlecan	One chain, dystroglycan
Tenascin	Four isoforms, integrin α:2, 5, β:1, 3

A variety of other proteins that form the ECM including nidogen or entactin, perlecan, tenascin, and osteonectin are listed in Table 11.3. Most characteristics of developing and mature tissues can be linked to the components of these basement membranes since they play such a fundamental role in compartmentalizing tissues and signaling behavior.

Box 11.2 Beyond the basics

From the discussion above, the reader may come to a conclusion that the ECM is a diverse network of matrix proteins, proteoglycans, and glycosaminoglycans that support multiple cell functions and provide the complete organization of a tissue. However, recent evidence has shown the ECM's role goes far beyond this. The ECM has been implicated in diverse effects on cell behavior including adhesion, survival, proliferation, migration, and differentiation. Strong evidence shows the action of specific components of the ECM interacting with cell surface receptors. In the next section, the classical cell surface receptor family of the integrins will be introduced. Yet, there are many other interactions between the ligand binding domains of the ECM and cell surface receptors.

Box 11.2 (*cont.*)

The term **matrikine** describes small fragments that have been proteolytically degraded from the ECM and possess bioactivity. In the past 20 years, these small fragments have emerged as key players in the wound-healing response. For example, the fragments of collagen IV provide a chemical signal to neutrophils. On the surface of these cells, receptors show affinity for the X–Gly–X–X–Pro–Gly (XGXXPG) patterns in collagen IV as well as elastin and laminin. This and similar mechanisms allow for the chemotractant phenomenon to permit immune cells to home-in on damaged matrix components. ECM fragments have also been linked with a variety of immune responses including the following:

- host ability to direct bacterial detection and elimination through induction of phagocytic function, and
- secretion of interferon from T cells to direct adaptive immunity.

These pro-immune responses from ECM fragments have been linked to genetic activation of many proteases including the matrix–metalloproteinases which direct ECM degradation. Clearly, the ECM's role goes far beyond cell and tissue support and is involved in numerous feedback systems that can direct functions as diverse as inflammatory and immune response.

11.2 Extracellular matrix mimics

From the previous discussion, it should be apparent how diverse and abundant are the connections of the ECM. As these networks provide the direct connection between cells and tissues, they represent the perfect target for biomaterial mimics that seek to bridge the divide between native and synthetic materials. Shown in Table 11.4 are a select few examples of products finding use as 3D cell-culture scaffolds, surface coatings, sealants, and fixatives currently on the market.

11.3 Cell interactions with non-cellular substrates

Most of the ECM proteins discussed had recognition domains specific for cell interaction. These interactions are made possible through affinity of the matrix

Table 11.4   A variety of trademark matrix products simulating or directly mimicking native tissue matrix components and functions used in tissue culture, biomaterial coatings, sealants, fixatives, and grafting

Product name	Matrix	Use(s)
Algimatrix	Algenic acid	Tissue culture matrix
Amniograft	Human amniotic membrane graft	Tissue healing
Artiss, Tisseel	Fibrin matrix polymer	Tissue sealant
Corgel	Hyaluronic acid gel	Injectable tissue matrix
Extracel	Glycosan	Tissue culture platform
Hyalbrix	Hyaluronic acid gel	Injectable tissue culture
Imedex Bio	Collagen gel	Tissue healing matrix
Integra	Collagen–glycosaminoglycan matrix	Implant matrix
Matigel	ECM reconstitution	Tissue culture platform
PriMatrix	Fetal bovine collagen	Matrix for dermal reconstruction
Sergis, Oasis	Small intestinal submucosa	Wound repair matrix
TrelX	Demineralized bone matrix	Grafting platform

with a variety of receptor families, which includes the integrins, immunoglobulins, cadherins, and selectins. These membrane-spanning proteins represent a much more selective attachment mechanism between cells and native matrix or biomaterial surface adsorbed matrix. This section will examine the major family of the integrins and their ability to form large focal adhesions on substrates.

The integrin family of receptors is a central control system for converting extracellular connections into intracellular metabolism. As identified at the beginning of this chapter, it is the environment that contributes to a cell's role, and the integrins are the main connection between the cell and its environment. Classically, integrins have been represented as two membrane spanning polypeptides which are non-covalently linked. They are classified according to their expression of α and β chains. Each chain contains a large globular head region on the extracellular spanning side and a long leg portion spanning the membrane. Recent evidence has demonstrated that the globular head region bends toward the plasma membrane in its inactive configuration but extends outward toward the matrix when ECM connections are made, as shown in Figure 11.9.

An incredible variety of cell ECM connections are possible due to the 18 different α chains and 8 β chains available for combinations. Currently, research

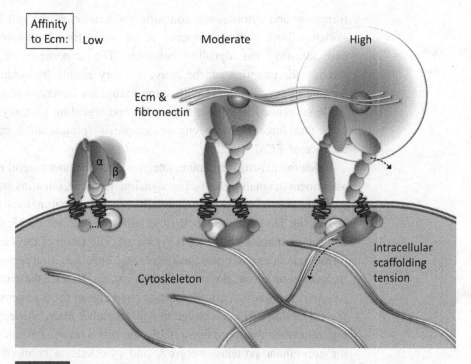

Affinity to Ecm: Low — Moderate — High

Ecm & fibronectin

α β

Intracellular scaffolding tension

Cytoskeleton

Figure 11.9

A representation of integrin binding. The heterodimeric integrin molecule goes through a conformation change upon binding to its corresponding ECM motif, originally positioned toward the plasma membrane, the extracellular globular proteins extend and selectively binds to protein causing an additional change in conformation in its tail regions which "signals" the binding inside the tail. Subsequently, the tail region allows binding of sub-membrane proteins to initiate cell signaling machinery.

has found approximately 26 expressed pairing combinations, with most cells exhibiting an assortment of different integrins. Communications through these proteins operates in both directions, that is, from the inside-out and outside-in. An outside-in communication is as simple as what an integrin pair binds. For example, the high binding to collagens or fibronectins can signal to the cell the type of ECM interaction and additional stimuli based on the affinity and intensity of the interaction. These interactions are indicated by the strength of the receptor α or β dimer binding to their target ligand. As such, the outside-in signal provides the feedback needed to determine cell metabolism and phenotypic expression of proteins in order to guide cell motility, proliferation, maturation, attachment, and chemotaxis.

Correspondingly, inside-out signaling allows the direct control of how many, what type, and where integrins will be expressed in the membrane. This is accomplished by the transcriptional and translational control of protein formation and cytoskeleton compartmentalization of the cell by its actin cyto-skeleton. Both pathways operate at the same time, continuously monitoring, and adjusting the signaling behavior. The importance of ECM signaling through the integrins can be conveyed very simply by comparison of normal adhesion dependent cells with some malignant cancers which can grow even while suspended in culture media. Several signaling pathways exist in normal cells that force them to engage apoptosis (programmed cell death) in the absence of ECM signaling.

Inside the plasma membrane, integrins operate under careful regulation with an assortment of small proteins. For signaling to occur, integrins are anchored with a large complex of proteins, such as talin, vinculin, paxillin, focal adhesion kinases, and tensin. The relative strength of an adhesion becomes highly dependent on how this sub-membrane clustering is organized. The result of this careful organization of multiple integrins is the focal adhesion. Most adhesion dependent cells do not make uniform connections with their substrates. Instead, these cells create specific locations for their integrin connections and cluster them in dense spots known as focal adhesions. The advantage to clusters rather than diffuse contacts is readily apparent when the speed of assembly and disassembly is considered. Maintaining the subcellular proteins, integrins, and cytoskeleton connections at individual locations gives the cell the ability to quickly move or begin mitosis through changes in the actin–myosin organization. As shown in Figure 11.10, focal contacts are easily observed *in vitro* by the application of fluorescence microscopy to integrin visualization.

In addition to ECM focal contacts, there is another extremely strong and organized binding called the hemidesmosome. This highly specialized binding mechanism is found in native tissues at the base of epithelial binding with its underlying membrane components. Both focal contact and hemidesmosome contacts are made possible by membrane-spanning integrins. However, in contrast to actin filaments in focal contacts, keratin filaments, also known as intermediate filaments, provide the connection in hemidesmosomes. Details of the intermediate filaments are discussed in Chapter 3. In addition, hemidesmosomes embed their integrins in a dense protein layer of plectin, and their globular domains typically are anchored with laminin. An additional transmembrane protein (BP180) is also present in these dense contacts. Because hemidesmosome contacts are found in abundance at tissue junctions such as the epidermal to dermal transition, they represent an attractive target for biomaterial surface mimics due to the more stable signaling behavior they mediate.

A major goal of modern biomaterial research is to recruit the native organiza-tion of ECM proteins onto the biomaterial surface and result in the correct cell integrin mediated signaling behavior. This gives the biomaterial researcher the

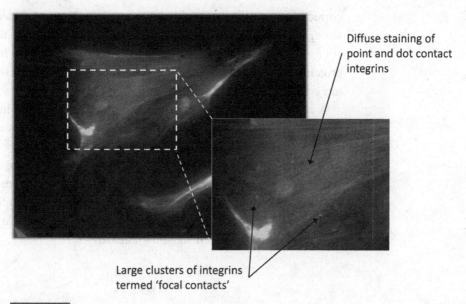

Diffuse staining of
point and dot contact
integrins

Large clusters of integrins
termed 'focal contacts'

Figure 11.10

Fluorescence microscopy image of an attachment dependent human palatal mesenchyme cell on a titanium substrate. Observe the cell spreading over the surface with distinct clustered integrins in focal contacts in connection with the actin cytoskeleton and nucleus, 200 × stained with Integrin-Cy3, FITC-phalloidin, and DAPI-Dilactate respectively.

ability to customize a surface to guide cell attachment, proliferation and differentiation. Many direct studies have been performed through characterization of how native adhesion proteins (plasma, fibronectin, vitronectin, serum, etc.) bind to material surfaces. For example, the groundbreaking work of Pierschbacher and Ruoslahti in 1984 allowed the identification of the Arg-Gly-Asp (RGD) peptide sequence known now for its integrin binding capacity in fibronectin.[1] This work began a large movement to identify the minimum required motifs necessary for an entire range of receptor–ligand binding behavior and have since become utilized in modern biomaterials to improve cell interactions. This immediately begs biomaterial scientists to ask the following questions.

- What is wrong with traditional biomaterial surfaces?
- Why is the field increasingly so reliant on surface modifications?

In the context of discussion, the importance of the affinity between receptor and substrate has been emphasized. Now consider the clean biomaterial, without any protein, amino acid, or molecule occupying its surface. When this material first encounters a biological fluid and has random protein absorption, binding characteristics can cause proteins to adsorb in many different conformations. The result could be that the peptide sequences necessary for receptor binding are not presented correctly or at all present for cell signaling. Thus, the ideal material which strongly binds the ideal protein may still not signal well because of protein conformation. For this reason, control over material protein binding in terms of protein composition, density, and conformation are all necessary for successful integration.

A common misconception held when first studying these integration mechanisms either in biomaterial testing or in native tissues is to assume high specificity from these interactions. Receptor-mediated adhesions do have high selectivity to their native target protein, but these interactions are selective and are not necessarily always specific. The majority of receptor connections are reversible indicating a relative affinity between proteins and are rarely covalently bound. Because of this "relative" affinity, mismatch connections are possible with a variety of receptor activity levels and signaling intensity.

11.4   Biocompatibility testing and techniques

Several of the most common *in vitro* cell testing techniques are reviewed in Chapter 13 and include evaluation of cell morphology, proliferation, viability, metabolic and differentiation activity, as well as tissue immunostaining. Additionally, in Chapter 3, a review of the basic biology and direct cell communication are outlined. In this section, we will review testing techniques to explore specific examples of cell to biomaterial contacts.

Techniques of *in vitro* experimentation can be placed into three classifications. *Cell culture* studies use primary or immortalized cells, which are typically from the same initial source and are isolated and grown without the early presence of an ECM. This represents one of the most controllable environments for cell–biomaterial exploration but also has major limitations. In contrast, *tissue culture* utilizes intact cell populations within a matrix and can include both single cell or mixed cell populations. Finally, *organ culture* employs full intact organs typically obtained from fetal or adult animals which may be perfused with a nutrient supply to test biomaterial interactions as close as possible to their designed use. *Organ culture* represents the most accurate testing platform for biomaterial studies but quickly encounters many difficulties in duplicating experiments, ethical and practical costs of harvesting, and ultimately cannot model the exact circulatory or biochemical signaling properties of native *in vivo* experimentation.

As the focus of this chapter is the interactions at a cellular scale, *cell culture* is the primary method for exploring questions in cell to ECM studies. Before highlighting several of these techniques, it is important to consider the major limitations of any *in vitro* assays. The major restrictions are:

- the absence of endocrine and paracrine signaling, and
- the diffusion limitation of nutrients and cellular waste by-products which would be supplied and removed via local circulation and lymphatic systems, respectively.

These restrictions are of major concern since in mammalian organisms cells are no farther than two to three hundred micrometers from a capillary. On a cellular scale, this limits diffusion-related transport to an estimated 10–15 cell layers in culture where cells are expressing an ECM component. However, diffusion in culture flasks is among the least efficient form of transport, and studies are seldom conducted exceeding one confluent layer of cells. In the absence of an ECM, most cells demonstrate a phenomenon called "contact inhibition," whereby cells will be inhibited from replication when they reach a defined distance from a neighboring cell.

Other significant differences that limit *in vitro* simulation of *in vivo* conditions include the absence of mechanical strains that are so prevalent in the cardiovascular and musculoskeletal systems. Mechanical deformation has been well linked with matrix formation and essential functions of these cells. In contrast, absence of mechanical deformation is often marked by cellular de-differentiation toward less committed lineages such as fibroblast-like phenotypes. The very substrate that cells are cultured on has significant implications on behavior and function of cellular processes, and this substrate includes the treated tissue culture plates. Additionally, the absence of cellular diversity in culture severely limits conclusions drawn from *in vitro* studies. Most cells *in vivo* are in close proximity to a range of cell types with necessary cell to cell communication needed for proper function. Co-culture studies have partially addressed this concern by using cell culture plate inserts. Two distinct cell types may be cultured in the same medium in co-culture studies but are separated to prevent one population from taking over all available surface area for growth.

Despite these limitations, cell biomaterial interactions can provide a wealth of data, particularly in screening early biomaterial modifications or to help elucidate specific mechanisms to improve integration. As in all scientific processes, care must be taken to refine the questions asked and to frame these questions in such a way so as not to exceed the limitations of the particular testing type. Universal considerations of *in vitro* tests include:

- identifying the most appropriate cell type to replicate the corresponding *in vivo* environment,
- carefully considering the volumes of medium in an *in vitro* assay to simulate the expected dosage, leaching of chemicals, and/or degradation of the biomaterial, and
- selection of proper controls for comparison which may include a well-established industry standard, existing biomaterial therapy, and/or natural material.

Table 11.5   Excitation and emission peaks for several of the most popular fluorophores

Fluorophore	Excitation peak (nm)	Emission peak (nm)
Cyanine, Cy2	350	450
Fluorescien, FITC	488	518
Indocarbocyanine, Cy3	550	570
Tetramethyl rhodamine, TRITC	550	570
Rhodamine red-X, RRX	570	590
Indodicarbocyanine, Cy5	650	670

11.4.1 Immunostaining techniques for studying cell–ECM interactions

By far, the workhorse of cell and molecular biology for studying protein interactions is the technique of immunostaining. Antibodies have highly specific recognition domains and can be custom engineered to bind with nearly all proteins fragments with known affinities. By tagging an antibody directly with a detection label (radioactive, fluorescent, and luminescent) or by tagging it with a secondary antibody with label, it is possible to visualize the precise location of the target in a cell. If quantification is desired, rather than use a visual label, an enzyme can be attached that has known reactivity to determine how much of the target is present, and hence the technique Enzyme Linked Immunosorbant Assay (ELISA) with steps shown in Figure 11.11. In this assay, after the antibody has bound its antigen (target), the complex is linked with an enzyme and finally provided with its substrate which will typically convert from an optically clear material to an optically opaque (colorimetric) or fluorescence substance at a known rate for detection in a plate reader by spectroscopy. Selection of primary and secondary antibodies should be carefully considered before beginning experiments. Best practices and entry into this field are achieved by contacting product specialists at major manufacturers of antibody and testing techniques such as Jackson Laboratories, Cell Signaling Technology, and Abcam. When considering multiple labeling experiments, a significant source of error due to false-positives can occur when primary and secondary antibodies cross react and overlap their emission spectrums into different channels. In Table 11.5, traditional absorption (excitation) and emission peaks are shown for several of the most popular fluorophores.

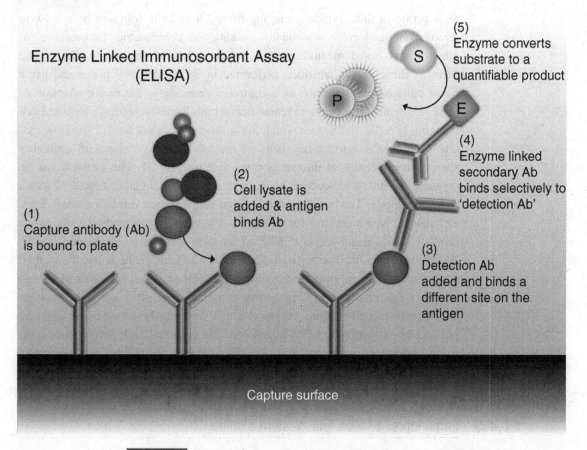

Figure 11.11

Overview and steps of a modern ELISA technique. The microplate is (1) coated with a capture antibody specific for a known protein; (2) lysate solution is applied and antigen is bound to the capture antibody; (3) a detection antibody is applied binding to a different location of the antigen; (4) an enzyme linked secondary antibody is bound to the detecting antibody; (5) a substrate is added and converted by the bound enzyme into a colorimetric or fluorescent product for detection and quantification.

11.4.2 Profiling a cell line for its ECM binding characteristics

Several commercially available products allow researchers to quickly identify the matrix proteins that a cell or tissue type particularly targets for adhesion. Companies such as Millipore, Invitrogen, BioPioneer, and others make cell adhesion assay plates which are coated with various ECM proteins such as laminin, fibronectin, collagen, integrin, and isoforms. These cell adhesion assay plates can be incubated with a cell population of interest to determine the number of cells adhering to each particular protein. Most of these tests involve incubation for a set period of time, typically ranging from 2 h to 24 h, followed by a wash to remove unbound cells, and finally staining via colorimetric, fluorescence, or luminescence-based methods to count the number of remaining cells on the surface. These assays are often performed in 96 or 384 well plates and use a tissue culture-treated surface as a control. Interestingly, the recent adoption of these simple tests has begun to replace some traditional fluorescence activated cell scanning (FACS) techniques which are a more traditional tool of the flow cytometer. In FACS test methods, cells are incubated in suspension with antibodies specific to a receptor of interest, for example integrin β4. The antibody can be tagged with either a radioactive, fluorescent or luminescent label directly or with a second antibody. The bound protein-labeled cell is passed through a small diameter channel with a detector for the signal of interest. In this method, cells can be profiled for as many receptor proteins as there are detection channels in the system. However, the start-up costs for such instruments are high, and the method is limited to cells in suspension culture. In contrast, adhesion assay plates are relatively cost effective and can also be combined with other assays to elucidate the strength of cell binding. As an example, cell attachment strength could be profiled by exposing the ECM-bound cells to controlled high gravity conditions using a laboratory centrifuge and quantifying the total cell number as a function of mechanically applied force.

11.4.3   Immunoprecipitation and Western blotting

Many lines of investigation require the researcher to further identify and screen particular proteins for study. The technique of immunoprecipitation takes a solution of proteins, most often an extraction from cellular and tissue volumes, and precipitates a specific protein by using a highly selective antibody for binding. While many technique variations exist, one of the most common technique variations is protein complex immunoprecipitation (Co-IP). In this variation, a protein that may be part of a larger cluster of proteins is targeted to precipitate all members of the complex out of solution to help identify what the unknown members of the complex may be. Precipitation by this technique is accomplished either directly or indirectly. Using the direct method, antibodies are first bound to a solid substrate such as agarose particles or superparamagnetic microbeads. In contrast, the indirect method first allows binding of antibody with antigen in solution that are then bead-bound due to high affinity. Both methods result in a bound complex which can then be isolated from the lysate solution by either centrifuge or magnetic field. After multiple washes, the bound proteins are finally eluted from their solid support for analysis by sodium dodecyl sulfate polyacrylamide gel electrophoresis (SDS-PAGE) with gel staining, cutting of individual protein bands, and potential sequenced by MALDI-mass spectrometry. Correspondingly, the eluted samples can be once again probed by Western blotting. In the Western blot, proteins that have been separated by SDS-PAGE are transferred to a nitrocellulose or polyvinylidene difluoride membrane. Following a blocking step to limit antibody reactions with the membrane itself, the proteins, now separated by charge or molecular weight, can be probed for specific antibodies and quantified.

11.5 Summary

With increasing understanding of cell–substrate interactions, it should not be discounted that new biomaterials will exploit native cell connections to a level not yet observed. The extracellular domain begins directly beyond the plasma membrane where a variety of cell to ECM interactions occur. The major proteins of the ECM, such as collagen, laminin, fibronectin, and proteoglycans, interact directly with the cell by means of linker receptor proteins such as integrins. Cells compartmentalize these receptors into highly complex focal contacts or hemidesmosomes and maintain affinity and quantity control over cell signaling. It seems very probable that future biomaterial surfaces could mimic membrane chemistry itself with tailored adhesion molecules that exploit known biochemistry. Toward this aim, modern biomaterial engineering will utilize the knowledge of cell ECM interactions based upon how native proteins and polypeptides are incorporated into biomaterial design.

Reference

1. Pierschbacher, M. D. and Ruoslahti, E. (1984). Cell attachment activity of fibronectin can be duplicated by small synthetic fragments of the molecule. *Nature*, **309**, 30–33.

Suggested reading

• Perrimon, N. and Bernfield, M. (2001). Cellular functions of proteoglycans – an overview. *Cell & Developmental Biology*, **12**, 65–67.
• Adair-Kirk, T. and Senior, R. (2008). Fragments of extracellular matrix as mediators of inflammation. *Int. J. Biochem. & Cell Biol.*, **40**, Issues 6–7 (June–July), 1101–1110.

- Arnaout, M. A., Mahalingam, B. and Xiong, J.-P. (2005). Integrin structure, allostery, and bidirectional signaling. *A. Rev. Cell Dev. Biol.*, **21**, 381–410.
- Moser, M. *et al.* (2009). The tail of integrins, talin, and kindlins. *Science*, **324**, 895.
- Dutta, R. C. and Dutta, A. K. (2001). Comprehension of ECM-cell dynamics: a prerequisite for tissue regeneration. *Biotechnol Adv.*, **28**(6), 764–769. (Epub 2010 Jun 18.)

Problems

1. Compare and contrast the roles of ECM proteins; collagen and proteoglycan. List several of the collagen types and their distributions where they are found.

2. Explain the implications of cross-linking in structural proteins such as collagen in the ECM with particular focus on the aging of tissues. Consider the environment directly around an implant that has limited turnover of tissue (less remodeling) due to a lack of blood flow. What would you expect to see in terms of protein cross-linking and how might this affect tissue biomaterial integration in time?

3. Mutations and resulting disease states represent a form of change to functionally competent tissue. Argue for or against the argument that ECM which forms next to a biomaterial but has structurally non-native organization can be classified as diseased.

4. Receptor binding motifs such as RGD have highly selective recognition for cell adhesion. Outline a strategy for incorporating a cell adhesive recognition domain to a biomaterial. Address the problems surrounding domain quantity, type, and affinity in your discussion.

5. From our discussion of ECM fragmentation, the concept of a matrikine was introduced. Briefly describe what a matrikine is, and how it can allow for cell homing to a target area. What would be the complications of unintended tissue fragmentation surrounding a biomaterial?

6. Concisely describe the differences between focal contacts and hemidesmosomes. What would be the implications of excessive formation of either at a biomaterial surface taking into consideration normal cell functions such as motility and mitosis?

7. Identify the most common classifications for *in vitro* studies.

12 Drug delivery systems

Goals

After reading this chapter the student should understand the following.
- The fundamentals and importance of drug delivery systems.
- How various different types of controlled drug delivery systems work.

Therapeutic drugs play a significant role in the therapy of almost all medical problems. From the treatment of common ailments such as headaches, and colds, to the reduction of fever, fighting infection, reducing cholesterol and blood pressure, and treating cancer, drugs play a major role in medicine. The systems used for delivering drugs take on a variety of forms. These include simple oral systems such as tablets, capsules, and syrups, transdermal systems like ointments and patches, and intravenous delivery using suspensions or nanoparticles.

In order for a drug to provide a therapeutic effect, the concentration of the drug in the blood plasma has to be above the minimum effective level and below the level that may be toxic. The difference between the minimum effective level and the toxic level is called the therapeutic index. As shown in Figure 12.1a, when the drug is conventionally administered as a single dose and is metabolized quickly inside the body, the level of drug in the blood plasma immediately increases followed by an exponential decrease. In such a conventional administration of the drug, the time frame over which the drug concentration is above the minimum effective level may not be long enough to produce a significant therapeutic effect in a single dose. Although this situation can be improved by increasing the amount of dose, this quickly raises the drug level to the toxic region. The alternative is to administer doses at regular intervals (Figure 12.1b). However, this method too has limitations, such as:

- the drug level in blood plasma is irregular and fluctuates with a high ratio of peak-to-valley concentrations, and
- the patient compliance is often poor.

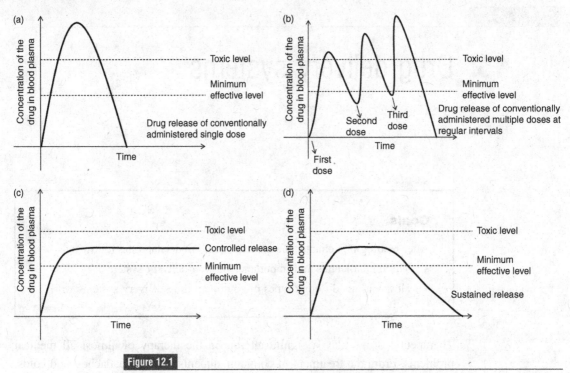

Figure 12.1

Schematics of drug release profiles of (a) conventionally administered single dose, (b) multiple doses administered at regular intervals, (c) controlled release, and (d) the sustained release.

Given the problems associated with the administration of drugs as a single dose and at regular intervals, there is a great need for controlled release systems, which can deliver drugs at a constant rate. In these systems, called zero-order release systems, the rates of absorption and of elimination of a drug in the body are equivalent. In such systems, a constant drug level in the blood plasma is maintained within the therapeutic index over the duration of use (Figure 12.1c). Since controlled release is the preferred choice among different drug delivery systems, there has been great interest in the development of novel controlled drug delivery systems.

- Controlled drug delivery – drug level in blood plasma is maintained within the therapeutic index for a long period of time.

This chapter will focus on the basic concepts involved in drug delivery and the different types of controlled drug delivery, which can be broadly classified under the following five categories:

- diffusion controlled systems,
- water penetration controlled systems,
- chemically controlled systems,
- responsive systems, and
- particle based systems.

12.1 Diffusion controlled drug delivery systems

There are two major types of diffusion controlled drug delivery systems, namely (a) membrane controlled reservoir systems, and (b) monolithic matrix systems.

12.1.1 Membrane controlled reservoir systems

In the membrane controlled reservoir system, the drug is contained in the core of the device (reservoir) and is covered by a thin polymer membrane (Figure 12.2a). The drug is delivered from the device through diffusion across the polymer membrane. The rate of drug release is controlled by the thickness, physiochemical characteristics, and porosity of the polymer membrane.

If the polymer membrane is non-porous, then the flux (movement) of the drug across the membrane is defined by Fick's first law of diffusion. As defined in Chapter 3 (Eq. (3.2)), the flux of the drug across the membrane can be expressed as:

$$J = -D\frac{dc}{dx},$$ (12.1)

where J is the flux per unit area (g/cm^2-s), dc/dx is the concentration gradient, and D is the diffusion coefficient of the drug in the membrane (cm^2/s).

Since the concentration of the drug in the membrane is difficult to obtain, Eq. (12.1) can be rewritten as Eq. (12.2) using partition coefficients to describe the ratio of the drug concentration in the membrane that is in equilibrium with the drug in the surrounding medium.

$$J = \frac{DK\Delta C}{l},$$ (12.2)

where K is the partition coefficient, ΔC is the difference in drug concentration in solutions on either side of the membrane, and l is the thickness of the membrane.

If the polymer membrane is porous, then the diffusion mainly occurs through the liquid-filled pores of the membrane, as shown in Figure 12.2b. In this instance,

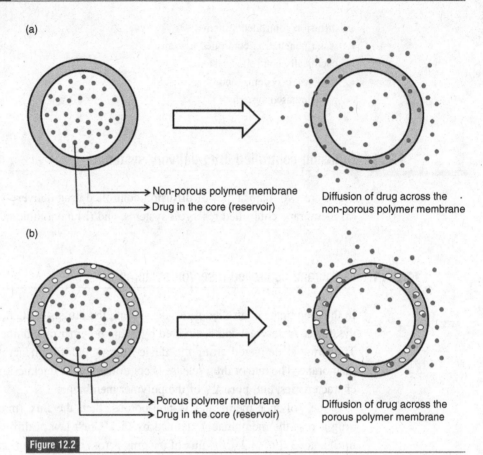

(a)

→ Non-porous polymer membrane
→ Drug in the core (reservoir)

Diffusion of drug across the
non-porous polymer membrane

(b)

→ Porous polymer membrane
→ Drug in the core (reservoir)

Diffusion of drug across the
porous polymer membrane

Figure 12.2

Schematics of diffusion of a drug in membrane controlled reservoir systems: (a) non-porous polymer
membrane, and (b) porous polymer membrane.

the diffusion primarily depends on the composition of the liquid and the flux of the
drug across the membrane can be defined by Eq. (12.3):

$$J = \frac{\varepsilon D K \Delta C}{\tau l},\tag{12.3}$$

where ε and τ are the porosity and tortuosity of the membrane, respectively.

12.1.2   Monolithic matrix systems

In the monolithic matrix system, the drug is either dissolved or dispersed in the
polymer matrix and is released through diffusion. The driving force for the drug
release is the drug concentration gradient between the matrix and the surrounding

environment. The polymer matrix has to be non-biodegradable so that the actual diffusion of the drug will not be affected by the physical changes in the matrix.

> **Diffusion controlled drug delivery**
>
> - Membrane controlled system – drug in the core is surrounded by a polymer membrane.
> - Monolithic matrix system – drug is dissolved or dispersed in the polymer membrane.
> - Drug release occurs by diffusion in both the cases.

In systems in which the drug is dissolved in the polymer matrix, the drug is loaded *below* its solubility limit (Figure 12.3a). As shown in Eqs. (12.4) and (12.5), derived by Baker and Lonsdale, early time approximation and late time approximation describe the release of the drug when it is dissolved in the polymer:

$$\text{early time approximation :} \quad \frac{dM_t}{dt} = 2M_x \left[\frac{D}{\pi \, l^2 \, t} \right]^{\frac{1}{2}} \tag{12.4}$$

$$\text{late time approximation :} \quad \frac{dM_t}{dt} = \frac{8DM_x}{l^2} \exp \frac{\pi^2 \, D \, t}{l^2}, \tag{12.5}$$

where M_x is the total amount of drug dissolved in the polymer, M_t is the amount of drug released at time t, D is the diffusion coefficient, and l is the thickness of the slab from which the drug is delivered.

In systems in which the drug is dispersed in the polymer matrix, the drug is loaded *above* its solubility limit (Figure 12.3b). Equation (12.6), as derived by Higuchi, describes the release of the drug when it is *dispersed* in the polymer:

$$\frac{dM_t}{dt} = \frac{A}{2} \left[\frac{2D \, C_s \, C_o}{t} \right]^{\frac{1}{2}}, \tag{12.6}$$

where C_s is the solubility of drug in the polymer matrix, C_o is the total concentration of the dissolved and dispersed drug in the matrix, and A is the area of the slab.

12.2 Water penetration controlled drug delivery systems

There are two types of water penetration controlled drug delivery systems, namely (a) osmotic pressure controlled drug delivery systems, and (b) swelling controlled drug delivery systems. The drug delivery rate in the osmotic pressure controlled

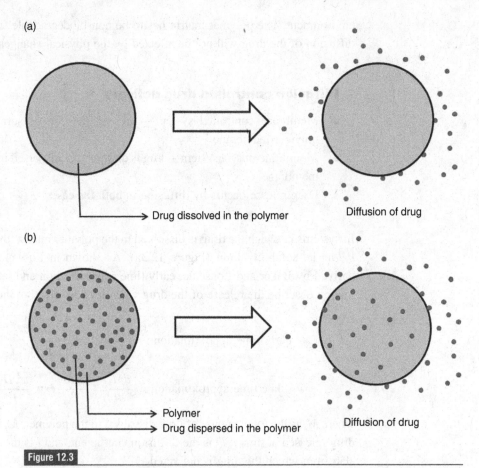

(a)

Drug dissolved in the polymer   Diffusion of drug

(b)

Polymer
Drug dispersed in the polymer   Diffusion of drug

Figure 12.3

Schematics of diffusion of a drug in monolithic matrix systems: (a) drug dissolved in polymer matrices, and (b) drug dispersed in polymer matrices.

drug delivery system is controlled by the osmotic pressure, that is, the pressure that is generated when the water flows from a higher concentration to a lower concentration across the semi-permeable membrane. In the swelling controlled drug delivery system, the drug delivery rate is controlled by the swelling of the polymer due to water penetration.

12.2.1   Osmotic pressure controlled drug delivery systems

This system primarily consists of an osmotic pump.[1] The pump has two compartments which are separated by a movable partition. One of the compartments is filled with an osmotic agent and the other is filled with the drug that is to be delivered. One side of the osmotic agent compartment is covered by a semi-permeable membrane and all the other sides of the pump are rigid with a laser drilled orifice at the side of the drug compartment. When the pump is placed into an aqueous system, the water from the surroundings diffuses into the osmotic agent compartment through the semi-permeable membrane. This causes an increase in volume of the osmotic agent compartment and exerts pressure on the partition which pushes the drug out through the orifice.

The volume of water passing across the semi-permeable membrane is defined by Eq. (12.7):

$$\frac{dV}{dt} = \frac{A}{l} L_p \left[\sigma \Delta\pi - \Delta P \right], \tag{12.7}$$

where dV/dt is the volume flux, A and l are the area and length of semi-permeable membrane, respectively, L_p is the membrane permeability coefficient, σ is the reflection coefficient, and $\Delta\pi$ and ΔP are the osmotic pressure difference and hydrostatic pressure difference across the membrane, respectively.

The drug delivery rate of this system is defined by Eq. (2.8) as:

$$\frac{dM}{dt} = \frac{dV}{dt} C, \tag{12.8}$$

where $\frac{dM}{dt}$ is the mass rate of delivery of the drug and C is the concentration of the drug in the solution which is pumped out of the orifice.

12.2.2 Swelling controlled drug delivery system

In the swelling controlled drug delivery system, the drug is dispersed or dissolved in a hydrophilic polymer when it is in a glassy state. Since the glassy state of the polymer is hard and rigid, as discussed in Chapter 6, the diffusion of the drug is very slow. However, when such a polymer is placed in an aqueous environment, it swells due to the penetration of water molecules into the matrix. As the water infusion causes the glass transition temperature of the polymer to be lowered below the ambient temperature, the drug diffuses out of the swollen rubbery polymer matrix. A schematic of the mechanism of drug release from swelling controlled systems is shown in Figure 12.4.

Water penetration controlled drug delivery

- Osmotic pressure controlled system – rate of drug delivery is controlled by osmotic pressure.
- Swelling controlled systems – rate of drug delivery is controlled by swelling of the polymer.

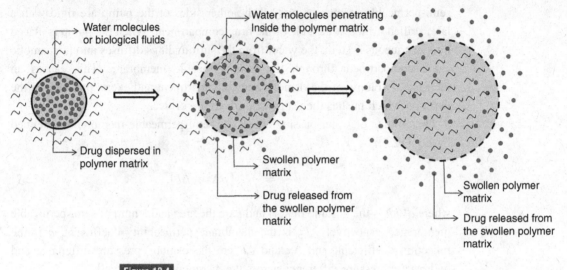

Figure 12.4

Schematic of drug release from swelling controlled systems.

12.3   Chemically controlled drug delivery systems

Chemically controlled drug delivery systems are classified under the following two categories: (a) polymer–drug dispersion systems, and (b) polymer–drug conjugate systems. In polymer–drug dispersion systems, the drug is uniformly dispersed or dissolved in a biodegradable polymer, and the drug release mainly occurs due to the degradation of the polymer under physiological conditions. With reference to drug delivery applications, the two terms commonly used to define the degradation of polymers are bioerosion and biodegradation. Although there is no clear differentiation between these two terms, bioerosion generally refers to either the dissolution of polymers or surface degradation, whereas biodegradation refers to the actual bulk breakdown of polymers. In polymer–conjugate systems, the drug is covalently attached to the backbone of the polymer, and the drug release mainly occurs through the cleavage of polymer–drug bonds under physiological conditions.

12.3.1   Polymer–drug dispersion systems

A variety of biodegradable polymers such as poly(lactic acid), poly(glycolic acid), poly(lactic-co-glycolic acid), polyanhydrides, polyphosphazenes, poly-ε-caprolactones, poly(orthoesters), poly(phosphoesters), polycyanoacrylates, polycarbonates, and polysaccharides have been extensively investigated for drug delivery applications. Some non-biodegradable polymers such as polysiloxanes and polyacrylates have also been explored for this purpose. Biodegradation occurs when the polymer backbone is broken down into low molecular weight compounds by hydrolysis or enzymatic actions. Ester and anhydride bonds are readily hydrolyzed by water molecules while amide bonds are easily cleaved by certain proteolytic enzymes. The types of degradation products depend on the nature of the polymer used, and it is advantageous if the products are water soluble and easily excreted from the body.

In polymer–drug dispersion systems, the drug is delivered from the polymer either by surface erosion or bulk degradation. As shown in Figure 12.5a, surface erosion occurs if the rate of hydrolysis is significantly greater than the rate of water penetration into the polymer. Polyanhydrides exhibit such a behavior. The anhydride bonds are hydrolytically cleavable while the polymer backbone is hydrophobic and resists water penetration. Zero-order release kinetics can be obtained with surface eroding polymers, provided the surface area of the device remains constant throughout the delivery period. However, zero-order release kinetics represent an ideal scenario, and in most instances, the drugs released are due to a combination of both erosion and diffusion. Diffusion is especially high for hydrophilic drugs since they have a strong affinity towards aqueous physiological environments.

As shown in Figure 12.5b, bulk degradation occurs if the rate of hydrolysis is significantly lower than the rate of water penetration. Poly(lactic acid) (PLA) and poly(glycolic acid) (PGA) exhibit such a behavior. The backbone of the polymer chain undergoes hydrolysis, and the degradation occurs uniformly throughout the device. Since the drug can easily diffuse out of the device when it is degraded uniformly, the drug release is a combination of diffusion and biodegradation controlled mechanisms.

12.3.2 Polymer–drug conjugate systems

In polymer–drug conjugate systems, also known as pendant-chain systems, the drug is covalently attached to the backbone of the polymer chain (Figure 12.6). The drug is delivered when the chemical bond between the polymer and the drug is cleaved by hydrolysis or by enzymatic actions. These polymeric drug carriers are also known as polymeric prodrugs. The drug can also be attached to the polymer through a spacer molecule such as an amino acid, and the drug is delivered when the spacer molecule is cleaved. One significant advantage of this system is that it provides control over the release of highly hydrophilic drugs. Since such drugs

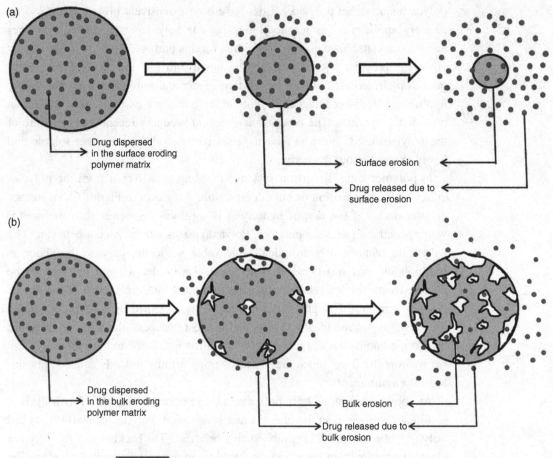

(a)

Drug dispersed
in the surface eroding
polymer matrix

Surface erosion

Drug released due to
surface erosion

(b)

Drug dispersed
in the bulk eroding
polymer matrix

Bulk erosion

Drug released due to
bulk erosion

Figure 12.5

Schematics of drug release by (a) surface erosion, and (b) bulk erosion.

readily diffuse into the system, a covalent attachment of the drug to the polymer provides more stability to the drug. However, care must be taken to prevent the alteration of the drug's chemical structure during the chemical attachment.

Chemically controlled drug delivery

- Polymer–drug dispersion systems – drug is dissolved or dispersed in a biodegradable polymer and drug release occurs due to biodegradation of the polymer.
- Polymer–drug conjugate systems – drug is chemically attached to the polymer and drug release occurs by cleavage of the chemical bond between drug and polymer.

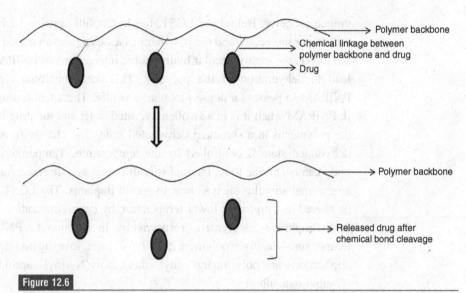

Figure 12.6

Schematic of drug release from polymer–drug conjugate systems.

12.4 Responsive drug delivery systems

The drug delivery systems which respond to external stimuli such as temperature, pH, solvents, ultrasound, electric field, and magnetic field are called responsive delivery systems or smart delivery systems. Depending on the type of stimuli and the nature of the polymer, the responses include hydration or dehydration, dissolution or precipitation, swelling or collapse, hydrophilicity or hydrophobicity, phase separation, and degradation.

12.4.1 Temperature-responsive drug delivery systems

Temperature-responsive drug delivery systems include polymers which undergo physical changes due to changes in temperature. Poly(N-isopropyl acrylamide) (PNIPAM) is a commonly used temperature-sensitive polymer. The lower critical solution temperature (LCST) of PNIPAM is 32 °C. The polymer is soluble in aqueous solutions below the LCST. However, PNIPAM becomes insoluble when the temperature is increased above LCST. This behavior of PNIPAM is in contrast with most polymers where the solubility generally increases with an increase in temperature. Such behavior makes the PNIPAM an interesting material for temperature controlled drug delivery. The interactions of hydrophilic and hydrophobic groups of PNIPAM with water molecules are considered to be responsible for this unique property. Below the LCST, the hydrophilic groups of PNIPAM interact with water molecules and the polymer possesses a random coiled structure. Above the LCST, the intermolecular hydrophobic interactions of PNIPAM dominate and lead to dehydration of the polymer. This conformational change allows the PNIPAM to possess a dense globular structure. Therapeutic drugs can be loaded in PNIPAM when it is in a swollen hydrated state, and the drug is delivered when the polymer is in a shrunken dehydrated state. The change from a hydrated to a dehydrated state is controlled by the temperature. Temperature changes in the body can occur due to an internal stimulus such as a diseased state of the body or an external stimulus such as heat triggered implants. The LCST of PNIPAM can be altered to a higher or lower temperature by copolymerization with hydrophilic or hydrophobic copolymers, respectively. In addition to PNIPAM, the other temperature-sensitive polymers that have been investigated for drug delivery applications are poly(methyl vinyl ether), poly(N-vinyl caprolactam), and poly (N-ethyl oxazoline).

The external stimuli in responsive drug delivery systems include the following

- Temperature.
- pH.
- Solvents.
- Ultrasound.
- Electric field.
- Magnetic field.

12.4.2   pH-responsive drug delivery systems

pH-responsive drug delivery systems utilize a group of polymers which undergo physical changes due to a change in the pH of the surrounding environment. The polymers may swell or collapse depending on the surrounding pH. This behavior is usually exhibited by the polymers with ionizable functional groups. The polymers with acidic functional groups such as $-COOH$ and $-SO_3H$ deprotonate at basic pH and acquire negative charge. Similarly, the polymers with basic functional groups such as $-NH_2$ protonate at acidic pH and acquire positive charge. A polymer with molecules which repel each other due to similar charges, exists in a swollen state. However, when the pH changes, the charged functional groups are neutralized and the polymer undergoes conformation change resulting in a collapsed state. Such changes involving swelling and collapsing of polymer are used to control the delivery of drugs. Poly(acrylic acid) and chitosan are some of the commonly used pH-sensitive polymers.

12.4.3   Solvent-responsive drug delivery systems

This injectable drug delivery system involves precipitation of polymer in a poor solvent.[2] This system is typically prepared by dissolving a water-insoluble biodegradable polymer in a water-miscible solvent. When the polymer is mixed with drugs and injected into a physiological system, the solvent diffuses out of the system and the water diffuses in, which leads to precipitation of polymer as a solid implant containing drug. The polymers which have been investigated for solvent-responsive drug delivery systems include poly(lactic-co-glycolic acid) copolymer dissolved in N-methylpyrrolidone, poly(lactic acid) dissolved in a mixture of benzyl benzoate and benzyl alcohol, poly(lactic-co-glycolic acid) copolymer dissolved in glycofurol, poly(lactic-co-glycolic acid) copolymer dissolved in benzyl benzoate, and sucrose acetate isobutyrate dissolved in N-methylpyrrolidone.[2]

12.4.4   Ultrasound-responsive drug delivery systems

Ultrasound can be used to control the delivery of drugs. When biodegradable polymers are used as drug carriers, ultrasound can significantly increase the rate of biodegradation as well as drug release. A variety of biodegradable polymers including poly(lactic acid), poly(glycolic acid), poly[bis(p-carboxyphenoxy)] alkane anhydrides and their copolymers with sebacic acid have been used as drug carriers for ultrasound responsive drug delivery systems.[3] The agents that are released using this system include p-nitroaniline, p-amino-hippurate, bovine serum albumin, and insulin. The integrity of these molecules is found to be intact after the exposure of ultrasound energy. In the case of non-biodegradable polymer-based drug carriers, ultrasound can increase the diffusion of drugs from the polymer matrix. The increase in diffusion rate could be caused by an increase in the temperature cause by the ultrasound vibrations. Ultrasound has also been used to increase the transport of drug across the skin. The delivery rate can be easily controlled by manipulating a number of ultrasound parameters such as frequency, power density, duty cycle, and sonication time.

12.4.5   Electrically responsive drug delivery systems

In the electrically responsive drug delivery system, an electric field is applied externally to control the delivery of drugs. Polyelectrolytes (polymers with ionizable groups) are commonly used for such delivery systems. When the electric field is applied to an electroresponsive gel, it changes its shape typically by bending or shrinking. The change in physical behavior allows the drug to be delivered at a constant rate. The magnitude of the applied current, the duration of the electrical pulses, and the interval between pulses are some of the parameters that can be altered to control the delivery of drugs from electroresponsive gels. A variety of natural polymers such as chitosan, hyaluronic acid, chondroitin sulfate, and synthetic polymers such as partially hydrolyzed polyacrylamide and polydimethylaminopropyl acrylamide gels have been used for this application. When partially hydrolyzed swollen polyacrylamide gels are placed between platinum electrodes and immersed in a 50% acetone–water mixture, the gels undergo volume transitions depending on the voltages applied.[4] The applied electric field not only produces a force on the hydrogen ions (H^+) to move them towards the cathode, but it also produces a force on the negatively charged acrylic acid groups in the polymer to be pulled towards the anode. The pull creates uniaxial stress along the gel axis with greater stress being experienced at the anode and lower stress being experienced at the cathode. This stress gradient deforms the gel. This process is reversible since the gel goes back to its original swollen shape when the electric field is removed.

An interpenetrating polymer network composed of poly(vinyl alcohol) and poly(acrylic acid) also shows electrosensitive behavior.[5] When the polymer network is placed between a pair of electrodes, it bends based on the applied electric field. The bending angle and the bending speed increase with the applied voltage and the amount of poly(acrylic acid) which contains negatively charged ionic groups. This polymer network shows step-wise bending behavior based on the applied electrical stimulus. Similar behavior is also shown by poly(vinyl alcohol) and chitosan interpenetrative polymer network hydrogel systems.[6]

12.4.6   Magnetic-sensitive drug delivery systems

Magnetic-sensitive drug delivery systems are also externally controlled drug delivery systems. Initially, the magnetic beads and the drug to be delivered are dispersed in a polymer matrix. Normally, the drug is released by diffusion. However, in the presence of an externally applied magnetic field, the diffusion of the drug is significantly increased.

12.5 Particulate systems

Particulate systems can be grouped into three categories, namely (a) polymeric microparticles, (b) polymeric micelles, and (c) liposomes.

12.5.1 Polymeric microparticles

Microparticles typically have a particle size of 1 μm to 1000 μm in diameter. Polymer-based microparticles used for drug delivery can be broadly classified into microcapsules and microspheres. Since microcapsules contain therapeutic drugs in the core that is surrounded by a rate-controlling polymer membrane, they can be grouped under membrane-controlled reservoir systems. The drug is released from the microcapsules through diffusion and an extended zero-order release is possible. The main limitations of microcapsules include:

- non-economical manufacturing methods, and
- the occurrence of burst release if there is a rupture in the outer membrane.[2]

- Polymeric microparticles – microcapsules and microspheres.
- Microcapsules – drug in the core surrounded by a polymer membrane.
- Microspheres – drug is dispersed in the polymer membrane.
- Drug releases by diffusion in both cases.

In contrast to microcapsules, microspheres can be prepared by a simpler method known as microemulsion polymerization. No outer membrane is present in microspheres, and the drug is uniformly distributed in the polymer matrix. The drug is released from the microspheres by diffusion. The limitations of microspheres are difficulties in achieving zero-order release and controlling the particle size distribution. However, microspheres have found applications in human use because of their inexpensive preparation method and the reduced chance for burst release.

- Self-assembly of amphiphilic block copolymers in aqueous solution produce polymeric micelles.
- Hydrophobic segments of block copolymer form the core while hydrophilic segments form the shell.

12.5.2 Polymeric micelles

Polymeric micelles are prepared from amphiphilic block copolymers composed of hydrophilic and hydrophobic segments. These polymers self-assemble in an aqueous solution in such a way that the hydrophobic segments form the core and the hydrophilic segments form the shell surrounding the core. These micelles can only be formed above the critical micelle concentration (CMC). Below the CMC, these polymers exist as individual polymer molecules. The unique core–shell structure of micelles paves the way for using such systems in drug delivery applications. The core of the micelles contains therapeutic drugs, and the shell interacts with the surrounding solvent and protects the drugs. Polymeric micelles provide stability to the drugs by preventing them from premature degradation before they reach the target site. In addition, poorly water soluble hydrophobic drugs can be solubilized by polymeric micelles leading to increased bioavailability. The narrow particle size distribution (10–100 nm) of polymeric micelles is an added advantage. Polymeric micelles can passively target tumors due to the enhanced permeability and retention (EPR) effect exhibited by tumors. The two main factors that contribute to the EPR effect are as follows.

- A tumor's microvasculature is highly disorganized with discontinuous or absent endothelial cell lining. Hence, they are hyperpermeable (leaky) and are easily accessible to nanoparticulates.
- The poor lymphatic drainage in a tumor results in retention of the delivery vehicles in the interstitial space of the tumor.

Polymeric micelles and liposomes (discussed in the next section), with an average diameter of 50–100 nm, can take advantage of the EPR effect and passively target tumors. A variety of poorly water soluble drugs such as paclitaxel, indomethacin, amphotericin B, adriamycin, and dihydrotestosterone have been successfully delivered using polymeric micelles. Poly(ethylene glycol) is extensively used for making hydrophilic components in block copolymers. For making hydrophobic components, polymers such as poly(propylene oxide), polycaprolactone, poly(lactic acid), poly(ortho esters), and poly(aspartic acid) have been investigated. Pluronic tri-block copolymers of poly(ethylene oxide)$_x$–poly(propylene oxide)$_y$–poly(ethylene oxide)$_x$ are commonly used to make polymeric micelles and are extensively investigated for drug and gene delivery applications.

12.5.3 Liposomes

Liposomes are made from phospholipids, which are a major component in all cell membranes and form lipid bilayers. Phosphatidylcholine (PC) is a class of phospholipids that is extensively used for making liposomes. The PC molecules consist of a hydrophilic polar head group (choline group) and hydrophobic tails (two fatty acids). When PC molecules are placed in an excess of water, these molecules arrange themselves in an ordered manner to produce closed spherical structures called liposomes or vesicles. These structures consist of an aqueous core surrounded by a hydrophobic membrane. Depending on the manufacturing conditions and chemical composition, liposomes can be formed with one or more concentric bilayer membranes. Liposomes can be generally classified into small unilamellar vesicles (SUV) and large multilamellar vesicles (LMV) depending on their size and lamella (a flat plate-like structure that is formed from lipid bilayers before converting into spherical vesicles). The diameter of SUV and LMV varies from 25 nm to 100 nm and 100 nm to several microns, respectively. Liposomes are hollow, and the inside contains liquids in which they were prepared. In this way, liposomes can encapsulate water soluble solutes in their aqueous inner core. In addition, these structures are capable of entrapping lipophilic drug molecules in their lipid bilayer membranes. Liposomes with sizes greater than 200 nm have short circulation time in the blood since they are rapidly cleared by the reticuloendothelial system. To overcome this short circulation time, liposomes have been coated with poly (ethylene glycol) (PEG), which has a water-retaining property. These PEG-coated liposomes are commonly referred to as stealth liposomes. Targeted drug delivery can also be achieved by conjugating appropriate ligands on the surface of liposomes corresponding to target cell receptors.

- Self-assembly of phospholipid bilayers in aqueous solution produces liposomes.
- Both polymeric micelles and liposomes contain drug in the core and passively target tumors due to EPR effect.
- Targeted drug delivery is possible by immobilizing ligands on the surface of the liposomes to target specific cell receptors.

12.6 Summary

Controlled drug delivery is the most preferred method for delivering drugs because the drug level in the blood plasma is maintained within the therapeutic index for a very long period of time. Controlled drug delivery can be achieved by different systems including diffusion controlled, water penetration controlled, chemically controlled, responsive controlled, and particulates. Diffusion controlled drug delivery systems can be broadly classified into membrane controlled and monolithic matrix systems. In membrane controlled systems, the drug is contained in the core of the device and is surrounded by a thin polymer membrane. In monolithic matrix systems, the drug is dissolved or dispersed in the polymer membrane. In both the membrane controlled and monolithic matrix systems, the drug is released by diffusion. Water penetration controlled drug delivery systems can be broadly classified into osmotic pressure controlled and swelling controlled systems. The two different types of chemically controlled drug delivery systems are polymer–drug dispersion systems and polymer–drug conjugate systems. In polymer–drug dispersion systems, the drug is dispersed or dissolved in a biodegradable polymer and the rate of drug delivery is controlled by the biodegradation of the polymer. In polymer–drug conjugate systems, the drug is covalently attached to the backbone of the polymer chain and released due to the cleavage of polymer–drug chemical bonds under physiological conditions.

Responsive drug delivery systems are smart systems that respond to external stimuli including temperature, pH, solvents, ultrasound, electric field, and magnetic field. The drug delivery from responsive systems is controlled by applying and removing external stimuli to drug-containing smart polymers. Particulate systems can be grouped under polymeric microparticles (microcapsules and microspheres), polymeric micelles, and liposomes. Microcapsules contain drugs in the core surrounded by a rate-controlling polymer membrane, whereas microspheres contain drugs that are uniformly distributed in the polymer matrix. The drug is released from the microparticles by diffusion. Polymeric micelles are produced by the self-assembly of amphiphilic block copolymers in an aqueous solution. The hydrophobic segments of the block copolymer form the core, whereas the hydrophilic segments form the shell surrounding the core. The core of the micelle contains the drug, while the shell protects the drug from premature degradation. Liposomes are produced by the self-assembly of phospholipid bilayers in an aqueous solution. Liposomes can contain hydrophilic drugs in their aqueous core. These structures are also capable of entrapping lipophilic drug molecules in their lipid bilayers. Liposomes and polymeric micelles can passively target tumors because of their enhanced permeability and the retention effect. Additionally, targeted drug delivery can also be achieved by immobilizing ligands on the surface of these liposomes to target specific cell receptors.

References

1. Theeuwes, F. and Yum, S. I. (1976). Principles of the design and operation of generic osmotic pumps for the delivery of semisolid or liquid drug formulations. *Ann. Biomed. Eng.*, **4**(4), 343–353.
2. Heller, J. and Hoffman, A. S. (2004). Drug delivery systems, in Ratner, B. D., Hoffman, A. S., Schoen, F. J. and Lemons, J. E., editors, *Biomaterials Science: An Introduction to Materials in Medicine*. London, Elsevier Academic Press, pp. 628–648.
3. Bawa, P., Pillay, V., Choonara, Y. E. and Toit, L. C. D. (2009). Stimuli-responsive polymers and their applications in drug delivery. *Biomedical Materials*, **4**(2), 022001.
4. Tanaka, T., Nishio, I., Sun, S. T. and Nishio, S. U. (1982). Collapse of gels in an electric field. *Science*, **218**(4571), 467–469.
5. Kim, S. Y., Shin, H. S., Lee, Y. M. and Jeong, C. N. (1999). Properties of electroresponsive poly(vinyl alcohol)/poly(acrylic acid) IPN hydrogels under an electric stimulus. *J. Appl. Polymer Sci.*, **73**(9), 1675–1683.
6. Kim, S. J., Park, S. J., Kim, I. Y., Shin, M. S. and Kim, S. I. (2002). Electric stimuli responses to poly(vinyl alcohol)/chitosan interpenetrating polymer network hydrogel in NaCl solutions. *J. Appl. Polymer Sci.*, **86**(9), 2285–2289.

Suggested reading

- Lee, P. I. and Good, W. R. (1987). Overview of controlled-release drug delivery, in Lee, P. I. and Good, W. R., editors, *Controlled-Release Technology Pharmaceutical Applications*. Washington, D.C., American Chemical Society, pp. 1–13.
- Hadgraft, J. and Guy, R. H. (1987). Release and diffusion of drugs from polymers, in Illum, L. and Davis, S. S., editors., *Polymers in Controlled Drug Delivery*. Bristol, Wright, pp. 99–116.
- Veronese, F. M. and Caliceti, P. (2002). Drug delivery systems, in Barbucci, R., editor, *Integrated Biomaterials Science*. New York, Kluwer Academic/Plenum Publishers, pp. 833–873.
- Heller, J. and Hoffman, A. S. (2004). Drug delivery systems, in Ratner, B. D., Hoffman, A. S., Schoen, F. J. and Lemons, J. E., editors, *Biomaterials Science: An Introduction to Materials in Medicine*. London, Elsevier Academic Press, pp. 628–648.
- Loo, Y. and Leong, K. W. (2006). Biomaterials for drug and gene delivery, in Guelcher, S. A. and Hollinger, J. O., editors, *An Introduction to Biomaterials*. Boca Raton, CRC Press, pp. 341–367.
- Bajpai, K., Shukla, S. K., Bhanu, S. and Kankane, S. (2008). Responsive polymers in controlled drug delivery. *Progr. Polymer Sci.*, **33**(11), 1088–1118.
- Bawa, P., Pillay, V., Choonara, Y. E. and Toit, L. C. D. (2009). Stimuli-responsive polymers and their applications in drug delivery. *Biomed. Mater.*, **4**(2), 022001.

- Schmaljohann, D. (2006). Thermo- and pH-responsive polymers in drug delivery. *Adv. Drug Delivery Rev.*, **58**(15), 1655–1670.
- Mathiowitz, E. and Kreitz, M. R. (1999). Microencapsulation, in Mathiowitz, E., editor, *Encyclopedia of Controlled Drug Delivery*. New York, John Wiley & Sons, pp. 493–546.

Problems

1. What are controlled drug delivery systems? Why are they so important?
2. What are the characteristics that control the drug release in membrane controlled reservoir systems?
3. What is the driving force for drug release in monolithic matrix systems? Can a polymer matrix be biodegradable in such systems?
4. How does an osmotic pump work in controlled drug delivery systems?
5. For which system does the glass transition temperature of the polymer play a vital role in controlling the delivery of drugs?
6. What is the significant advantage and limitation of polymer–drug conjugate systems?
7. What is the role of poly(N-isopropyl acrylamide) (PNIPAM) in controlled drug delivery system?
8. Give examples of pH-sensitive polymers used in controlled drug delivery systems and explain the drug delivery mechanism.
9. What are the parameters that can be altered to control the delivery of drugs in ultrasonically controlled drug delivery systems?
10. What are the advantages of microspheres over microcapsules?
11. How can polymeric micelles passively target tumors? Explain.

13 Tissue engineering

Goals

After reading this chapter the student will understand the following.

- Basic fundamentals of tissue engineering.
- Different cell types pertinent to tissue regeneration.
- Typical scaffold fabrication techniques.
- Techniques used to evaluate scaffolds, cells growing on scaffolds, and neo-tissue.

Can cells be used as living materials to engineer organs and tissue? Over the past several decades there has been increasing interest within the biomedical field to develop methodologies to restore the function of damaged tissue or organs without the use of long-term implants. This has led to the advent of the field of tissue engineering, which is often described as "an interdisciplinary field that applies the principles of engineering and life sciences toward the development of biological substitutes that restore, maintain, or improve tissue function or a whole organ."[1] Initially tissue engineering was considered a sub-field of biomaterials but has now evolved into its own distinct area. Nevertheless, although the role of biological sciences has significantly increased in tissue engineering, the field has stayed closely related to biomaterials.

Box 13.1

- Every year thousands of human lives are lost due to a lack of organs available for transplantation. Successful tissue engineering can solve this problem by re-growing the patients' own organs.
- Successful tissue engineering can also potentially provide skin for burn victims and repair nerves and restore function to those paralyzed.

The basic principle of tissue engineering involves the use of living cells to repair or re-grow a tissue or organ damaged by disease or trauma. For most applications, these cells need a carrier as well as a structural or mechanical support to be able to function normally, and to proliferate and form tissue. These supports are called scaffolds and are fabricated from biomaterials. The basic tissue engineering methodology involves harvesting appropriate cells from the body and seeding them on a three-dimensional porous scaffold structure or construct. These cell–biomaterial constructs are then placed in the correct environment (*in vitro* or *in vivo*) to encourage the formation of tissue. The cells can be differentiated cells such as osteoblasts for bone growth, chondrocytes for cartilage, or endothelial and smooth muscle cells for vasculature. The use of precursor or stem cells is also a possibility and is the subject of many scientific studies.

Box 13.2

- Often the terms regenerative medicine and tissue engineering are used interchangeably.
- Tissue engineering is different from cloning where the goal is to reproduce a copy of a whole organism.

Polymers, ceramics, metals, and natural materials have all been considered as candidate scaffold materials for various tissue engineering applications. Regardless of the type of biomaterial or cells used, these two form the most essential components of tissue engineering.

In this chapter we will first cover some of the basics about tissue engineering before identifying the most important cell types and ways to preserve them. The properties of cell delivery systems or scaffolds are then discussed followed by a description of common techniques to manufacture them. The chapter further describes cell–scaffold combinations and their culture, how to assess properties of cells and new tissue and concludes with addressing some of the challenges facing the field.

13.1   Tissue engineering approaches

13.1.1   Assessment of medical need

The first step in tissue engineering is the identification of the type of tissue that needs to be repaired or regenerated and the extent of damage. This is a very important step as different tissue types can require very different strategies and techniques. For example, load-carrying tissues such as bone, ligaments, and articular cartilage (present in all the major joints) may require scaffolds that are capable of sustaining stresses. On the other hand, load-carrying capability may not be critical for scaffolds used in tissue engineering of the liver or breast. Additionally, even in the family of load bearing musculoskeletal tissues, there are significant differences. Bone is vascularized and experiences compressional and torsional loading. Ligaments are loaded primarily in tension. Articular cartilage is avascular and is loaded primarily in compression and shear. These tissues will require uniquely different tissue engineering strategies.

The type of tissue defect also plays a role in determining the best tissue engineering based therapy. For example, a partial thickness defect in articular cartilage does not repair itself spontaneously. However, if the defect is full-thickness and the underlying sub-chondral bone is exposed, then there may be some spontaneous repair activity. Tissue engineering the repair of full-thickness cartilage defects involves repairing both a vascularized tissue (bone) and an avascular tissue (cartilage) and their interface. Similarly, repairing a bony defect in the skull (non-load-bearing) may require different strategies than those used for regenerating bone in a femur.

13.1.2 Selecting a tissue engineering strategy

Depending on the clinical need, one of three tissue engineering approaches can be used:

- scaffold-based,
- cell-based, and
- cell-loaded scaffolds.

In the scaffold-based approach, a scaffold without any cells may be used to fill the tissue defect. Native cells can then infiltrate this scaffold from the surrounding tissue. Biochemical molecules can be attached to or released from the scaffold to encourage cell ingress, cell attachment, and to enhance cell function. These molecules could be chemotactic or mitogenic growth factors. Depending on its location, a variety of cells types can migrate into the scaffold. This approach can be an advantage if the migrating cells are the kind required for the regeneration of that specific tissue. However, the presence of cells of a highly proliferative nature such as fibroblasts can lead to the formation of scar tissue instead of the desired tissue.

In the cell-based approach, cells may be introduced into the defect without a scaffold or matrix. The cells can be at different levels of development – stem cells (embryonic or adult) or differentiated cells. This approach can be successful when the cells are introduced into an area where there is existing matrix to support them.

An example could be the introduction of healthy myocytes into the myocardium after it has been partially damaged by a heart attack.

> • Most current tissue engineering methods use a combination of cells and biomaterial-based scaffolds.
> • Scaffolds serve as a carrier for cells.
> • Scaffolds also provide a 3D framework and structure for the cells to proliferate, produce extracellular matrix and generate tissue.

The third option is to use a scaffold preloaded with cells. In this cell–scaffold combination approach, biological signals can also be delivered using the scaffold as a vehicle.

When a scaffold with or without cells is placed into a tissue defect site, a classic wound healing reaction occurs. There is initial bleeding, hemorrhaging, and the formation of a fibrinous clot. The release of biological signals carried by the blood results in an inflammatory response and the arrival of macrophages and other types of "cleanup" cells. This leads to the gradual resorption of the clot and the formation of granulation tissue, which is eventually replaced by vascularized unspecific scar tissue. Based on the microenvironment present, this scar tissue may or may not be fully replaced by the specific tissue desired. A challenge in tissue engineering is to prevent the formation of scar tissue by providing a conducive environment so that the new tissue formed is of the specific kind needed and is functional over the long term. Strategies that encourage desired cellular attachment and proliferation on scaffolds while discouraging other types of invasive cells can provide a step in the right direction.

13.2 Cells

Tissue engineering uses cells as therapeutic agents or as living engineering materials. This concept is not entirely new because cells have been used as therapy in blood transfusions or in bone marrow transplants for a long time. However, the idea of using cells as engineering building blocks in conjunction with a biomaterial-based carrier is unique. Cells for this purpose can be obtained from a variety of sources:

- autologous (cells are obtained from the donor for re-implantation in the donor),
- allogenic (cells are obtained from another individual of the same species),
- syngenic (cell donor is genetically identical, such as a twin or a clone), and
- xenogenic (cells are obtained from a different species).

Autologous cells do not usually cause problems regarding immune response or rejection and are most often used for tissue engineering. However, extracting the cells requires an additional surgical procedure which may cause morbidity in the patient and carries the risk of infection and pain. Other issues include the possibility that the patient may not yield an adequate supply of cells. Additionally, the harvested cells may have to be multiplied *in vitro* before implantation, thereby introducing a delay in the treatment. Recent advances in scientific understanding and new protocols have led to techniques that can induce cells harvested from bone marrow or fat tissue to differentiate into cell types that can potentially regenerate bone, cartilage, or nerves.

The average human body contains about 100 trillion cells. These cells are present in a variety of types and the body has more than 200 types of fully mature cells. Some common types are listed in Table 13.1.

13.2.1   Stem cells

Instead of using tissue specific cells, the use of stem cells is also a possibility for tissue engineering applications. Stem cells are cells that have not yet differentiated terminally and have the potential to give rise to different types of cells. The three important characteristics that distinguish stem cells from other cell types are that (1) they have the ability to self-renew through cell division for a long period of time, (2) they are non-specific cells, and (3) they can be induced to transform into tissue or organ-specific cells under the right physiologic or laboratory conditions. In some organ systems, stem cells are continuously dividing to perform normal repair of the tissue.

Embryonic stem cells are pluripotent cells derived from embryos and can differentiate into any cell type. These cells have great potential in tissue engineering applications. Human embryonic stem cells used in scientific studies are derived from embryos created for reproductive purposes using *in vitro* fertilization methods. Issues related to the ethics of using human embryonic stem cells are the subject of an ongoing national debate in the United States.

- Depending on their source cells can be classified as:
  autologous,
  allogenic,
  syngenic,
  xenogenic.
- Stem cells have the ability to differentiate into different types of cells.

Table 13.1   Cell types

Connective tissue
Osteoblasts (bone)
Chondrocytes (cartilage)
Fibroblasts
Endothelial cells (blood vessels)

Nervous tissue
Neurons
Schwann cells
Glial cells
Astrocytes

Epithelia
Secretory cells (hepatocytes, β-cell)
Absorptive cells (e.g. intestine)
Ciliated cells (e.g. trachea)

Metabolism and storage cells
Adipose cells (white and brown fat)
Hepatocytes

Blood and immune system
Monocytes
Macrophages
Erythrocytes (red blood cells)
Megakaryocytes (platelet precursor)
Lymphocytes (immune system)
Neutrophils
Eosinophils
Basophils
T cells
B cell

Muscle
Skeletal
Cardiac
Smooth

Adapted from reference 2.

- Based on how many types of cells they can form, stem cells can be classified as:
  multipotent,
  pluripotent.

Adult stem cells are undifferentiated cells found among the differentiated and specialized cells of different tissues and organs. They can differentiate to form the specialized cell types present around them. They are not very common and typically have a small population (e.g. 1 in 100 000 in bone marrow). Also known as *somatic stem cells*, adult stem cells are multipotent and can form many different types of cells, but are restricted to their lineages. For example, hematopoietic stem cells found in bone marrow can differentiate into all types of blood cells. Another type of stem cell present in bone marrow is the mesenchymal stem cell which can form bone, cartilage, and fat. Stem cells present in the brain can form all three major cell types in the brain – astrocytes, oligodendrocytes and neurons.

Recent scientific advances have shown that adult stem cells can differentiate into cells that are different from their own lineage. For example, brain stem cells can produce blood cells. This phenomenon is known as *transdifferentiation*. Additionally, adult stem cells can be reprogrammed to behave like embryonic stem cells via gene manipulation. These cells are known as *induced pluripotent stem cells* (IPSC).

13.2.2   Biopreservation of cells

Tissue engineering depends on the availability of cells, both for conducting laboratory experiments and for clinical applications. In some cases, it may be possible to harvest the cells from the person in need, close to the time of clinical use. At other times, it would be necessary to harvest and store the cells for use at a later time. For example, cells may be harvested from the placenta at the time of birth and preserved for possible use in the future by the same person. The need for cryopreservation of cells is becoming increasingly important as the field of tissue engineering grows.

> - An approach to preserve cells for tissue engineering applications is to freeze them.
> - Ice formation during the cooling process can be damaging to the cells due to related cell shrinkage and change in electrolyte concentration.
> - Cryoprotectants can be added to the sample to prevent cell shrinkage and an increase in electrolytes.

Research has yielded various protocols for cryopreservation of cells. For example, red blood cells have been successfully preserved and used clinically for the past 50 years. Successful cryopreservation protocols usually involve an optimum cooling rate and a cryoprotectant. However, a cryopreservation technique that works for one cell type may or may not work for another. There are still many cell types which cannot be successfully preserved.

It is critical that, during the cooling process, ice is not formed within the cell. In slow freezing methods, the cells are cooled in a gradual fashion and ice nucleates and forms outside the cells. As the water turns into ice, it releases the solutes in it, thereby raising the osmolality of the extracellular solution. This causes the cells to contract or shrink. The cells continue to shrink with decreasing temperature, but no ice forms within the cells because there is very little water. At a certain temperature, the intracellular solution enters into a glassy phase. The danger is that because of the excessive cell shrinkage and the change in electrolyte concentrations, the cells may die. To overcome this issue, cryoprotectants such as glycerol are added. After entering the cells, these molecules dilute the electrolytes and also reduce the shrinkage.

The cooling rate is a critical parameter but varies from cell type to cell type. For some stem cells which have relatively impermeable membranes, the cooling rate may be as low as 1°C per minute. For highly permeable cells, the optimum rate can be as high as 1000 °C per minute. The optimum rate is defined as the fastest cooling rate possible that would not form intracellular ice. Based on this definition, water transport equations, and the assumption that ice will not form if the cell had less than 5% of its initial water when it reaches −30 °C, a theory has been developed that uses the cell membrane permeability and the cell's initial surface-to-volume ratio to predict the optimum cooling rate for a particular cell type.

Ice formation within the cell can also be avoided by *vitrifying* the cell or making the entire sample enter a glassy phase without letting ice form either inside or outside the cell. This can be achieved by very rapidly freezing the sample at rates such as 1 000 000 °C per second which causes the water to enter a glassy phase without forming ice. Although this is a viable technique when only a few cells are involved, it is difficult to scale up for larger samples. An alternate technique is to use large amounts of cryoprotectants so that the water will enter a glassy phase even at slow cooling rates. Since no ice is formed anywhere in the sample, the cells do not shrink. This technique eliminates the need for a fast rate of cooling. At high cooling rates, uniformity of cooling is essential when large samples are to be cryopreserved, and this is difficult to achieve. With the high cryoprotectant technique, the actual cooling rate is not important, and slow rates can be used to maintain uniformity and preserve large samples. However, the addition and removal of the high concentration of cryoprotectants can cause cell damage due to stress and toxicity.

In recent years, progress has been made in the area of anhydrobiotic preservation. This technique preserves the cells in a dried, but not frozen state. The cell membranes are first stabilized by using non-toxic sugars such as sucrose and trehalose, followed by drying the cells using air and freeze drying. Another advantage of this method is that the storage is at or near room temperature, thus eliminating the need for liquid-nitrogen storage and the expenses associated with it.

13.3 Scaffold properties

Over the past three decades, a variety of biomaterials have been used for scaffold fabrication, and a multitude of fabrication techniques have been developed. Scaffolds can be porous structures or gels. No one material, fabrication technique or scaffold design is appropriate for all applications. However, there are some essential properties that most scaffolds should possess to be successful, including the following.[3]

- The biomaterial(s) used should be biocompatible for the specific application. It should be noted that although a material may be biocompatible in general, it may not be optimum for every application.
- The material should have the ability to be remodeled *in vivo* or be biodegradable so that it does not retain its original form over the long term and is gradually dissolved or incorporated in the tissue.
- The biodegradation or remodeling should be synchronized with tissue regeneration so that the scaffold initially does provide adequate support to the cells but does not interfere with the regenerative process or tissue function over the long term.
- The scaffold should have high permeability to enable adequate diffusion of nutrients for the cells and the removal of waste products.
- The scaffold porosity should be sufficiently high to allow for the ingress of cells and provide the cells space to proliferate and form the extracellular matrix (ECM).
- The pore size of the scaffold should be optimum for the cells in use.
- Cells respond biochemically to applied stresses and thus the scaffold or cell carrier should possess mechanical properties that closely mimic those of the surrounding native tissue so that the cells are subjected to the correct stress microenvironment.
- The scaffold or carrier should ideally encourage and promote the formation of ECM.
- The surface of the scaffold should be conducive to cell attachment and should facilitate normal cell functions. This may occur naturally due to the inherent properties of the biomaterial or may require appropriate surface modification.
- The scaffold should have the ability to act as a carrier for biomolecules such as growth factors to stimulate the repair process.

13.4 Fabrication techniques for polymeric scaffolds

A large variety of materials and fabrication techniques have been used to create carriers for cells used in tissue engineering. The fabrication techniques depend on the type of biomaterial used and its physical and chemical properties. Some of the more common techniques are described below.

13.4.1 Solvent casting and particulate leaching

In this widely used technique, the polymer is first dissolved in an organic solvent. Particles of a water soluble material, generically called the *porogen*, are then mixed into the solution (Figure 13.1). Examples of the porogen include common salt and sugar. The solvent is then removed via evaporation and the resulting polymer–particle composite is immersed in water to dissolve and extract the porogen particles. The particles leave behind empty spaces thereby creating porosity. The size, shape, and quantity of the particles determine the pore size, porosity, and interconnectivity of the resulting foam or sponge-like scaffolds.

Other variations of this technique include the use of heat to melt the polymer before the addition of the porogen, thereby eliminating the use of an organic solvent. Alternatively non-water soluble porogens may be used. In such cases, a non-aqueous solvent has to be used to extract the porogen.

Figure 13.2 shows an example of polylactic acid scaffolds made using solvent casting and particulate leaching technique. Although this is a simple technique for fabricating scaffolds, limitations of this technique include the inability to precisely control the architecture of scaffolds and the variance of pore size and interconnectivity of the pores.

13.4.2 Electrospinning

As shown in Figure 13.3, electrospinning involves the dissolution of the polymer in an organic solvent and the ejection of the solution through a fine needle. A metal receptor, which can be a foil, plate, or rotating mandrel, is placed so that there is an air gap between it and the needle. When a high voltage is applied between the needle and the receptor, the applied voltage exerts forces on the surface of the polymer solution which exceed the surface tension. As a result, a charged polymer jet is ejected from the needle and is deposited on the receptor in the form of micrometer- or nanometer-scale (diameter) fibers (Figure 13.4).

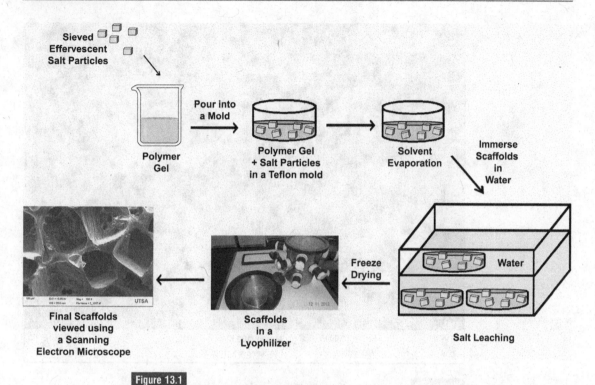

Figure 13.1

The salt leaching method for making scaffolds.

The solution concentration, speed of ejection, the gap distance, and the applied voltage all play a role in determining the fiber diameter and configuration. The architecture of structures generated by electrospinning is random in nature and cannot be closely controlled.

13.4.3 Solid free form fabrication (SFFF)

The solid free form fabrication (SFFF) process involves the formation of 3D-structured layers using rapid prototyping (RP) manufacturing. The object is first modeled using computer-aided design which is rendered into a series of thin, stacked virtual cross-sections by the software. Starting from the bottom, the RP machine lays down a layer of material corresponding to a cross-section. Based on the additive principle of laying materials in layers, the scaffold is created by laying subsequent layers over the preceding layers. Each layer is attached to the preceding layer by using either a solvent or heat-induced adhesion. The advantage of the SFFF process is that closely controlled and repeatable, reproducible architectures

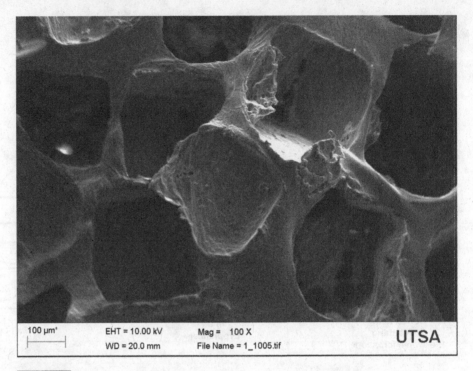

100 μm   EHT = 10.00 kV   Mag = 100 X
          WD = 20.0 mm   File Name = 1_1005.tif   UTSA

Figure 13.2

Micrograph of polylactic acid scaffold made by using a salt leaching technique.

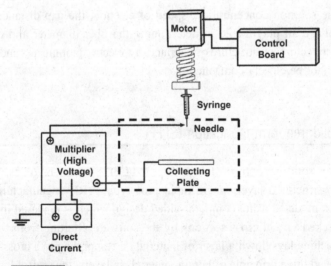

Figure 13.3

Schematic showing the electrospinning apparatus.

EHT = 5.09 kV
WO = 22.5 mm
Mag = 1.00 K X
File Name = 12wt125e14Kv_0.tif
10μm'
UTSA

Figure 13.4

Micrograph of polycaprolactone scaffold showing electrospun fibers.

can be produced for scaffolds. In general, there are several different techniques that can be categorized under the umbrella of SFFF, including:

- fused deposition modeling,
- selective laser sintering,
- stereolithography, and
- 3D printing.

Fused deposition modeling (FDM) involves the use of a heated nozzle to extrude a polymeric fiber. Figure 13.5 shows an example of a polylactic acid (PLA) scaffold that is fabricated using FDM. The fiber is deposited in specific patterns, in a horizontal plane on a platform. The platform is then lowered by a distance of one fiber diameter and the next layer of fibers is subsequently deposited. The process is repeated until the scaffold is completed.

In *selective laser sintering*, an infrared laser is used to melt a thin layer of polymer powder at selected points so that it adheres to the layer below. After a layer is sintered, a new layer of powder is added, and the process is repeated to create a 3D structure.

Similar to selective laser sintering in concept, *stereolithography* involves a highly focused ultraviolet (UV) beam used to photopolymerize a liquid,

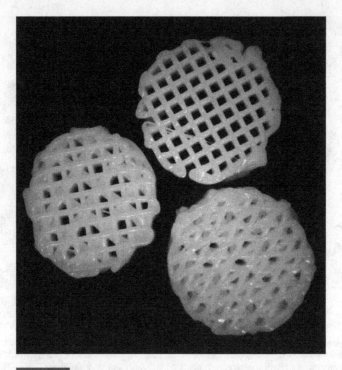

Figure 13.5

Micrograph of an FDM fabricated PLA scaffold.

photocurable monomer at precise spots. After each layer is cured, a new layer of liquid monomer is added, and the process is repeated to create a 3D structure.

In *3D printing*, the process first involves the spreading of a layer of a powder material. Figure 13.6a shows the platform inside a 3D printer where the layer of powder material is being laid. This powder could be a polymer or a ceramic or a composite. An inkjet print head is then used to precisely deposit a binder on the powder material. The scaffold is created by repeating this process of laying down layers of the powder followed by the deposition of binder from the inkjet print head. As shown in Figure 13.6b, complex structures can be built using a 3D printer, with control of shape and size.

13.5   Fabrication of natural polymer scaffolds

Natural polymers such as collagen and chitosan are also used extensively as materials for fabricating scaffolds. Usually, scaffolds fabricated from these materials do not have high mechanical properties but do possess chemical structures that are closer to the native tissues.

(a)

(b)

Figure 13.6

(a) Platform inside a 3D printer where the powders are spread; and (b) a complex structure built using a 3D printer.

Collagen is often used to make scaffolds, and Figure 13.7 shows a typical collagen scaffold. A variety of techniques can be used to make collagen scaffolds. For example, a collagen solution in HCl can be neutralized using ammonia gas from an ammonia solution in a closed chamber. This yields a white collagen gel with a honeycomb structure, which can be washed and slowly lyophilized.[4] Collagen suspensions in low pH solutions can be neutralized by using a number of chemicals to generate collagen gels and scaffolds. In some cases, cells can be mixed in with the collagen and then frozen to yield cell–collagen composite scaffolds. Composite scaffolds of chitosan and collagen can be produced by controlled freezing and lyophilizing mixed solutions of these natural polymers.

Composite scaffolds of chitosan with hydroxyapatite can be fabricated by adding calcium and phosphate solution to a chitosan solution in mild acid as schematically depicted in Figure 13.8. The resulting mixture is then frozen, followed by neutralizing it with a basic agent to yield a chitosan–hydroxyapatite composite porous scaffold. Modifications to such fabrication have included the use of a syringe pump to form bead-like chitosan–hydroxyapatite scaffolds, as shown in Figure 13.8b.

Natural tissue can also be used as scaffolds. Small intestinal submucosa (SIS) is an extracellular matrix derived from porcine small intestines. It is processed to make it acellular and is used in studies in an attempt to create de novo abdominal wall, urinary bladder, tendons, and blood vessels. Human-derived acellular dermal matrices can also be used as scaffolds. Since the epidermis and cells in donated human skin can cause an immune response, the donated skin is processed to remove these cells, and the processed skin is used as a scaffold for hernia repair and breast reconstruction.

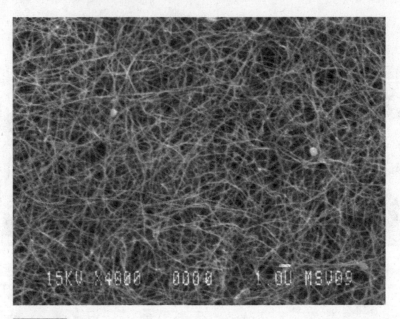

Figure 13.7

Collagen scaffold. (Courtesy of Sebastien Meghezi and Diego Mantovani, Laval University, Quebec City, Canada.)

(a)

(b)

Figure 13.8

(a) Process for making chitosan and hydroxyapatite composite scaffolds;[5] and (b) scanning electron micrograph of a bead-like chitosan–hydroxyapatite scaffold made with the use of a syringe pump. (Courtesy of Joel Bumgardner, University of Memphis, TN.)

13.6   Fabrication techniques for ceramic scaffolds

13.6.1   Template sponge coating

In the template sponge coating technique, a highly porous and interconnected sponge made of a polymer, such as polyurethane, is first impregnated with a slurry of hydroxyapatite (HA) or other calcium phosphate (CaP) materials. An initial heat treatment at a sufficiently high temperature burns off the polymer, leaving behind a very highly porous and interconnected structure. Subsequent sintering of the ceramic at high temperature (>1000 °C) aids in strengthening the scaffold. Figure 13.9 shows an example of a calcium phosphate scaffold made by using the template sponge coating technique.

13.6.2   Non-sintering techniques

In some scaffold fabrication approaches, high temperature sintering has to be avoided because the scaffolds may be made of composites containing polymers or heat sensitive biomolecules such as growth factors. In such cases, super-cooled

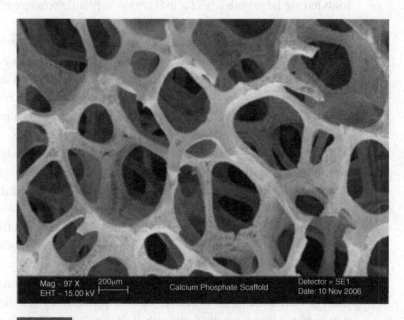

Figure 13.9

Scanning electron micrograph of a porous calcium phosphate scaffold made by using a polymer sponge method showing a bone like structure.

liquids of hydrogen peroxide or sodium phosphate ice are used to form a solution and then bubbled to create porosity as the super-cooled phase evaporates. Although the resulting scaffolds usually possess good porosity, they lack high mechanical strength.

13.7   Assessment of scaffold architecture

The architecture of a scaffold can be a critical factor in determining its success in tissue growth. For example, scaffolds with high surface area-to-volume ratio provide a large area for cell attachment. Scaffolds with high porosity provide the 3D empty volume needed for cell proliferation and tissue formation. Additionally, for tissues that are vascularized, a porous and interconnected structure facilitates the formation and infiltration of blood vessels.

The measurement of porosity is an important aspect of characterizing scaffolds. A common technique is the use of mercury intrusion porosimetry where the pore volume is determined by forcing mercury into the void space under pressure. Porosity due to closed pores cannot be detected and measured by this technique. Additionally, the mercury intrusion method tends to be less accurate with scaffolds having large pore sizes (> 800 μm) as very little pressure is required to drive the mercury into the pores.

Gravimetry can also be used to calculate the porosity. If the external dimensions of the scaffold can be accurately measured, its volume (v) can be calculated. The scaffold can then be weighed to yield its mass (m). Its overall density (d) can be calculated as $d = m/v$. If the density of the biomaterial (d_p) is known, then the porosity, p, of the scaffold is given by $p = 1 - d/d_p$. Gravimetry does not distinguish between closed and interconnected pores.

Micro-CT is another technique that can be used to measure both pore size and porosity. As shown in Figure 13.10 the micro-CT image contains 3D volumetric information. As the resolution of such machines has increased in recent years, so has the reliability and accuracy of their measurements.

A high permeability in scaffolds is conducive for successful tissue regeneration as it aids in the efficient diffusion of nutrients and the removal of waste products. It should also be noted that although a high porosity usually indicates a high permeability, this positive correlation is not always true. Porosity is the measurement of void space in the structure, whereas permeability is a measure of the ease with which a fluid can flow through it. A scaffold with high porosity may be covered by a thin film or skin of material rendering it impermeable. This is similar to disposable cups used for hot drinks that have internally porous walls to

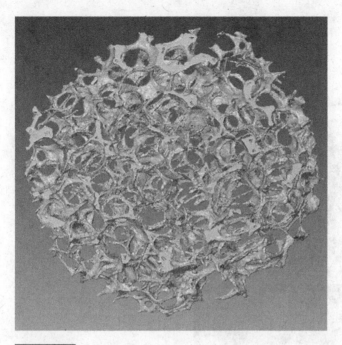

Figure 13.10

Micro-CT of a calcium phosphate scaffold.

provided insulation but have impermeable layers on both the inside and outside walls so as to prevent the liquid from leaking.

- The architecture of a scaffold can be characterized by measuring:
  porosity,
  permeability, and
  pore size.

The permeability of a scaffold can be measured by forcing a liquid, such as water, through the scaffold under a known constant pressure. The amount of liquid that passes through the scaffold in a given amount of time can be measured and Darcy's law can be used to calculate the permeability $k = QL/(hAt)$, where Q is the quantity of discharge, L is the length of the sample in the direction of the flow, A is the cross-section of the sample, h is the hydraulic head and t is the time. Based on the above principles, Figure 13.11 shows an apparatus that is used for measuring the permeability of scaffolds.

High porosity, large pore size, and interconnected pores all lead to high permeability. Permeability is also affected by *tortuosity*, which is a measure of

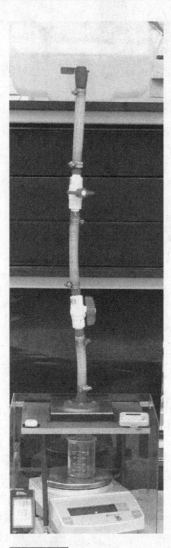

Figure 13.11

Picture of a permeability measuring apparatus.

the twisted path a fluid has to take to traverse from one end of the scaffold to the other end. Unlike scaffolds with a straight line path, the fluid that flows through a tortuous scaffold has to change paths every time it encounters an obstruction created by the biomaterial. Thus tortuosity is a reflection of the hindrance to fluid flow. The higher the hindrance, the higher will be the tortuosity and the lower will be the permeability. In general, a low permeability can lead to lower diffusion rates within the scaffold.

13.8 Cell seeded scaffolds

13.8.1 Cell culture bioreactors

Cells seeded on scaffolds can be cultured *in vitro* under static or dynamic conditions. Under static conditions, the cell seeded scaffolds are immersed in the appropriate media containing the nutrients needed by the cells, and there is no movement of the media or the scaffolds. Under dynamic conditions, the media can be drawn through the scaffold using a pump. Alternatively, dynamic conditions can include suspending the scaffold in mechanically agitated media or tumbling the scaffold in a container containing the media. Cell culture experiments performed under dynamic conditions are conducted in a *bioreactor*. In general, bioreactors are devices where complex biological processes can be enabled under close control of environmental conditions, such as temperature, pressure, pH, and fluid flow.

Box 13.3

- Bioreactors are an essential component of *ex vivo* tissue engineering.
- Compared to static conditions, bioreactors improve the quality of tissue being engineered.
- Bioreactors can improve reproducibility in engineered tissue because they operate under closely controlled conditions.

Dynamic conditions in a bioreactor generally improve the diffusion of nutrients to the cells in the interior of the scaffold and the removal of metabolic waste released by the cells as well as the flushing of degradation products eluting from the scaffold itself. There are several categories of bioreactors, including:

- spinner flask (SF) bioreactors,
- rotating wall vessel (RWV) bioreactors,
- hollow fiber (HF) bioreactors, and
- direct perfusion (DP) bioreactors.

Spinner flask (SF) bioreactors usually have the scaffolds suspended from the top of the flask, which contains the growth medium (Figure 13.12a). A rotary mechanism is used to stir the media, resulting in enhanced mixing of oxygen and nutrients in the medium thus making them more readily available to cells resident on the scaffold. The enhanced mixing can also cause the breakdown of boundary layers at the surface of the scaffold–cell construct. In addition to the

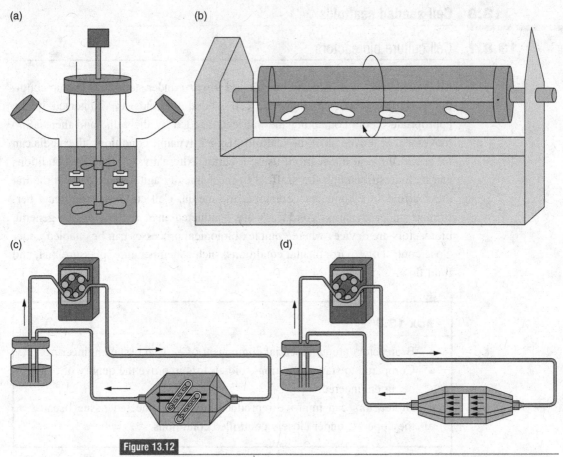

Figure 13.12

Different types of bioreactor shown schematically: (a) spinner flask (SF), (b) rotating wall vessel (RWV), (c) hollow fiber (HF), and (d) direct perfusion (DP).

delivery of oxygen and nutrients, improved mass transport also allows a more efficient removal of the waste products. However, turbulent eddies can form in this system, and these can be detrimental to the growth of such tissues as cartilage.

In *rotating wall vessel (RWV) bioreactors*, the scaffolds are tumbled about a central horizontal axis as shown in Figure 13.12b. The rotation generates a laminar flow and the relative motion between the scaffold and the fluid imposes a shear force on the scaffold. By adjusting the speed of rotation, the shear forces, the centrifugal force, and the gravitation force can be balanced, causing the scaffold to stay in a state of free-fall through the media. RWV reactors result in a significant amount of mass transport and enhance the diffusion of nutrients, oxygen, and waste products. Cartilage tissue grown in such bioreactors has been shown to be both biochemically and biomechanically superior than that grown in spinner flask or static bioreactors. RWV bioreactors have also been used to grow cardiac and bone tissue with good outcomes.

Hollow fiber (HF) bioreactors consist of hollow fibers or tubes with perforated walls to enable perfusion as depicted in Figure 13.12c. They are often used to culture cells such as hepatocytes that are highly metabolic in nature. These cells are cultured on the outside surfaces of the hollow fibers or tubes.

Direct perfusion (DP) bioreactors have a flow of medium going through the scaffold as shown in Figure 13.12d. This forced flow results in very good diffusion and mass transport, not only in the outer layers of the scaffold, but also through the internal pores of the scaffolds. Laboratory studies have shown that direct perfusion enhances the proliferation, differentiation, and mineralization of osteoblasts. Other cell types such as chondrocytes and cardiomyocytes have also been reported to do well under direct perfusion. Although direct perfusion helps with diffusion of nutrients into the scaffold and the egress of metabolic waste and scaffold degradation products, a high flow rate can also wash out newly deposited extracellular matrix components. Thus, it is critical that an optimum, balanced flow rate is maintained. Flow rates should be controlled by using software and computational models so that flow fields and shear stresses within the scaffold can be predicted. Both the flow fields and shear forces inside a scaffold will have much greater variability if the pores within the scaffold are not uniform and pore sizes vary a lot.

It is becoming increasingly known that mechanical forces imposed on cells elicit biochemical responses through a phenomenon known as mechanotransduction. Although the different aspects of this process are not fully understood, various scientific studies have shown that mechanically loading the scaffold–cell construct during the tissue generation process can be beneficial for tissues such as cartilage, bone, and blood vessels. In general, bioreactors can be designed to impose static or cyclic loading on the scaffold. The cyclic loading can vary in frequency, magnitude, and waveform.

13.8.2 Cell seeding

Various techniques can be used for seeding scaffolds with cells *in vitro*. These techniques can be broadly classified as either static or dynamic. Static techniques include the drop seeding technique wherein a drop of cell suspension is delivered onto a scaffold, and the cells infiltrate the structure due to gravity. Although this technique has been widely used, it results in relatively lower seeding efficiency (low number of cells) as well as an uneven distribution of cells within the scaffold. This cell seeding technique is also operator dependent, adding to the variability in its outcome.

The seeding efficiency is improved if the cells are added to the media in a spinner flask, which has scaffolds suspended in it. Agitation of the fluid causes seeding through random interactions between the cells and the scaffold. However, the cells are not very uniformly distributed in the interior of the scaffolds as most of the fluid flow remains limited to its exterior regions.

Alternatively, the liquid containing the cells could be made to flow through the scaffold under positive or negative pressure. Such a technique provides the cells the opportunity to attach to the scaffold as they traverse through the pores in its structure. Direct perfusion bioreactors can be used for this purpose, yielding high seeding efficiency and more uniform cell distribution compared to the other techniques. The ability for a large number of cells to be uniformly distributed in the 3D scaffold is highly desirable as this leads to improved tissue formation. Added benefits of cell seeding in direct perfusion bioreactors include:

- the process can be computer controlled,
- the process can be made reproducible, and
- the cell culture can take place in the same vessel.

If the scaffolds are implanted in the body without any cells, then a natural seeding process will take place from the adjacent tissue or blood. In this case, however, the lineage of the cell type on the scaffold is not assured, and it is possible that the scaffold may be invaded by cells that do not generate the desired tissue. The surface of the scaffold can be chemically altered so that only a specific cell type is likely to adhere. Additionally, the scaffold can be designed to release chemo-attractants to draw a certain type of cells into the scaffold.

13.8.3 Growth factors

Cell proliferation, differentiation, and tissue formation can be enhanced by the addition of growth factors. Growth factors are small proteins that serve as signaling molecules for cells. *In vivo*, these growth factors are secreted by cells, and they reach their targets either by diffusion (*paracrine*) or by being released in the blood stream (*endocrine*). In the *autocrine* system, the cell self-signals itself. A large variety of growth factors have been discovered and isolated (Table 13.2), including Vascular Endothelial Growth Factor (VEGF), the family of Bone Morphogenetic Proteins (BMP), Insulin-Like Growth Factor (IGF), and Transforming Growth Factor Beta (TGF-β). These factors promote activities such as the proliferation of chondrocytes, mineralization by osteoblasts, and blood vessel formation by endothelial cells. Growth factors are naturally present in the body but can be added to the media when cells are cultured on scaffolds *in vitro* for tissue

Table 13.2 Examples of growth factors

Growth factor	Activity
Bone Morphogenetic Proteins (BMP)	Cartilage and bone formation
Fibroblast Growth Factors (FGF)	Angiogenesis, wound healing, and embryonic development
Nerve Growth Factor	Axonal growth and branching
Insulin-Like Growth Factors (IGF)	Growth of muscle, cartilage, bone, liver, kidney, nerves, skin, and lungs
Transforming Growth Factor-ß (TGF-ß)	Cartilage and bone formation
Vascular Endothelial Growth Factor (VEGF)	Angiogenesis

engineering. However, to ensure that the growth factors are present when scaffolds are implanted *in vivo*, it is often necessary to incorporate and release them from the scaffolds themselves. The factors can be coated on the surface of the scaffold or can be mixed in with the biodegradable biomaterial for controlled release. Since growth factors are easily denatured by heat or strong chemicals, their addition to scaffolds limit the techniques available for scaffold fabrication. In general, the scaffold biomaterial, the fabrication method, and the coating technique can significantly influence the bioactivity and release kinetics of the growth factor. In most applications, a burst release (large amounts released over a very short time period) is not desirable and a long-term, steady release rate is preferred.

The spatial and temporal availability of growth factors during the formation of new tissue is critical to the success of tissue engineering. This is especially important when more than one cell type is involved in the tissue regeneration process. Polymeric scaffolds and delivery systems have the ability to regulate the simultaneous and temporal release of growth factors. However, to ensure the bioavailability of growth factors to the cells in sufficient concentrations, the loading of the factors on the scaffolds has to significantly exceed what is normally available in the body.

A different approach to delivering growth factors is to transfect the cells in the healing milieu. Using this approach, the transfected cells could up-regulate and produce the required concentration of growth factors. However, there are concerns that viral gene delivery vehicles can become pathogenic at high concentrations. Another drawback of this approach is that non-viral systems are inefficient at gene transfection. Thus, the cell transfection approach could be cytotoxic at high concentrations and ineffective at low concentrations. In addition to the concerns with the gene delivery vehicles, the success of gene therapy in tissue engineering has been limited by the short-term expression of the DNA-plasmid or adenoviral vectors.

Recent advances in bone tissue engineering using more than one growth factor indicate that it is more effective to release these growth factors from locations that are spatially removed from each other, such as opposite ends of the same scaffold. This is just one example of the complexity involved in effectively delivering growth factors. Designing an efficacious delivery system is just one of the many issues facing biomedical engineers involved in the optimization of processes for regenerating functional tissues. Other issues that are not fully understood include:

- the amount of growth factor(s) needed for a specific application,
- the optimum time points at which the growth factors should be released during the regeneration process,
- the duration over which the growth factor released should be available to the cells, and
- the sequence of emergence for the various growth factors involved in a tissue regeneration cascade

13.8.4 Mechanical modulation

In their normal physiologic environment, several tissue types in the body are subject to mechanical loading. Examples of such tissues include bone, cartilage, ligaments, muscles, blood vessels, and cardiac tissue. In the case of bone, normal development of skeletal tissue from its cartilaginous precursors requires loading. The maintenance of normal health for such tissues also requires mechanical loading as is exemplified by the loss of bone suffered by astronauts in zero gravity for prolonged periods of time and muscle atrophy seen in patients confined to beds. Thus, it is important that cells are exposed to physiologically relevant mechanical loading during tissue engineering, in addition to biochemical signals such as growth factors.

Scientific studies have shown that physiologic levels of mechanical loading not only increase the amount of proteoglycans and collagen type II present in tissue engineered cartilage but also result in better mechanical properties. Similarly, mechanical loading results in higher elastic modulus and more organized microstructure in newly regenerated bone tissue. Pulsatile pressures help to improve the microstructural composition and the mechanical properties of blood vessels. Although it is clear that mechanical stimulus is critical to engineering tissues that are normally subject to loading, there is still not a full understanding of how mechanical forces are sensed by cells and if only a few or all cell types are capable of sensing such forces.

As previously mentioned in this chapter, mechanical loading during *ex vivo* tissue engineering can be provided through the design of bioreactors. Cells can be exposed to different degrees of shear stress by varying the fluid flow rate in a bioreactor. Additionally, compressive stresses can be delivered by using hydrostatic pressure, whereas pulsatile loading can be imposed by using fluids, as in the case of blood vessels. Also, cyclic compressive or tensile loads can delivered to cells and new tissue via the scaffold by using specialized equipment.

13.9 Assessment of cell and tissue properties

In research leading up to the development of tissue engineering products, it is critical to measure and evaluate the properties of the main components of the tissue (cells and extracellular matrix) at various stages. Such evaluations start with *in vitro* studies in the laboratory, progressing to preclinical evaluations in animals, and finally ending with clinical studies in humans. An overview of some selected basic evaluation techniques is provided in the following sections.

13.9.1 Cellular properties

Cellular properties and cell function can be evaluated by using a variety of techniques. These include microscopy, radioactive and biochemical assays, DNA microarrays, and mechanical testing. Protein synthesis, metabolic activity, and signal transduction are cell functions that are commonly monitored.

Morphology

The shape and size of individual cells can be used as a qualitative assessment of the health of cell cultures. The morphology (structure and form) of cells can be influenced by a variety of factors including their direct interaction with the scaffold substrate, their reaction to degradation products eluting from the scaffold, signals from other cells, and cell density. Different types of cells exhibit different morphologies. For example, fibroblastic cells are elongated in shape and grow attached to surfaces. Epithelial-like cells are rounded or polygonal in shape, and grow in clusters or patches while attached to a surface. Cell morphology can be evaluated using optical or confocal microscopy techniques. Electron microscopy can provide a high level of detail but is a destructive test as it entails the fixing and coating of cells.

Cell numbers and proliferation

The cells seeded on a biomaterial should ideally proliferate and increase in number. To estimate the number of cells present, the cells adherent over a known area can be counted under a microscope and used as representative of the whole sample. However, this technique can lead to inaccurate counts if the cells are not evenly distributed over the entire substrate and the area chosen for measurements may not be representative of the whole sample.

Alternatively, cells can be removed from the substrate to form a cell suspension. They can then be counted using an electronic particle or cell counter such as the Coulter counter. The limitations of this method include the need for individual cell suspensions as cell aggregates would only be counted as one cell.

Cell numbers can also be determined using destructive assays which measure DNA content, protein content, or metabolic activity. These assays each require an independently developed reference standard curve, which provides the relationship between the measured quantity and cell numbers.

While a variety of techniques have been developed to assess cell numbers and function in culture, none has been developed specifically for cells cultured on 3D scaffolds. The tissue engineering community has thus adapted a few techniques generally used for cell culture studies in 2D for use with scaffold-based cell culture. However, it should be noted that these techniques have inherent shortcomings when applied to scaffolds because they rely on rates of diffusion and cell extraction, which have not been tested or optimized for 3D configurations.

Metabolic assays quantify the metabolic activity of cells without disrupting the metabolic process. If the metabolic rate can be assumed to be constant over the period of time of interest, then the measure of metabolic activity can be correlated to cell numbers, and these assays can be used as cell proliferation measures. Two commonly used assays in this category are AlamarBlue and Celltiter-Blue. As there is no need to destroy the cells or remove them from the scaffold for these techniques, long-term continuous studies with repeated measures can be conducted.

Both assays contain resazurin, a blue dye that is able to diffuse into the cells and become reduced to resorufin within the electron transport chain. The resazurin can be reduced in mitochondrial, cytosolic and microsomal enzymes. The resorufin as is then able to diffuse back out of the cell into the media. In resazurin, the mid-point potential, which determines whether a compound will be reduced in the electron transport chain, is large enough to accept electrons by all of the donors in the chain without interfering with the redox itself. Unlike resazurin, resorufin fluoresces and is pink in color, allowing it to be quantified by either fluorescent or colorimetric methods. By detecting the fluorescence or change in color, it is possible to detect the level of redox activity which may be used to extrapolate more general data regarding the cell response to an environment.

Using AlamarBlue or Celltiter-Blue for tissue engineering applications is relatively straightforward. The reagent is diluted in culture medium to $1\times$ concentration. Scaffolds are rinsed in phosphate buffered saline (PBS) and an equal amount of the $1\times$ solution is then added to each sample in a well plate, as well as a few empty wells for normalizing the color change. The plate is incubated for an optimized time period (1–4 h). The solution is then removed from the samples and aliquoted into either an opaque 96-well plate for fluorescent reading or a clear 96-well plate for absorbance reading. After reading, scaffolds are rinsed again with PBS. The medium is then re-introduced and scaffolds are placed back in the incubator in order to continue the study. The assay plates are read by using either an excitation of approximately 565 nm and reading an emission at approximately 590 nm for fluorescence, or measuring the absorbance at 570 nm using a 600 nm reference wavelength. The results are normalized by subtracting readings for the wells without samples from the readings with samples. A standard curve may be produced by assessing the metabolic activity of a series of known numbers of cells and adding a linear best fit line. However, it should be noted that the standard curve remains linear only up to a certain number of cells.

Some of the potential limitations of AlamarBlue and Celltiter-Blue are due to the fact that the cells remain intact on the scaffold (this quality can also be a great advantage of the assay). Because the cells remain viable on the scaffold and the assay itself was not designed for a 3D scaffold, it has been questioned whether cells at the center of a scaffold are accessed by the assay and accurately measured. If such a concern should arise, it is advisable to use a secondary, perhaps more destructive, assay to ensure that all of the cells are accounted for. The association of metabolism with cell growth permits the extrapolation of measures of metabolic activity to cell numbers when a standard curve is developed for reference. This correlation is not valid in the instance that the metabolic rate has changed but the cell number has not, or when the solution cannot diffuse properly through a scaffold.

The MTT and MTS assays are similar to the AlamarBlue or Celltiter-Blue assay in that they work within the electron transport chain to be reduced and give a quantitative colorimetric assessment of metabolic activity. The assays comprise a yellow tetrazolium compound, which is reduced to a dark purple formazan product. In tissue culture well plates, the tetrazolium compound is added to samples with fresh media and allowed to incubate for approximately 4 h. An SDS–HCl solution is then added and allowed to incubate for 4–18 h. Because it has a lower mid-point potential, the tetrazolium compound can neither be reduced by cytochromes nor allow electrons released by other donors to be passed to cytochromes, thus interrupting the electron transport chain and shutting down respiration. Unlike the AlamarBlue or Celltiter-Blue assay, a limitation for using MTT or MTS assay is the inability for investigators to perform multiple time point evaluations on a single sample. The released, solubilized formazan product is measured by an absorbance plate reader at 570 nm. While the MTS assay is reportedly more stable and straightforward than the MTT assay, both assays work on the same principle.

DNA quantification assays provide a more direct measure of cell numbers. The PicoGreen assay is developed to quantify small amounts of double stranded DNA (dsDNA). The assay contains a proprietary dye, which is highly specific to dsDNA and only fluoresces when the dye is attached to dsDNA. As each cell contains a very specific amount of dsDNA, this quantity can be correlated with cell numbers when applied to a standard curve. For tissue engineering applications, this assay relies on lysing the cells and removing all dsDNA content from the cells on the scaffold. This can be achieved by a series of freeze–thaw cycles which lyse the cells. The PicoGreen assay kit contains a PicoGreen reagent and a TE buffer consisting of 200 mM Tris–HCl and 20 mM EDTA at pH 7.5. The PicoGreen working solution is prepared by using a TE buffer to solubilize and protect the DNA and PicoGreen reagent. The TE buffer is first added to the samples and incubated, followed by the adding of the working solution and incubating at room temperature. The samples are read using a fluorescent plate reader to excite at 480 nm and read the emission at 520 nm. A standard curve should be prepared and measured at the same time as the samples to correlate the fluorescence with a cell number.

Although a DNA quantification assay is more accurate than a metabolic assay in determining cell numbers on scaffolds, this technique is destructive, which prevents the use of a single scaffold for multiple analyses. It should be noted that alternative brands, such as QuantiFluor™, appear to work similarly.

Like PicoGreen, the CyQUANT® assay attaches a fluorescent dye to nucleic acids to allow DNA quantification and to determine a cell number for proliferation studies. However, it is not exclusive to dsDNA. There are several variations of this assay available, some of which allow measurement of the DNA without lysing the cells. Unfortunately, in tissue engineering, placing an entire scaffold into a plate reader is not usually an option. Not only is the size and shape of scaffold a limiting factor but, like the metabolic assays, it is not certain that the assay will penetrate the scaffolds in a reliable manner. Therefore, using an assay such as the CyQUANT® that requires a freeze step and cell lysing is generally preferred. The CyQUANT® assay contains two components, a lysis buffer and the reagent, which are combined into a working solution before introducing them to the samples. Freezing at −70 °C then thawing the samples prior to introducing the working solution makes the lysing more efficient. After briefly incubating at room temperature, the samples can be read in a plate reader using 480 nm excitation and 520 nm emission. Fluorescence should be compared to a standard curve which correlates known cell number with fluorescence.

Cell proliferation rates can also be quantified by measuring the rate of DNA production. Fluorescent (bromodeoxyuridine, BrdU) or radioactive (^{3}H-thymidine) measurements of DNA synthesis can be made by using labeled DNA precursors that are incorporated into nucleic acids.

Cell differentiation

Differentiation of cells in *in vitro* cultures can be evaluated by measurements of biochemical changes and the secretion of biomolecules by the cells. For example, the early stage differentiation of osteoblasts can be monitored by measuring alkaline phosphatase activity and the late stage differentiation can be assessed by quantifying the secretion of osteocalcin, which is a non-collagenous bone protein. Differentiation can also be monitored by looking at expression profiles using DNA microarray technology.

13.9.2   Tissue properties

Immunostaining

Tissue properties can be assessed by studying the general visual appearance of the tissue. Additionally the cellular make up of the tissue can be determined under the microscope. The type of cells present, their distribution, and their orientation can all be relevant factors to assess which assays to use. Also important is the evaluation of the extracellular matrix. The amount and orientation of extracellular matrix molecules can be characterized using biochemical assays, immunostaining methods, and radioisotope labeling. For example, hematoxylin and eosin (H&E) is a commonly used stain that colors cell nuclei in a dark blue, and other cytoplasmic structures and extracellular components in various shades of red and pink. Alcian blue is used for the demonstration of glycosaminoglycans in cartilage, and the von Kossa stain colors mineral deposits in bone formation.

Mechanical testing

The mechanical behavior of tissue samples can be assessed using equipment generally used for testing engineering materials. These include tensile, compressive, shear, and indentation tests as described in Chapter 2. Soft tissues generally demonstrate both significant elastic and viscoelastic behavior, and so, the test techniques have to account for time related deformation. The mechanical properties of tissues are usually also dependent on strain rate. Special test standards for tissue and scaffolds are published annually by standard-making organizations such as the American Society for Testing and Materials (ASTM) International.

13.10 Challenges in tissue engineering

Although significant progress has been made on a variety of fronts in the field of tissue engineering, there are still only a few successes leading to commercial products. In addition to several regulatory hurdles, there are several scientific issues that still need solutions and these issues include:

- vascularization of the regenerated tissue,
- scale-up of scaffolds for large defects, and
- supply of appropriate growth factors.

The need for *vascularization of the regenerated tissue* remains a challenge. With some notable exceptions such as articular cartilage, most tissues have a network of blood vessels to provide nutrients and to remove metabolic waste. It is not merely sufficient to regenerate the tissue, but it is imperative to concomitantly regenerate the blood supply. *In vitro*, this may necessitate the use of co-cultures of two or more different types of cells to regenerate the tissue and blood vessels simultaneously. For scaffolds implanted *in vivo* without cells, it may be necessary to recruit more than one type of cell. For example, the tissue engineering of bone may require the recruitment of osteoblasts and endothelial cells.

The *scaling-up of scaffolds* studied in laboratories to clinically relevant dimensions requires more focus. Scaffolds tested in laboratories are usually small in size, on the order of a few millimeters. However, most organs in the human body or tissue defects caused by trauma or disease are larger than a few millimeters in size. Large scaffolds are often more challenging in providing adequate nutrients to cells in the interior of the scaffolds as a result of limited diffusion. As a result, the success of small scaffolds does not often correlate well when scaffolds are scaled up for large defect applications.

The *supply of appropriate growth factors* at relevant times and in relevant quantities still needs to be fully understood and accomplished. Natural healing in the human body involves a cascade of biomolecules signaling cells and orchestrating the healing process. For successful tissue engineering, it may be necessary at least to trigger the healing cascade by releasing appropriate growth factors at the suitable time.

13.11   Summary

Tissue engineering is a rapidly evolving field, with the goal of restoring function to damaged tissue or organs by regenerating the damaged tissue instead of replacing it with permanent implants. A common approach in tissue engineering involves the delivery of cells to the defect site using a scaffold. The scaffold functions both as a cell carrier and as a mechanical support and framework for the cells to function and produce tissue. Scaffolds have been fabricated from a variety of synthetic and natural biomaterials, and numerous fabrication techniques have been reported in the scientific literature. Although the properties and architecture of a scaffold depend on its specific proposed application, there are some basic characteristics that all scaffolds should possess. Experiments leading to the development of tissue engineering protocols and products involve the assessment of the scaffold, cellular function on the scaffold, and the properties of the neo-tissue. There are still significant challenges that need to be overcome, and these challenges include regenerating tissue that has blood vessels, diffusion of nutrients in scaled-up, clinically relevant scaffolds, and releasing growth factors from the scaffolds in a fashion that mimics the natural cascade.

References

1. Langer, R. and Vacanti, J. P. (1993). Tissue engineering. *Science*, **260** (5110), 920–926.
2. Palsson, B. and Bhatia, S. (2004). *Tissue Engineering*. Pearson Prentice Hall, ISBN 0–13–041696–7.
3. Agrawal, C. M. and Ray, R. (2001). Biodegradable polymeric scaffolds for musculoskeletal tissue engineering. *J. Biomed. Mater. Res.*, **55**(2), 141–150.
4. George, J., Onodera, J. and Miyata, T. (2008). Biodegradabe honeycomb collagen scaffold for dermal tissue engineering. *J. Biomed. Mater. Res.*, **87A**, 1103–1111.
5. Chen, J., Nan, K., Yin, S. *et al.* (2010). Characterization and biocompatibility of nanohybrid scaffold prepared via *in situ* crystallization of hydroxyapatite in chitosan matrix. *Colloids and Surfaces B: Biointerfaces*, **81**, 640–647.

Suggested reading

• Palsson, B. and Bhatia, S. (2004). *Tissue Engineering*. Pearson Prentice Hall, ISBN 0–13–041696–7.
• Lanza, R. P., Langer, R. and Vacanti, J., editors (2000). *Principles of Tissue Engineering*. Academic Press.

Problems

1. Describe the three basic tissue engineering approaches. What are the advantages and disadvantages of each?
2. Describe the biological processes that occur when a tissue engineering scaffold is inserted in the body. What type of tissue is not desirable as the final result?
3. Based on the source of cells, what are the different classifications for cells?
4. Which are some of the common type of bioreactors for *in vitro* tissue engineering? Compare and contrast these methods.
5. Describe two techniques commonly used for making polymeric tissue engineering scaffolds. Can these same techniques be used for ceramics? If not, then explain why.
6. List five important properties of a tissue engineering scaffold. If you were to be asked to tissue engineer a liver, which of these would be most important? Why?
7. If you had to perform an experiment where you had to measure cell proliferation, which assay would you prefer? Under what conditions is a metabolic assay advantageous?
8. What are the main challenges facing the field of tissue engineering?
9. Which of the following refer to the same scientific field: tissue engineering, cloning, regenerative medicine?
10. Briefly describe some of the complexities involved in the delivery of growth factors from a scaffold.

14 Clinical applications

> **Goals**
>
> After reading this chapter, students will understand the following.
> - Advantages of different biomaterials used in clinical applications.
> - Rationale for use of the different biomaterials in clinical applications.
> - Role of biomedical engineers in medicine.

In today's healthcare industry, synthetic and natural biomaterials are being introduced into the human body at an increasing rate. In the United States alone, several million biomaterial-based implants or devices are used for human patients each year. It is not uncommon to find that many of the biomaterials used today were first introduced for industrial applications other than medicine. In order to protect and promote public health, all biomaterials used by the medical device industry have to meet numerous stringent criteria required by regulatory agencies such as the Food and Drug Administration in the USA. These stringent criteria include material properties, preparation and sterilization of the materials, biocompatibility, and short-term and long-term issues related to the use of the material for a specific application. With these protections in place, reasons for the conventional use of biomaterials in clinics and hospitals include:

- treating in order to return to normalcy,
- repairing in order to reinstate organ function, and
- restoring in order to retain or modify organ shape or form.

Examples of biomaterials used for treatment include pacemakers and cardiac valves, which provide treatments for specific disorders in the body. Joint replacements and fracture fixation devices are examples of biomaterials that are used to repair fractured or diseased bone. Biomaterials used for restorations include dental implants or breast implants that are used for reconstructions, thereby providing esthetics and maintaining the psychological and social well-being of the patient. Today, the clinical uses of biomaterials have been extended to include patient needs, such as rehabilitation, comfort, and convenience. Such shift in paradigm has challenged clinicians, materials scientists, and biomedical engineers to explore new innovative materials, new processing methodology, and new surgical skills in providing therapeutic approaches to the treatment of end-stage diseases as well as providing elective restoration of chronically damaged tissue. Listed below are some uses of biomaterials in today's clinical applications.

14.1   Cardiovascular assist devices

In cardiac assist devices such as pacemakers, the cardiac pacing leads have to survive in the harsh endocardial environment. The biomaterials used for the leads must ideally be stable, flexible, possess adequate conductive and resistive properties, and provide endocardial contact within the heart. Figure 14.1 shows a photograph of three left-ventricular pacing/sensing leads. The leads are positioned in the heart and are connected to a battery-powered heart failure device such as a pacemaker. Implantation of the heart failure device and the leads allows the transmission of electrical signals to the heart for cardiac resynchronization therapy. Major components of the cardiac pacing leads include the electrode, the conductor, and the insulation. Each component of the pacing leads is critical to the functionality of the pacemaker. As such, material selection for each of the components is important.

Box 14.1

- Given the harsh endocardial environment, most pacing failures occur because of malfunction of the pacing leads.
- Electrodes made of other metals such as copper, zinc, nickel, and silver have been known to cause toxic reactions with the myocardium.

Current electrodes are made of metals, including platinum–iridium, Elgiloy (a composite alloy of nickel, chromium, cobalt, manganese, molybdenum, and iron), and carbon. However, because of the intense inflammatory response at the electrode–endocardial interface after implantation, complex electrode designs that gradually release steroids have been developed to markedly attenuate the inflammatory response. It is crucial that the ideal electrode should possess the following properties:

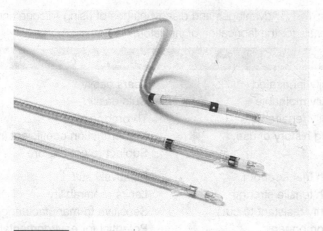

Figure 14.1

Photograph of EASYTRAK®, EASYTRAK®2, and EASYTRAK®3 left-ventricular coronary venous pacing/sensing leads. (Courtesy of Boston Scientific, Natick, MA, USA.)

- be biologically inert,
- be resistant to degradation,
- be resistant to corrosion, and
- cause minimal tissue inflammation and fibrosis at the electrode–endocardial interface.

Like the electrodes, the pacing lead conductor is a wire made of metal. The function of the pacing lead conductor is to conduct current from the pulse generator to the pacing electrode and to send cardiac signals from the electrode(s) to the sensing amplifier of the pulse generator. Ideally the pacing lead conductor should be non-corrosive, possess adequate tensile strength, and have the ability to reduce resistance to metal fatigue. Current conductor materials include non-corrosive alloys such as MP35N, which is a composite of nickel, chromium, cobalt, and molybdenum.

Insulating materials are used to separate the conductor coils. Biomaterials commonly used as insulators in cardiac pacing leads include silicone rubber and polyurethane. Properties of both silicone and polyurethane are discussed in Chapter 6. Being partly organic and partly inorganic, the main polymer chain for silicones consists of alternating silicon and oxygen atoms. As a biocompatible polymer, silicones possess a high degree of lubricity, excellent resistance to water, are chemically inert, and are capable of retaining most of their physical properties at elevated temperatures. Critical properties of polyurethane include its load-bearing capacity that is comparable to cast steel, exceptionally smooth surfaces, resistantance to colonization by fungal organisms, and excellent hydrolytic stability.

Table 14.1 Advantages and disadvantages of using silicone and polyurethane materials for the fabrication of insulators

Biomaterials	Advantages	Disadvantages
Silicone rubber	Easily fabricated Easily moldable Easily repairable Long history of use	Tears easily Cuts easily Thrombogenic Higher friction coefficient in blood Subject to cold failure
Polyurethane	High tear strength High tensile strength Highly resistant to cuts Thrombogenic Low friction coefficient in blood High abrasion resistance High compressive properties Less thrombogenicity	Relatively stiff Lacks repairability Sensitive to manufacturing process Potential for environmental stress cracking Potential for metal ion oxidation

Although there are some known problems associated with the use of silicone rubber as insulators, crack propagation and metal ion oxidation are two of the common causes of lead failure when the insulator is made of polyurethane. Lead failure occurs as cracks propagate through the full thickness of the polyurethane insulator due to environmental stress. Additionally, the production of hydrogen peroxide by inflammatory cells at the electrode–endocardial interface also results in a reaction between the polyurethane and oxygen, thereby resulting in oxidative degradation of polyurethane and subsequent failure of the lead. Table 14.1 summarizes the advantages and disadvantages for insulations made from silicone and polyurethane.

14.2 Cardiovascular stents

As shown in Figure 14.2, bare metal stents are tubular scaffolds that are commonly made from either 316L stainless steel (example: R Stent™ from OrbusNeich, Hong Kong), nitinol or NiTi alloy (example: S.M.A.R.T.® CONTROL® Iliac Stent System from Cordis Corporation, New Jersey, USA), or cobalt–chromium-based alloys (example: PALMAZ® BLUE™ Transhepatic Biliary Cobalt Chromium Stent from Cordis Corporation, New Jersey, USA). The properties of these metals are discussed in Chapter 5. From our understanding of biomaterials and their properties we know that different metals lead to differences in performance

(a) (b)

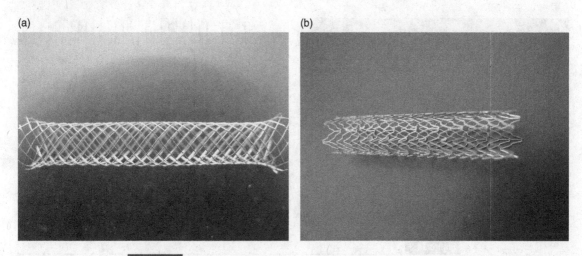

Figure 14.2

Photographs of metal stents with different strut designs.

and this greatly influences the design of the stent. For example, stents made from nitinol can be designed as self-expandable owing to the superelasticity of the material at body temperature, whereas cobalt–chromium-based stents can be designed to have thinner strut thickness compared with stainless steel stents without reducing radial strength and radiopacity. Additionally, the cobalt–chromium-based stents are less prone to biocorrosion when compared with stainless steel stents.

Used in patients to reduce coronary artery blockage, stents find application during revascularization procedures of coronary or peripheral arteries. Similar to the conventional balloon angioplasty, where an inflatable, small, sausage-shaped balloon is placed at the catheter's tip, the procedure for coronary stenting involves mounting a stent on a balloon catheter in a "crimped" or collapsed state. Figure 14.3a shows a photograph of the entire catheter, and Figure 14.3b shows an enlarged view of the "crimped" stent at the end of the catheter. However, unlike the inflation of the balloon to dilate an area of the arterial blockage during the conventional balloon angioplasty, the inflation of the balloon during the coronary stenting procedure allows the stent to expand from its crimped state, thereby pushing it against the lumen of the coronary artery and dilating the area of the arterial blockage. Once in place, the stent remains open as the balloon is deflated and removed. A benefit of coronary stenting over conventional balloon angioplasty is a significant reduction in restenosis. Restenosis is defined as a greater than 50% re-narrowing of a vessel diameter at 6 months' follow-up. This is caused by smooth muscle cells hyperproliferating followed by thickening, scarring, and remodeling of the region as a result of damage to the endothelial barrier at the site

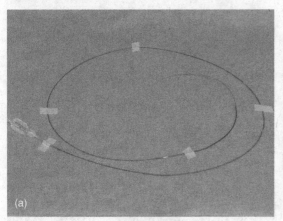

Figure 14.3

Stent on a catheter in a "crimped" or collapsed state. (a) Entire catheter. (b) Enlarged view at the end of the catheter showing the "crimped" stent.

of balloon inflation. Additional benefits of stenting include a reduction in the elastic recoil effect of the coronary artery and a reduction in neointimal hyperplasia, which is the immune system's reaction to the intrusion of angioplasty. As a result, stenting is now a well-established technique in endovascular interventions.

Although the use of bare metal stents reduced the concern for elastic recoil, in-stent restenosis and neointimal hyperplasia still persisted in some patients. Current advances in technology have allowed stent manufacturers like Boston Scientific Corporation (Natick, MA, USA) to market TAXUS® Express^2® Atom™ Paclitaxel-eluting Coronary Stent System, a drug eluting stent. Drug eluting stents are designed to slowly release a drug *in situ*, over a period of time, to pharmacologically reduce in-stent restenosis. Other drug-eluting stents include CYPHER® Stent (Cordis Corporation, New Jersey, USA), a stainless steel stent that releases 80% of sirolimus in 30 days. In many instances, biodegradable polymers such as poly(lactic-co-glycolic acid) (PLGA) are used as a matrix for drug-eluting coatings on current generation cardiovascular stents. Both the polylactic acid (PLA) and polyglycolic acid (PGA) are discussed in Chapter 6, and PLGA is a block copolymer of the PLA and PGA. Vast research and clinical activity related to stents has also provided ample knowledge on the current problems associated with the drug-eluting stents. Although the availability of these drug-eluting stents with a greater potential to eliminate restenosis is likely to change the field of interventional cardiology, there is a concern that the current generation of drug-eluting stents may be releasing non-uniform doses of drugs from the polymeric coatings. There are also concerns with allergic reactions to the stent coatings in a segment of the patient population. Thus current research in this area includes the investigation of other biomaterials to coat the stents. Research on methods to optimize drug delivery kinetics is also being performed to develop the next generation stents. Additionally, the use of stents is not limited to blood vessels. Besides uses in endovascular interventions, other stent applications have included lumen enlargement of hollow structures such as the tracheobronchal tree biliary duct and urinary tract.

14.3   Dental restoration

Guidance and standards for biomaterials used in dental restoration are found in publications of the American National Standards, American Dental Association, and the International Organization for Standardization. Used commonly for either block replacement or shell coverage, materials employed for the restoration of teeth include metals alloys (amalgams, gold, stainless steel, and cobalt–chrome), and ceramics (porcelain and alumina). Other uses of restorative materials in dentistry include polymers as sealants for surface lamination.

Examples of block replacement are fillings and inlays, and this type of restoration requires the use of a large volume of restorative materials. Supported on several sides by the containing tooth, the significant thickness of the restorative biomaterials used for block replacement helps to resist loads as well as to provide adequate strength. Additionally, since these restorative materials are in contact with the pulp and oral tissues, the ideal material should possess the following properties:

- be harmless when in contact with oral tissues,
- be bioinert or contain no diffusible components that may leach out of the material to stimulate local or systemic toxic response, and
- be free of sensitizing agents that may lead to an allergic response.

Examples of biomaterials used for block replacements include silver amalgam, resin composite, and glass ionomer cements. In many instances, the selection of material for this application depends on whether esthetic needs to be taken into consideration. For example, silver amalgam, which is not tooth-colored, is used as a filling for tooth cavities found in the posterior maxilla and mandible where esthetics is not a requirement. Figure 14.4 shows fillings of posterior teeth filled with amalgam.

Consisting principally of silver, tin, and copper, amalgam alloy is produced when mixed with mercury. Advances in amalgam research have resulted in variations in composition, and these amalgams are classified as either low or high copper amalgam as well as non-zinc or zinc containing amalgam alloys. The primary rationale for such variations in amalgam composition are:

- to minimize expansion and thus avoiding cracking of the tooth after the tooth cavity is filled,

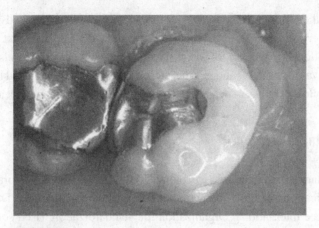

Figure 14.4

A dental patient with cavities in the oral cavity filled with amalgam. (Courtesy of Dr. Daniel C. N. Chan, University of Washington, School of Dentistry.)

- to increase corrosion resistance,
- to improve strength, and
- to improve the clinical outcome.

Although amalgam is an effective restorative material with a long clinical history, other restorative materials have been developed to improve the esthetics of the restoration. Such new restorative materials include resin composite and glass ionomer cements. Unlike dental amalgam, which is used mostly to restore teeth found in the posterior region of the mouth, resin composites and glass ionomers are more tooth-colored and are recommended for restoring anterior teeth. Shown in Figure 14.5 is a cavity that is filled with resin composite.

Although composite resin is esthetically pleasing to the patients, it exhibits the following problems:

- it shrinks during polymerization, resulting in gap formation between the composite resin and the tooth. The gap allows for bacterial invasion, plaque accumulation, and the formation of secondary caries or a carious lesion near an existing restoration,
- it wears excessively especially in load bearing areas, and
- it has a higher coefficient of thermal expansion for most commercially available composite resin compared to the tooth structure. This can possibly result in cracking of the tooth structure as the composite resin expands.

Aside from amalgam and composite resins, there are other restorative materials including glass ionomer, porcelain, cast metal alloys and ceramics. The glass ionomer is an adhesive restorative material with a favorable coefficient of thermal

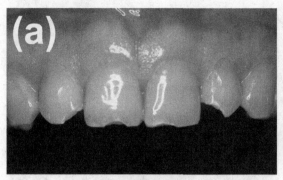 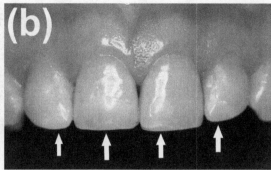

Figure 14.5

A dental patient with (a) chipped incisors, and (b) the incissors filled with resin composite, as indicated by arrows. (Courtesy of Dr. Daniel C. N. Chan, University of Washington, School of Dentistry.)

expansion. However, it is brittle and weaker than the other directly placed restorative materials and is recommended only for the restoration of class 3 and 5 lesions. Other materials such as porcelain, cast metal alloys and ceramics are significantly more expensive than the directly placed resin or amalgam because of the additional treatment time required for the completion of restorations. Additionally, porcelain is significantly harder than tooth structure and thus wears the opposing tooth.

Box 14.2

- Tooth caries are classified into six classes, depending on their size and location.
- A class 3 lesion is found in the proximal surfaces of central and lateral incisors, and cuspids or canines, whereas a class 5 lesion is found in the gingival third of the facial or lingual surfaces of the anterior or posterior tooth.

Similar to block replacements, shell coverage is made from either cast metals or ceramics. Examples of shell coverage are porcelain crowns and gold crowns. Such crowns are usually used in applications to completely cap a fractured tooth or are used in conjunction with dental implants. Shown in Figure 14.6a is a patient with a fractured central incisor in the maxilla, and Figure 14.6b shows the tooth being completely restored with a porcelain crown.

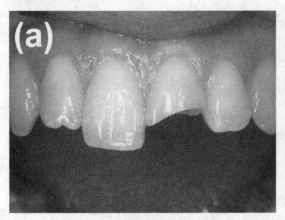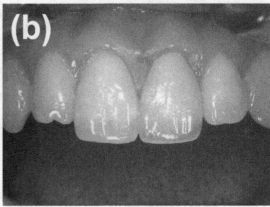

Figure 14.6

A dental patient with (a) a fractured incisor in the maxilla, and (b) the fractured incisor capped with a porcelain crown. (Courtesy of Dr. Namsik Oh, Inha University Hospital, South Korea.)

14.4 Dental implants

Unlike dental restorative materials, dental implants are used to restore tooth function for partially edentulous and totally edentulous patients. Today, almost all of dental implants used are root-form endosseous implants. These implants are similar to an actual tooth root. Because they are implanted in the bone, biocompatibility and functionality of the implants to support dental prostheses such as crowns or dentures are key to the success of the implants. Figure 14.7a shows an implant being placed in the right mandible of a patient, and Figure 14.7b shows three implants placed in the left mandible of the same patient. Figure 14.7c is a radiograph of the same patient showing crown restorations after successful osseointegration between the implants and maxilla bone had been achieved.

With today's knowledge of biomaterials, all commercially available dental implants are made of either titanium or titanium alloys. Clinical success rates of the placed implants have exceeded 90%. In many instances, the high clinical success of the placed implants is aided by improved implant designs and implant surface modifications for maximizing bone–implant osseointegration. As shown in Figure 14.8, alterations of implant design have included changes in the pitch of screws, tapering of screw threads as well as tapering of the implants to mimic the root of a tooth.

Aside from implant design, it is also common to see dental implant vendors modifying their implant surfaces using techniques such as plasma spraying, sandblasting, or acid etching (Figure 14.8). Plasma spraying to deposit either titanium or hydroxyapatite on dental implant surfaces is common, and this process is

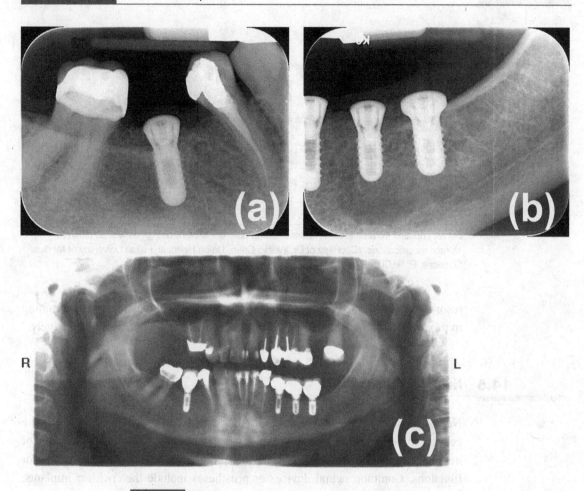

Figure 14.7

Radiographs of (a) an apical X-ray on the right mandible of a patient showing placement of one root-form implant, (b) an apical X-ray on the left mandible of the same patient showing placement of three root-form implants, and (c) a panoramic X-ray of the same patient showing implants restored with crowns after successful osseointegration. (Courtesy of Dr. Tony Dacy, Private Practice, Boerne, TX.)

typically performed after the implant is sandblasted to enhance the coating–implant adhesion strength. The principle of plasma spraying is discussed in Chapter 9. A rationale for the use of titanium particles as a medium for the plasma spraying process is that roughness of the implant surface is enhanced and thus allows for maximum bone–implant osseointegration. In contrast, modification of the implant surfaces through plasma spraying of hydroxyapatite allows for a more bone-like material to be implanted. Additionally, the osteoconductive property of hydroxyapatite also permits more rapid bone formation and implant fixation. Other common surface modifications for implants include blasting of

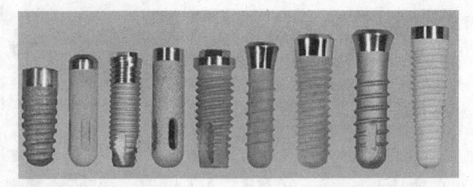

Figure 14.8

Dental implants from different vendors showing different implant design as well as different surface modifications. (Courtesy of Dr. Weihui Chen, Union Hospital, Fujian University of Medical Sciences, P. R. China.)

resorbable-blasted media (RBM) and the combined use of blasting and acid etching to produce RBM and sandblasted large acid etched (SLA) implants, respectively.

14.5 Neural prostheses

Neural prostheses are used to provide short bursts of electrical impulses to the central or peripheral nervous system. These impulses aid in the stimulation of neurons, which ultimately result in the production of sensory and/or motor functions. Common neural devices or prostheses include the cochlear implants, brain implants, bladder management stimulators, as well as devices that aid patients in their walking and grasping.

All neural prostheses are made up of four common components, namely a stimulus generator, a battery-powered or external alternating current power supply, electrodes, and a user-control interface. Metal electrodes can either be percutaneous, transcutaneous, or implanted to stimulate the nerves. Unlike percutaneous or implanted electrodes, transcutaneous electrodes are surface electrodes that are in contact with the skin. These transcutaneous electrodes are non-invasive, easy to apply, and are inexpensive. Although higher intensity signals are required due to the impedance of the skin, some nerves are too deep to be stimulated by transcutaneous electrodes. Percutaneous electrodes are thin wires that are inserted through the skin and into muscular tissue. Unlike percutaneous electrodes that are temporarily placed into muscular tissues, implanted electrodes are permanently placed, generally requiring a lengthy and invasive surgical process. These electrodes are made of metals such as tungsten, iridium, platinum, gold, and stainless steel, and advances in technology have allowed for improvement of neural signals by coating the electrically conductive electrodes with biocompatible polymeric coatings of alginate hydrogel and nanotubes made of poly(3,4-ethylenedioxythiophene) (PEDOT).

14.6   Opthalmology

In the field of ophthalmology, most implants are fabricated from synthetic polymers or ceramics. These implants include orbital implants, keratoprostheses and intracorneal implants, sclera buckles, glaucoma filtration implants, and intraocular lenses. Intraocular implants can also be designed to deliver therapeutic drugs into the eye, thereby overcoming many barriers that inhibit effective treatment of eye diseases conventionally treated with eye drops or by oral or intravenous treatment. It is critical that biomaterials selected for these applications are primarily based on their biomechanical properties and biofunctionality. Two applications of biomaterials in the field of ophthalmology are described below.

The loss of an eye due to trauma, tumor, or end-stage ocular disease such as glaucoma, or diabetes can be devastating to patients at any age. The diseased or injured eye is typically removed by enucleation (removal of the eyeball) or evisceration (removal of the contents inside the eye, including cornea, iris, lens, vitreous, and retina, and leaving behind the sclera or shell of the eye). Such surgeries result in an unsightly sunken depression of the eyelids into the eye socket that is typically not esthetically pleasing and may have a major impact on the patient's self-image, self-confidence, and self-esteem. Orbital implants have been used to restore the esthetics of the eye socket and these implants can be fabricated from a wide range of biomaterials, including glass, cork, ivory, silicone, and aluminum. Glass was used as early as 1885 and is still one of the most common non-integrated implants used today. In addition to glass, other common non-integrated orbital implants can also be made from poly(methyl methacrylate). These implants do not have direct or indirect integration with the orbital structures. Motility of these implants is achieved through surface tension at the conjunctival–implant interface. Although non-integrated implants made from glass range in size from 14 mm to 20 mm, surgeons typically place a larger size implant to fill the anophthalmic socket and to avoid volume loss, as well as subsequent unfavorable cosmetic deformities. In general, the biomaterials used for non-integrated orbital implants should be:

- inexpensive,
- well-tolerated, with few complications, and
- technically not complicated to fabricate.

Box 14.3

- The high density porous polyethylene implants are most popular for enucleation and evisceration.
- Implant sizing is important, since smaller implants tend to displace or migrate, whereas larger implants may result in tension on the conjunctival wound and implant exposure.

Since there is a strong desire for these implants to permit cellular ingrowth and vascularization, the porosity of orbital implants becomes critical to allow integration with the extraocular muscle. These porous implants are termed as biointegrated orbital implants. The integration of the muscle as well as the innervations of the muscles allow the biointegrated orbital implants to move in coordination with the healthy eye. This ability to move the orbital implant results in restoring the esthetic appearance of the eye and the surrounding tissue as well as preserving some degrees of normal eye motility. At the present, most of the biointegrated orbital implants are fabricated from coralline hydroxyapatite, synthetic hydroxyapatite, aluminum oxide, or porous polyethylene. Figure 14.9 shows an example of a biointegrated orbital implant made from hydroxyapatite.

Although not directly related to the properties of the hydroxyapatite (HA), many of the HA orbital implants are limited by increased cost as well as

Figure 14.9

Bio-Eye® orbital implant made from hydroxyapatite. (Reproduced with permission from Integrated Orbital Implants, Inc., San Diego, CA, USA.)

Table 14.2 Commercially available orbital implants currently in the market

Biomaterial	Product	Manufacturer
Coralline HA	Bio-Eye®	Integrated Orbital Implants Inc., San Diego, CA, USA
Bovine HA	Molteno M-Sphere	IOP Inc., Costa Mesa, CA, USA
Synthetic HA	FCl₃ HA Implant	FCI Ophthalmics Inc., Issy-Les-Moulineaux, Dedex, France
Synthetic HA	Hydroxyapatite Sphere	Altomed Limited, Boldon, Tyne and Wear, UK
Aluminum oxide	Bioceramic Implant	FCI Ophthalmics Inc., Issy-Les-Moulineaux, Dedex, France
PMMA	Acrylic (Lucite) Sphere	Altomed Limited, Boldon, Tyne and Wear, UK
Silicone	Silicone Orbital Implants (Eye Sphere)	FCI Ophthalmics Inc., Issy-Les-Moulineaux, Dedex, France
Silicone	Silicone Sphere	Altomed Limited, Boldon, Tyne and Wear, UK
Porous polyethylene	Medpor™	Porex Surgical Inc., Cooledge Park, GA, USA

complications such as implant exposure, conjunctival thinning, socket discharge, pyogenic granuloma formation, implant infection, and persistent pain or discomfort. Additionally, direct suturing of HA implants to extraocular muscles is not possible. As a result, many of these complications may be related to surgical implantation technique, implant wrap selection, and host factors. These experiences suggest that the ideal orbital implant should exhibit the following characteristics:

- easy to insert and manipulate,
- should not be prone to infection,
- should not be prone to extrusion,
- should be easily anchored to surrounding structures,
- should be cost effective, and
- should not elicit fibrous tissue formation.

Complications and costs associated with the use of HA orbital implants have resulted in the introduction of alternative implant materials. Examples of current commercially available orbital implants are shown in Table 14.2.

Unlike orbital implants that are used to restore the esthetics of the eye socket, implantation of intracorneal lenses in the central corneal stroma is utilized to correct conditions such as myopia and hyperopia. Common biomaterials used for intracorneal lenses are PMMA, hydrogel, and polysulfone. Although PMMA is considered biocompatible and possesses a range of good optical and mechanical properties, transport of nutrient across the cornea is disrupted with the use of solid PMMA lenses. Hydrogel lenses, such as Vue+ from ReVision Optics™, Inc. (Lake Forest, CA, USA) have a water content of greater than 70% and are used to improve nutrient flow. However, hydrogel lenses possess a lower refractive index (refractive index of 1.37) compared to PMMA lenses (refractive index of 1.49). Additionally, lens migration and invasion of epithelial cells have been reported as side effects in patients. In contrast, polysulfone lenses have a relatively high refractive index of 1.63. As a result, polysulfone lenses can be made extremely thin, thereby facilitating the implantation surgery. Since polysulfone is impermeable to aqueous solutions and may disrupt nutrient flow, these lenses are fenestrated with a network of holes 10 mm in diameter. While these holes allow fluid movement but are too small for cellular invasion, optical quality of the lens is not compromised.

14.7   Orthopedic implants

Most orthopedic implants are made of stainless steel, cobalt–chrome alloy, or titanium or titanium alloys. These orthopedic devices include the knee, total hip, and finger prostheses as well as intramedulary nails, and fracture fixation plates and screws. In instances where the prostheses possess articulating surfaces, polyethylene polymers are used for such surfaces. Additionally, the ball of the total hip prostheses is commonly made of cobalt–chrome alloys or a ceramic, which is more wear resistant compared to titanium or titanium alloys. Figure 14.10a shows a Synergy™ hip implant (Smith and Nephew, Inc., Orthopaedic Reconstruction and Trauma, Memphis, TN, USA) with femoral heads made from alumina. Like dental implants, modifications of orthopedic implant surfaces are also performed to enhance osseointegration. These modifications include roughening of surfaces as well as the use of plasma-sprayed hydroxyapatite. As a representative of implant surface modifications, the proximal end of the stem shown in Figure 14.10b is modified with titanium bead coatings, followed by hydroxyapatite coating.

Other new metal alloys introduced in the market include the OXINIUM™ that is used in the Smith and Nephew's Synergy™ hip implants and the Legion™ CR total knee systems. Consisting of a metal alloy with an oxidized zirconium surface, OXINIUM™ is marketed as a material that has excellent fracture toughness and provides wear resistance without being brittle. Shown in Figure 14.11, the femoral head of the hip implant as well as the knee femoral component are made of

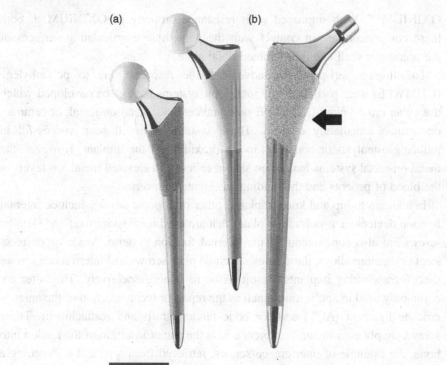

Figure 14.10

Synergy™ hip implant with (a) alumina femoral heads, and (b) porous bead coating followed by hydroxyapatite coating (arrow) on the proximal end of the stem. (Reproduced with permission from Smith and Nephew, Inc., Orthopaedic Reconstruction and Trauma, Memphis, TN, USA.)

Figure 14.11

Photographs of the (a) Synergy™ total hip, (b) magnified view of femoral head in the acetabular cup, and (c) Legion™ CR total knee systems. These photographs showed components made of OXINIUM™. (Reproduced with permission from Smith and Nephew, Inc., Orthopaedic Reconstruction and Trauma, Memphis, TN, USA.)

OXINIUM™. With improved wear resistance property of OXINIUM™, both these components are in contact with the polyethylene articulating surfaces of the acetabular shell and the stemmed tibial plate.

To mitigate peri-prosthetic adverse tissue response due to polyethylene (UHMWPE) wear particles, new total joint systems have been developed which use either cross-linked UHMWPE-on-metal/ceramic, metal-on-metal, or ceramic-on-ceramic articulating couples. These systems have all been successful in reducing osteolysis or bone loss in the proximity of the implant. However, the metal-on-metal systems have been shown to result in elevated metal-ion levels in the blood of patients and this finding has created concerns.

In addition to hip and knee implants, other orthopedic devices include internal fixation devices such as fracture plates that are used for fracture fixation. Metallic screws are also components of the internal fixation systems. Made of stainless steel or titanium alloys, these screws include plate screws and interference screws used for anchoring implants or soft tissue to bone, respectively. The latter are commonly used in applications such as the repair or reconstruction of the anterior cruciate ligament (ACL) and for bone–tendon grafts and reattachment. These screws are placed with an interference fit at the site of insertion of the tendon into bone. An example of interference screws, retrieved from a patient undergoing a revised ACL reconstruction, is shown in Figure 14.12.

Aside from the use of metals such as titanium and stainless steel, internal fixation devices can also be made from absorbable polymers such as poly(L-lactic

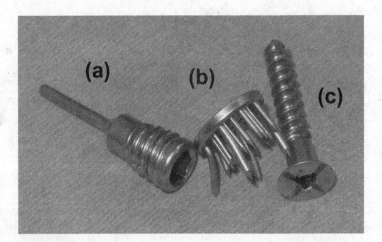

Figure 14.12

Interference screws such as the cannulated Bone Mulch™ Screw (a) for femoral tunnel fixation, a WasherLoc™ tibial fixation device (b), and a cancellous compression screw (c) that are made of titanium alloy (Arthroteck®, a Biomet, Inc. Company, Warsaw, IN, USA). These are used for fixing of graft construction during the restoration of the anterior cruciate ligament.

acid) (PLA), poly(glycolic acid) (PGA), PLA–PGA copolymers, polydioxanone (PDO), and polyglyconate. These fixation devices include pins, screws, and plates for fracture fixation. Additional absorbable devices include interference screws for repair of the ACL, staples for meniscal repair, and suture anchors, tacks, and staples to aid in soft tissue fixation to bone.

14.8 Renal

The kidneys are critical for removing metabolic end-products or solutes and excess water from the blood. Non-treatment of failed kidneys eventually leads to death due to the accumulation of the solutes in the blood. As such, the potential for fatality due to kidney failure requires the patients to either undergo renal replacement therapy or kidney transplantation in order to survive. Efficacy of dialysis therapy is dependent on the properties of the semi-permeable membranes, which play a key role in the removal of these solutes and water from the blood. Ideally, these membranes should:

- remove a broad range of metabolic end-products,
- be biocompatible,
- be resistant to change during sterilization, and
- not induce adverse reactions with the blood or administered drugs during the therapy.

There are different types of dialysis membranes used for dialysis therapy. The first generation membranes are made from regenerated cellulosic membranes. Cellulose is a natural linear polysaccharide material and its properties are discussed in Chapter 8. These membranes are highly hydrophilic, with low water and solute permeabilities (small pore size), a relative thin wall thickness of 5–15 μm, and an increase in the degree of C3 complement activation. An example of a commercially available dialysis membrane is Cuprophan® (Membrana GmbH, Wuppertal, Germany). Popularity for the use of Cuprophan® has decreased over the years owing to poor biocompatibility and its poor removal of middle molecules and low molecular weight proteins. In this application, the definition for biocompatibility refers to the non-adhesion or activation of blood cells, including non-adsorption or transformation of proteins and non-proteinaceous substances by the membrane. Efforts to improve the biocompatibility of cellulose membranes have resulted in the use of modified cellulosic membranes. Through acetylation, these modified cellulosic membranes do not bind to the C3 molecules, thereby decreasing the degree of complement activation. An example of dialysis membrane made from modified cellulosic membrane includes Hemophan® (Membrana GmbH, Wuppertal, Germany).

> **Box 14.4**
>
> - The C3 molecule is an abundant serum protein.
> - Being the most abundant complement protein, the C3 molecule is capable of being activated during hemodialysis, depending on the types of membrane.
> - Activation of the C3 molecule results in proteolytic cleavage of the molecule into two biologically active fragments, namely C3a and C3b.
> - High concentration of C3a fragments can result in the patient experiencing severe allergic reaction.

Dialysis membranes can also be made from synthetic polymers such as polyacrylonitrile, a copolymer of acrylonitrile and sodium methallyl sulfonate known as AN69, polysulfone, poly(methyl methacrylate), and polyamide. Although these membranes are more porous, their hydrophobic properties allow for an increase in capacity for protein adsorption. Although membranes made from these synthetic polymers have greater wall thickness of at least 20 μm, these membranes also exhibit increased biocompatibility, which accounts for their increasing popularity over the years.

14.9 Skin applications

The skin functions as a protective barrier against the environment. It prevents infection as well as the loss of water and electrolytes from the body. As a result, skin loss due to wounds, such as severe burns and skin ulcers, requires patients to undergo skin replacement. In many instances, such loss of skin can lead to an unacceptable quality of life, disfigurement, and death. Biomedical engineers and scientists have utilized a variety of different approaches in an attempt to replace damaged skin. These approaches include culturing of epidermal autografts *in vitro* and the use of biologically active tissue-engineered scaffolds. Although a variety of biomaterials have been used to treat skin loss, the ideal biomaterials must:

- provide an effective temporary barrier,
- promote healing,
- minimize scarring.

Cultured epidermal autograft products, such as Epicel® (Genzyme Biosurgery, Cambridge, MA, USA) and Holoderm® (Tego Science, Korea) have been indicated for severe burn patients with deep dermal or full thickness burns comprising a total body surface area of greater than or equal to 30%. Keratinocytes are derived from full-thickness biopsy of the patient's undamaged skin and cultured *in vitro*. These cultured grafts are typically available to patients after 15–21 days of culture and delivered in a spill resistant package containing a nutrient medium. Additionally, the transfer and ability to dispose the confluent epithelial cell sheets play a critical role in achieving a successful skin graft. Cellular support for such cell transfer could include a thin biomaterial film that would facilitate direct culture, whereby cells could adhere and proliferate. These films have been fabricated from collagen, hyaluronan, chitin, fibronectin, and fibrin molecules. Ideal biomaterials for such use should possess the following materials and clinical characteristics:

- be biocompatible,
- be easy to handle,
- have short transfer time,
- be flexible in order to allow total contact between the cultured epithelial autograft and the receiving site,
- be thin and permeable to prevent any risk of complete occlusion of the wound,
- facilitate the escape of exudates from the underlying tissues, and
- be biodegradable.

The combination of cultured epidermal autografts with biomaterials has resulted in several products such as Apligraf® (Organogenesis, Inc., Canton, MA, USA). Made of an upper layer containing keratinocytes, the lower layer of Apligraf® is composed of collagen and fibroblast.

Other advances in biomedical engineering have also allowed the development of temporary skin substitute dressings. These are used clinically when the surgeon encounters a lack of sufficient donor sites to perform skin autografts, has difficulty evaluating the depth of excision, has difficulty in controling bleeding, has to perform massive excision of the infected skin, or has determined that the general condition of the patient is poor. Ideally, these temporary skin substitute dressings should:

- adhere to the wound,
- control loss of water,
- be biocompatible,
- be flexible,
- be durable,
- be stable on various wound surfaces,
- provide a barrier to bacterial infection,
- be easy to apply and remove,
- be easily available and stored,

- be cost-effective, and
- possess hemostatic efficiency.

Biomaterials that are used as skin wound dressings have included silicone, polyvinyl chloride and polyurethane membranes, cotton gauze bonded to silicone, nylon mesh bonded to silicone, and porous copolymer layers of L-lactide and ε-caprolactine with a poly(ether urethane) membrane dressing. These dressings are broadly classified by the FDA as either non-interactive or interactive dressings. Non-interactive wound dressings, such as Oasis® Wound Matrix (Cook Biotech, Inc., West Lafayette, IN, USA), do not integrate into the body tissue, but are placed on the wound for the primary purpose of protecting the wound. Indicated for all wound types except for third degree burns, non-interactive dressings are removed from the wound site after several weeks and prior to wound therapy or grafting. In contrast, interactive wound dressings are long-term skin substitute or a temporary synthetic skin substitute that directly or indirectly interacts with the body tissues to promote healing. Commercially available interactive skin substitute dressings include Dermagraft® (Advanced BioHealing, Inc. LaJolla, CA, USA), Biobrane® (Smith and Nephew Wound Management, Hull, UK), and Integra® Dermal Regeneration Template™ (Integra LifeSicence Corporation, Plainsboro, NJ, USA).

In the Dermagraft® dressing, human fibroblasts are seeded onto polyglactin mesh scaffolds. These fibroblasts proliferate and secrete dermal collagen, matrix proteins, growth factors and cytokine to create a dermal substitute. In the case of Biobrane®, the dressing is made of silicone film with a nylon fabric partially imbedded into the film. The presence of the nylon fabric presents a complex 3D structure, allowing for the interaction with collagen and clotting factors in the wound site, and thereby allowing the dressing to adhere better until epithelialization occurs. The adhesion of the dressing also forms a durable protective layer at the wound, forming a barrier to exogenous contaminations and decreasing the risk of infection. Unlike Biobrane®, the Integra® Dermal Regeneration Template™ is made up of two layers, with the upper layer being made from a medical-grade, flexible silicon sheet and the bottom layer being composed of a matrix of interwoven bovine collagen and glycosaminoglycan. Placed on a wound, this two-layered design allows the silicone to mimic the top, epidermal layer of skin which helps to close the wound and prevent loss of fluid. The bottom matrix mimics the fibrous pattern of the bottom layer of skin, permitting blood vessels and other cells to grow, and thereby regenerating a new layer of skin as the collagen resorbs. After placement for about 14–21 days, the upper layer of silicone can be removed, allowing the graft of the patient's skin to be applied to the wound area.

14.10   Summary

The clinical applications described above represent some of the uses of today's biomaterials. In many instances, the development of implants or devices with improved design, biofunctionality, and surface properties is done for a specific clinical application. Fundamentally, the development of any implant or device for clinical application will require the knowledge of material properties and how these properties influence tissue response after implantation. Additionally, it is not uncommon to utilize a combination of different materials and drugs to achieve the optimal performance of the device for a specific application. Examples of such utilization of combined materials in current medical devices include total hip prostheses whereby titanium, cobalt–chrome alloys and ultrahigh molecular weight polyethylene are used to fabricate the different components. As observed in stents, advances in science and technology have always allowed alterations of existing technologies to improve and optimize the material properties and design of devices. Overall, in an effort to improve the performance of a device in a specific clinical application, engineers, scientists, and device manufacturers should take into account the available manufacturing technologies as well as advances in new materials in their overall design. In addition to the new technologies and materials, engineers and scientists should, at the same time, keep in mind the ease of the manufacturing process, limitations of the materials, cost, and biocompatibility.

Additional reading

- Bhatia, S. K. (2010). *Biomaterials for Clinical Applications*. Springer, ISBN 9781441969194.
- Rakhorst, G. and Ploeq, R. (2008). *Biomaterials in Modern Medicine: The Groningen Perspective*. World Scientific Publishing Company, ISBN 9812709568.

Problems

1. What role do the FDA and other regulatory agencies play in the use of biomaterials for clinical applications?
2. How are the current biomaterials used in the clinics?
3. What role does silicone rubber and polyurethane play in the cardiac pacing leads?
4. What are the advantages and disadvantages of using silicone rubber and polyurethane in the cardiac pacing leads?

5. What is the advantage of using cobalt–chromium stents when compared to stainless steel stents.

6. List the current problems associated with drug-eluting stents.

7. Why is the composite resin preferred by patients over the use of amalgams when restoring a tooth cavity?

8. What are the current problems associated with the use of composite resins?

9. What is the rationale for the use of plasma spraying titanium or hydroxyapatite (HA) on dental and orthopedic implant surfaces?

10. What are the four common components of neural prostheses?

11. What are the differences between non-integrated and biointegrated orbital implants?

12. Why is an increase in the degree of C3 complement activation not favorable for patients using regenerated cellulosic membranes for hemodialysis?

13. What does biocompatibility of the membranes mean in renal hemodialysis applications?

14. What are the ideal properties that temporary skin substitute dressings should possess?

15. What should the biomedical engineers keep in mind when developing devices for a specific clinical application?

Index